STOREY'S GUIDE TO

GROWING
ORGANIC
VEGETABLES
& HERBS
FOR MARKET

Site & Crop Selection

Planting, Care & Harvesting

Business Basics

Keith Stewart

Storey Publishing

*The mission of Storey Publishing is to serve our customers by
publishing practical information that encourages
personal independence in harmony with the environment.*

Edited by Deborah Burns and Sarah Guare
Art direction by Cynthia N. McFarland
Book design by Kent Lew
Text production by Jennifer Jepson Smith

Cover photography by © Валерий Военный/iStockphoto.com (salad greens), © Daniel Bendjy/
iStockphoto.com (mesclun leaf), © Digital Paws, Inc. / iStockphoto.com (tomato), © Julián
Rovagnati/iStockphoto.com (garlic), © Konrad Kaminski/iStockphoto.com (basil),
© Le Do/iStockphoto.com (cilantro), © sunstock/iStockphoto.com (red pepper), © Susan
Trigg/iStockphoto.com (peapod right), and © YinYang/iStockphoto.com (peapod left)
Illustrations and graphics by © Elayne Sears, except for: courtesy of the author, 41 and 219;
Alison Kolesar, 225, Charles Joslin, 368 right, 369, 371, 374, 375, 376, 378, 379, and 381;
© Flavia Bacarella, 8; Ilona Sherratt, 42, 44, 45, 58, 59, 80, 201 bottom, 224 (adapted from
USDA), and 281–287; Kent Lew, 199; and Regina Hughes, 462, 463, and 465

Indexed by Christine R. Lindemer, Boston Road Communications

Storey Publishing
210 MASS MoCA Way
North Adams, MA 01247
www.storey.com

Storey Publishing is committed to making environmentally
responsible manufacturing decisions. This book was printed in the
United States on paper made from sustainably harvested fiber.

Printed in the United States by McNaughton & Gunn, Inc.
10 9 8 7 6 5 4 3 2 1

LIBRARY OF CONGRESS CATALOGING-IN-PUBLICATION DATA

Stewart, Keith, 1944–
   Storey's guide to growing organic vegetables & herbs for market / by Keith Stewart.
      p. cm.
   Guide to growing organic vegetables & herbs for market
   Growing organic vegetables and herbs for market
   Includes index.
   ISBN 978-1-60342-571-1 (pbk. : alk. paper)
   ISBN 978-1-61212-007-2 (hardcover : alk. paper)
   ISBN 978-1-60342-907-8 (e-book)
   1. Organic gardening. 2. Farmers' markets. 3. Vegetables—Organic farming. 4. Herbs—Organic farming
   I. Title. II. Title: Guide to growing organic vegetables & herbs for market. III. Title: Growing organic veg-
      etables and herbs for market.
   SB453.5.S84 2013
   631.5'84—dc23
                              2012032604

Acknowledgments

Farming is among the most basic and essential of human activities. It has been with us in various forms for a long time. This book is essentially a compilation of the accumulated knowledge and experience of others. I am therefore greatly indebted to all those farmers and agriculturalists, past and present, who have lighted the way to a more healthful and sustainable food system.

Beyond this broad indebtedness, I want to thank two individuals who read the entire first draft of the book, chapter by chapter, and generously offered their professional input and advice, which was a great help: Maire Ullrich, Agricultural Program Leader and Vegetable Crops Educator for Cornell Cooperative Extension; and Brian Caldwell, fellow farmer and Farm Education Coordinator for the Northeast Organic Farming Association of New York. I would also like to thank Abby Seaman, Vegetable IPM (Integrated Pest Management) Coordinator at Cornell University; and Bob Schindelbeck, Extension Associate in the Department of Crops and Soil Science at Cornell University, both of whom provided valuable input.

A farm is no better or more productive than the sum of its workers. I owe a very large thank you to all those who have lived on my farm and labored beside me in the fields over many years. I hope these individuals, many of them good friends to this day, gained something from working with me. I know I gained much from working with them.

My thanks also to Deb Burns at Storey Publishing for getting me rolling and encouraging me along the way, and to Sarah Guare for her editing expertise. Both were a real pleasure to work with. Finally, thanks to my wife, Flavia Bacarella, who willingly endured the demands that book writing places on a spouse and was always my first reader.

Contents

Part IV. The Crops We Grow

Part V. Harvesting and Marketing

Preface

I never expected this book to get as big as it is. Like a field of vegetables, blessed with good soil and sufficient sun and rain, it just kept on growing. There was always more useful material to present, more advice to offer. The challenge was knowing when to stop. Now that it's done, I wouldn't want to leave the impression that any existing or aspiring farmer needs to have all this information neatly stored in his or her cerebral filing cabinet. I certainly don't. If you asked me to tell you the optimum spacing for French tarragon, or the diameter of a clay particle, or the best temperature for green-sprouting potatoes, I'd say, "Give me a minute, and I'll look it up." I might even look it up in my own book. But that's not to say that I wouldn't have other books on hand, because, in truth, there's far more to know than is contained in these pages.

My goal in writing the book has been threefold:

- To compile a broad swath of material of practical value to small- and mid-scale organic growers, especially those in the early stages of their farming careers

- To weave into the narrative some of my own story and experience as a farmer

- To encourage the reader to look at the big picture — to view the farm as a living entity, an ecological unit, rather than merely a unit of production

The last of the three goals is seldom spelled out directly in the book's many pages. But if I have done my job well, it will be a theme that emerges here and there between the lines. I hope so, because it is this broader ecological perspective that is at the core of the practice and philosophy of organic farming.

To think like a true farmer is to think about many things. It is to think about what makes crops grow, what constitutes healthy soil, and what it takes to develop a market for your produce, keep customers happy, and run a successful business. It is to think about, then develop, a farm plan as well as a plan to dodge nature's weather-related curveballs and her pests and diseases. And it is to never lose sight of what it takes to be a good steward of the land, not just for yourself but for future generations as well.

For those who are persuaded of its value and willing to take hoe in hand, the kind of farming described herein is a life's work and then some. It is not a job in the conventional sense, though it can and certainly should be a viable way to make a living. In my mind, the well-managed, diversified farm using organic methods and functioning within a community or region is an agent for positive change, an example of a different way of relating to one another and the world in which we live. It is a model for a more sustainable planet.

Operating a diversified, organic farm will bring challenges and heartbreaks by the score — no question about it. But it can also bring great satisfaction, and it can put you in touch with elemental forces that, if you have eyes to see and a holistic turn of mind, will cast a richer and warmer hue on this green planet of ours.

"In the cycles of farming, which carry the elemental energy again and again through the seasons and the bodies of living things, we recognize the only infinitude within reach of the imagination."

Wendell Berry, *The Unsettling of America*

"Whether you will or not
You are a King, Tristram, for you are one
Of the time-tested few that leave the world,
When they are gone, not the same place it was.
Mark what you leave."

Edwin Arlington Robinson, "Tristram"

PART I

IN THE BEGINNING

Chapter 1

··

Thinking About
Becoming a Farmer?

I MADE THE SWITCH from city living and the corporate world to becoming a farmer in my early forties. It was the best move I ever made. Not a day has passed since then on which I doubted the rightness of my choice. Though I had virtually no experience, the idea of farming had been brewing inside me for a few years. I was not happy going into the office every day and playing the part of a company man. I hankered for rolling fields, open space, and blue sky. I remembered my parents' vegetable garden in New Zealand and the fresh, wholesome food that came from it, and I remembered my Uncle Roger's sheep farm where I passed many a youthful summer's day. I wanted to hear the sound of birds and feel the pulse of nature. And I wanted to reclaim my body and follow my environmental inclinations. Thoughts of getting back in shape, doing good physical work, being my own boss, and growing healthy food kept rattling around inside my head.

It was the mid-1980s. Small farms were disappearing at an alarming rate, often being merged into larger ones. Industrial monoculture held full sway. "Get big or get out" was the motto — and don't spare the chemicals. In the age of Reagan, organic farming was a minor fringe movement viewed by most as a holdover from the radical and rebellious '60s and seldom accorded much respect. At the time, even to me, becoming a small farmer seemed like an impractical goal, if not a totally harebrained scheme. But I was drawn to it with an unusual energy and passion, the likes of which I had not known since my younger days. Quite simply, it was what I wanted to do, regardless of my chance of success. I have since learned that this kind of enthusiasm counts for a lot.

I read everything I could get my hands on by such authors as Wendell Berry and Gene Logsdon and several others less well known. I pored through numerous Rodale publications on sustainable farming and gardening. On weekends I visited friends in upstate New York and relished working on their land. In idle moments, I imagined a small, diversified farm with an abundance of life on it. I conjured up images of fruit trees, berry bushes, and rows of vegetables sparkling in the early sun.

As it turned out, many of my imaginings did become reality, though only after years of work. But from the beginning of my farming career, even in the face of my own inexperience and numerous setbacks, I truly enjoyed what I was doing, perhaps for the first time in my adult working life. Looking back, I can see it was a chancy undertaking, and had fortune not smiled on me in those early days, things might have turned out a little differently and not necessarily to my liking. Were I to go back as a younger man and start all over again, knowing what I know now, I would approach the agrarian life with a little more caution and advance preparation. But I'd go for it, nonetheless.

Twenty Points to Ponder

With the benefit of hindsight, I've assembled some points that I might have given more thought to 25 years ago. They are questions to turn over in your mind before you set out to become a farmer — to get a sense of how well suited you might be for the farming life.

The desire to escape from a lousy job or a domestic situation with which you're unhappy may be valid reasons to make changes in your life, but they are not a sufficient foundation upon which to build a farm. Neither is a love of nature or good food, though these are certainly not impediments. Every aspiring farmer should try to understand what he or she is getting into and avoid conjuring up an idealistic (and probably unrealistic) pastoral future. Farming is predominantly work — some of it hard work — day after day. For you to succeed, it has to be meaningful work, and you have to keep wanting to do it, year after year.

Many of these questions are somewhat loaded, and the preferred answers are rather obvious. But in case you're in doubt, each one is followed by expounding remarks that make my position clear. To make them a little more digestible, I've divided the points to ponder into three loose and somewhat overlapping categories:

- Deal Makers
- Deal Breakers
- Question Marks

If you come down on the right side of most of the Deal Makers, you're in good shape, but watch out for the Deal Breakers. The Question Marks will call for a bit more self-evaluation. Finally, by way of disclaimer, let me emphasize that in the world of farming, there are not many "right" answers. I've no doubt that other time-tested farmers might see things a little differently.

It's not likely that any one of us, including yours truly, will score high marks on all of these questions. I've not yet come upon a flawless farmer and am not sure I would want to. The idea is to get you thinking about what it takes to be a good farmer and how well suited you might be. If there are aspects of your nature that could be modified to better suit your agrarian goals, well, it's not too soon to start working on them.

Deal Makers

1. *Do you like being your own boss and calling the shots, or are you stressed out when you have to make too many decisions?*

Operating a small, diversified farm is tailor-made for anyone who likes to be his or her own boss and is comfortable calling most of the shots. If you have a hard time making decisions, it might be wiser to work on someone else's farm rather than try to run your own place.

2. *Do you like watching things grow? Are you fascinated by nature's ways, the changing seasons, the ebb and flow of life?*

A good farmer has an eye for the natural world. If growing plants doesn't turn you on and you're not especially fascinated by nonhuman living things and the roles they play on this planet, consider a different line of work.

3. *Do you enjoy working with nature, or are you happier when the natural world is in an ordered state and under control?*

Working cooperatively with nature is at the very heart of organic farming. If you prefer to control the natural world, you may be better suited to conventional farming with its arsenal of chemicals, though this is not a path I would recommend, nor is it one without its own considerable challenges.

4. *Do you enjoy working outside? Can you tolerate different kinds of weather? Are you a big fan of air conditioning?*

Farming, of course, is mostly conducted outdoors, so you must be okay with this. Sometimes it will be a little too hot for your comfort and sometimes too cold or too wet. Regardless of the weather, work needs to get done. If all-season field work is not your cup of tea, you might focus your energies on marketing farm products rather than growing them.

5. Are you in good shape? Do you enjoy physical work? Do you know how to pace yourself and use your body efficiently?

A farmer should be fit and physically capable and know how to use his or her body intelligently and efficiently. This type of knowledge can be learned on the job, but it would be helpful to have a modest supply of it before you start. Think of your body as a machine, like a tractor, that needs to be well maintained if it is to keep doing the work you will ask of it, year after year.

6. Can you balance a checkbook, keep accounts, and avoid spending more than you have?

Farming is a business, and every business needs to make a profit to keep going. Expenses must be offset by income. A farmer must keep a close watch on the bottom line and avoid slipping into debt. If you want a new tractor, it's probably best to wait until you can afford it.

7. If you're starting your own farm, do you have enough reserves of capital to weather a few lean years? Can you keep a part-time job that will bring in some income while you're learning the ropes?

Don't expect your farm to be profitable in its first year or two. It might happen, and of course, that is what you will hope for. But it would be safer to assume that it won't.

Deal Breakers

1. Are you reluctant to try something when you're not sure of the outcome? Do you feel crushed when a project you had high hopes for turns out badly?

Farming is not for the fainthearted or the risk averse. Uncertainty goes with the territory. Better to take your wins and losses in stride (you'll have plenty of both) and move on to the next challenge.

2. Do you hate getting dirt or grease on your hands and clothes? Do sweat, dust, oil, and — occasionally — a little blood make you uncomfortable?

All of this will happen. Just get yourself a bar of soap and some hard-wearing work clothes.

3. *Are you reluctant to admit that you've made a mistake? Will you stubbornly stay with a bad bet, or are you willing to cut your losses and move on?*

This one's a bit like throwing good money after bad. To a large extent farming is about the allocation of scarce resources — namely your own capital and labor and the labor of your employees. When a crop or project is not generating the expected income or desired result, you should take a very hard look at it and be willing to bring down the ax.

4. *Are you short on patience? (Many of us are, though we might like to think otherwise.) Do you like quick action and quick results? Or are you willing to plan ahead and take the long view?*

Organic farming is seldom about instant gratification. Many of the challenges you will face do not lend themselves to quick solutions. Plants do not grow on command. They develop at their own pace and in response to variables that are beyond human control, like the sun, the wind, and the rain. The wise farmer understands this. He or she takes an even stride, allows events to unfold, and cultivates a philosophic turn of mind. Then again, there will be times when quick and radical action is called for, and you need to be ready.

5. *Do you dread being bitten by a mosquito or stung by a bee? Are you afraid of snakes?*

Snakes are beautiful and beneficial to farmers; mosquitoes are food for bats and barn swallows; bees will pollinate your plants. None of these creatures is the enemy. They are part of the wild dance of life on an organic farm.

Question Marks

1. *Are you looking for a different life — or a different job?*

Farming is a life, not a job. It will consume most of your energy and place numerous and diverse demands on you. Don't expect much in the way of leisure time or traditional vacations.

2. *Are you a highly detail-oriented person? Do you always want to get things just right?*

Being too meticulous and detail oriented can slow you down. A few small weeds around a head of lettuce are not necessarily a problem. Not every bunch of basil has to be exactly the same size. If you're ill at ease with all the variability in nature, you might better focus your energies on some aspect of farming that does not call for a lot of hands-on work with plants — like sales and promotion, or managing community-supported agriculture (CSA) accounts.

3. *Are you a generalist or a specialist? Do you mind having lots of different things to do and often not quite enough time to get all of them done?*

Running a small diversified farm requires that you wear many different hats and feel comfortable in most of them. If a very full plate makes you nervous, operating your own farm may not be the best choice. Working on someone else's place, where you have a defined area of responsibility, could make more sense.

4. *Do you feel compelled to fix everything that malfunctions or breaks down? Do you think not doing so constitutes personal failure?*

Mechanical and fix-it-up skills are great to have, but a farmer must be careful not to spend too much time solving mechanical or other specialized technical problems. There's all that farming to do. Sometimes it's wiser and more cost-effective to hire a professional who has the experience and the right tools to repair what is broken more quickly than you can. This way you get to spend your time planting what needs to be planted or harvesting what needs to be harvested — which is how you make your money. That said, however, many larger farms with plenty of tractors and equipment are very happy to have a full-time maintenance and repair person on board.

5. *Are you an optimist or a realist? When you embark on a new undertaking, are you willing to entertain the worst-case scenario as well as the best?*

Optimism and realism are not mutually exclusive. Most farmers hope for good growing conditions, bountiful crops, and a booming market, but we understand that these three will not unite in flying colors for us every season. Envisioning only the best outcome just sets you up for disappointment. Sometimes you're going to get clobbered. Better to know this in advance and be ready with a backup plan.

6. *Are you willing to be an employer? Can you manage other people? Can you delegate responsibility?*

Many farmers, being self-starters and independent-minded individuals, may not excel in this department. Managing other people is a tricky business and can take some time to master. Most employees will not want to work as hard as you do, and almost all of them will want some time off — even if you don't take any for yourself.

A good approach is to try wearing your employees' shoes every now and then — metaphorically speaking. Balance your needs against theirs. Try to imagine how they feel about what you are asking them to do. Be organized, be fair, be firm, and never go back on your word. Admit error when the error is yours.

If you don't like the idea of being an employer, you can, of course, work alone or with a partner. This is a reasonable and perhaps correct approach in the beginning, but for most farmers, especially those who would like to see some growth in their business, working alone is not practical over the long haul. We'll talk more about a workforce in chapter 23.

7. How are your social skills? Can you see yourself selling vegetables to the public at a farm stand or a farmers' market or through a CSA, or do you shy away from social contact with people you don't know?

These days, most successful small farmers have learned to be direct marketers as well as farmers. The retail dollar is highly desirable. Selling wholesale or to an intermediary is seldom a good path to take. You'll have a leg up in your marketing if you enjoy interacting with other people and can show them that you genuinely appreciate their business. If marketing is not your strong suit, you might want to stick to growing and find someone to do the marketing for you.

8. Do you really, really want to be a farmer, and do you understand what it entails?

The more you want to be a farmer and the better you understand what you're getting into, the greater your chance of success.

Concrete Steps to Take

The questions posed above are mostly concerned with your skills, your interests, your likes and dislikes, and your individual disposition. Let's assume you've mulled them over and are not deterred. Now it's time to consider some of the more practical steps you can take to educate and prepare yourself for the farming life.

Attain a Formal Education

You could enroll in an agricultural program at one of the land grant universities that every state has, but this is probably not a good idea. These schools are more geared toward the large-scale agribusiness models that have prevailed in North America for the last 50 years or more. They'll teach you about management practices on farms that rely on chemical pesticides and fertilizer and grow large acreages of just one or two crops. No doubt, you can acquire some useful and specialized information at the land grant schools, but there are easier and cheaper ways to learn how to become an independent, diversified organic farmer.

There are now a few schools that focus on the small-scale, sustainable model. These might be a better bet than the land grant colleges, if you have your heart set on formal education and the money to spend on it. One of my former interns is looking into an apprenticeship program at the Center for Agroecology and Sustainable Food Systems, University of California, Santa Cruz. The cost is several thousands of dollars per session. Housing is not provided.

Fortunately, there are other approaches that may be more effective and definitely cost less. In fact, most of these "other" approaches will cost very little, or nothing at all, or will even pay you while you're learning. Following are some ways to immerse yourself in the world of small farms.

Get to Know a Few Farmers

Start shopping at farm stands and farmers' markets, if you haven't already, and try to get to know the farmers you would like to emulate. Don't attempt to engage them in conversation when they have a line of customers, are packing up at the end of a long day, or are obviously busy in some other way. At our market, there's often a lull in mid-afternoon, between about 2:00 and 4:00 P.M. That's a good time for someone to come by with questions. Another good time is when it's raining heavily and business has almost come to a standstill. Or simply ask the farmer when would be a good time to ask a few questions.

Formulate some questions. Approach this type of research (and it really is research) as you might a job interview. Tell the farmer that you're really interested in small-scale, organic agriculture and are thinking about becoming a farmer. You might then ask a few questions, such as:

- Where's your farm?

- How big is it?

- How long have you been farming?

- Do you find this to be a good market?

- Do you have other marketing outlets?

- What type of equipment do you use?

- Do you have a crew?

- What crops make the most money for you?

Compliment. In the midst of your questions, let the farmer know that you really appreciate what he or she does and that the food has tasted great — a little flattery seldom goes amiss.

Study your subject first. I find myself more willing to talk seriously with people who have already done their homework — for example, if they have read a few of the current relevant books (such as those of Eliot Coleman, Joel Salatin, Michael Pollan) or seen some of the movies that deal with food issues, sustainability, and small-farm life (*The Future of Food*; *Food, Inc.*; *The Real Dirt on Farmer John*; etc.). I'm less impressed when I sense that I'm the first stop on the road to becoming a farmer and that even the most basic concepts need explanation. This may seem a little ungenerous, but there's a reason for it: Talking with small farmers has lately become a popular, even fashionable, pastime, but most of us have only so much talking energy to expend in a given day. Often, I don't have the time for extended lightweight conversation and basic education.

Sign Up with Your Local Organic Farming Organization

Become a member of your state or local organic farming and certification organization. These institutions usually have educational and beginning farmer programs that offer technical advice and assistance. Most have monthly or quarterly newsletters, in either paper or electronic form, that will keep you abreast of all matters organic in your state or region. You'll read articles written by farmers explaining their practices, their successes, and their failures. You'll learn about upcoming field days on successful farms and workshops on subjects ranging from growing vegetables, herbs, grains, fruits, and berries to artisanal cheese and wine making to raising small and large livestock.

Attend Organic Farming Conferences

Many organic farming organizations hold annual conferences. Some of the better-known ones are the Pennsylvania Association for Sustainable Agriculture (PASA), the Midwest Organic and Sustainable Education Service (MOSES), the Ecological Farming Association (EcoFarm) in California, the Ohio Ecological Food & Farm Association (OEFFA), the Maine Organic Farmers and Gardeners Association (MOFGA), and the Northeast Organic Farming Association (NOFA). The conferences these organizations hold provide a treasure trove of information and the opportunity to network with like-minded individuals. A thousand or more people routinely attend these 2-, 3-, and even 4-day affairs. They usually have excellent keynote speakers and a tremendous selection of workshops, the majority of which are conducted by experienced farmers.

The NOFA–NY conference I attended in Saratoga Springs, New York, last year had more than 80 separate workshops to choose from over a 4-day period. Deciding which ones to go to was a challenge. I opted for a half-day session on growing in high tunnels, and several shorter workshops on such topics as choosing potato varieties; managing summer cover crops, beneficial insects, and biological pest controls; and growing brassicas in the Northeast.

At organic farming conferences the emphasis is on sharing practical, "how-to" knowledge — for beginners and experienced farmers alike. The price for attending is usually reasonable, the atmosphere uplifting and inspirational, and the meals extremely tasty — they're made with local, organic ingredients, many of them donated by organic farmers in the region. Any new or aspiring farmer (and experienced farmers, too) should make it his or her business to attend at least some of these excellent conferences.

Once you start getting involved, you'll discover that the organic farming community is a good community to be a part of. It is made up of men and women who genuinely believe in and strive for a saner and more sustainable food system, and who understand that their goals will be more readily achieved when they share information and help one another. Who can argue with that?

Subscribe to Small-Farm Publications

There are several magazines and other publications for *small* farmers who function outside the mainstream of conventional American agriculture. *Growing for Market* is one of my favorites. There's barely an issue of this monthly, or almost-monthly, publication that doesn't inform me of ways to improve my own farm operation or sow seeds for some future project. All the articles are written by farmers or nonfarmers who are engaged in their subject matter. The information provided is practical, detailed, and easy to follow.

Other useful publications include *The Natural Farmer, Small Farmer's Journal* (especially good for farmers who use animal traction), and *Acres USA*. These, like *Growing for Market*, provide practical information for small farmers wishing to follow an organic or sustainable path. Also, don't overlook local farming publications that can be found in most rural areas. These are good for finding used equipment and supplies, making contact with other farmers, and keeping abreast of what's going on in your neck of the woods. In Pennsylvania, *Lancaster Farming* is a good example.

Research Information Online

Cornell University's Northeast Beginning Farmers Project (see Resources) provides online courses, publications, videos, and advice tailored to those who are new to farming. Many other state agriculture departments and land grant colleges are beginning to lend a hand as well, as they become increasingly aware that small-scale and sustainable farming is experiencing a remarkable renaissance.

Other grassroots organizations that might help you along the way include Farm Hack–National Young Farmers' Coalition, the Greenhorns, and beginningfarmers.org.

Visit and Work on Organic Farms

This is a great way to get your feet wet. Contact farms that are of interest to you, and ask if they would be willing to put you up for a week or two in exchange for some help. Those who have available housing and are short on labor will often say yes. If you find that you're having a good time and the arrangement is mutually beneficial, you might stay longer. Many small market farms and CSAs take on seasonal apprentices every year.

Farm Lists. The New England Small Farm Institute in Belchertown, Massachusetts, sponsors a regional Farm Apprenticeship Placement Service coordinated by North East Workers on Organic Farms (NEWOOF), which publishes an annual list of farms seeking both paid and unpaid interns. Many young people have gotten their start in farming through the New England Small Farm Institute. Many organic certification organizations also publish lists of farms offering apprenticeships.

ATTRA. Appropriate Technology Transfer for Rural Areas — admittedly a bit of a mouthful — is a federally funded organization that provides a wide range of information for small and sustainable farmers. ATTRA, also known as the National Sustainable Agriculture Information Service (NSAIS), is managed by the National Center for Appropriate Technology (NCAT). On its ATTRA website, under "Learning Opportunities," you'll find an extensive list of farms in the United States that are looking for help. Each listing is accompanied by a description of the farm, what it has to offer, and what is expected in return. In most cases the farms will take on inexperienced workers and will provide lodging and a stipend in exchange for an agreed-on amount of labor. If this approach appeals to you, it's a good idea to contact several farms and narrow down your search to those that will provide the best overall environment and learning experience for you. Taking the time to visit farms before committing to working on them is a smart thing to do. This way you shouldn't be in for any big surprises.

WWOOF. The acronym WWOOF used to stand for Willing Workers on Organic Farms. More recently, it has morphed into Worldwide Opportunities on Organic Farms, but it still serves the same purpose. WWOOF is a loose network of national organizations that list farmers in many different parts of the world who are willing to take on workers, for short or longer periods of time. In most cases, no money changes hands. The host farmer provides food and lodging in exchange for help around the farm. The WWOOF list caters to itinerant young people who want to combine travel and farm work along with a nice dose of adventure.

If you have the time, spending an entire season on a suitable farm is the best way to get some serious experience. You'll learn a lot about how that farm operates and get a sense of whether you are suited to the farming life. Ten years ago, college students or recent graduates made up the majority of applicants for farm internships. These days, many more people in their late 20s, 30s, and even 40s and beyond, with diverse backgrounds, are discovering that working with an experienced grower with a good track record is a great way to immerse themselves in farming before making the final step to beginning their own operation. Some choose to work with more than one farmer and devote 2 or 3 years to this pursuit.

CHAPTER 1 RECAP

Twenty Points to Ponder

DEAL MAKERS

1. Do you like being your own boss?
2. Do you like watching things grow? Are you fascinated by nature's ways?
3. Do you enjoy working with nature, or are you happier when the natural world is in an ordered state and under control?
4. Do you enjoy working outside? Can you tolerate different kinds of weather?
5. Are you in good shape? Do you enjoy physical work? Do you know how to pace yourself and use your body efficiently?
6. Can you balance a checkbook, keep accounts, and avoid spending more than you have?
7. If you're starting your own farm, do you have enough reserves of capital to weather a few lean years? Can you keep a part-time job that will bring in some income while you're learning the ropes?

DEAL BREAKERS

1. Are you reluctant to try something when you're not sure of the outcome? Do you feel crushed when a project you had high hopes for turns out badly?
2. Do you hate getting dirt or grease on your hands and clothes? Do sweat, dust, oil, and — occasionally — a little blood make you uncomfortable?
3. Are you reluctant to admit that you've made a mistake?
4. Are you short on patience? (Many of us are, though we might like to think otherwise.)
5. Do you dread being bit by a mosquito or stung by a bee? Are you afraid of snakes?

QUESTION MARKS

1. Are you looking for a different life — or a different job?
2. Are you a highly detail-oriented person? Do you always want to get things just right?
3. Are you a generalist or a specialist? Do you mind having lots of different things to do and often not quite enough time to get all of them done?
4. Do you feel compelled to fix everything that malfunctions or breaks down? Do you think not doing so constitutes personal failure?
5. Are you an optimist or a realist? When you embark on a new undertaking, are you willing to entertain the worst-case scenario as well as the best?
6. Are you willing to be an employer? Can you manage other people? Can you delegate responsibility?
7. How are your social skills? Can you see yourself selling vegetables to the public at a farm stand or a farmers' market or through a CSA, or do you shy away from social contact with people you don't know?
8. Do you really, really want to be a farmer, and do you understand what it entails?

Looking for a Place of Your Own

YOU'VE DONE YOUR HOMEWORK: You've read a few relevant books, become a member of your state's organic certification organization, attended at least one organic farming conference, subscribed to two or three small-farm publications, talked with farmers in your area, put your hands in the soil and tried at least some backyard gardening, visited and perhaps worked on one or two successful farms that are practicing sustainable and organic methods, allowed the idea of farming to gestate in your mind for a suitable amount of time (preferably at least a year or two), and considered whether you are temperamentally suited to the undertaking. Now you're ready to act and eager to get a place of your own.

Or maybe you can't check off all of the above, just some of them, but you want to farm anyway. Fair enough. Give it a try. Let's assume you have enough money to actually buy a farm — that is, to put down a deposit and take out a mortgage. Here are some criteria to take into account as you set out to find a place of your own.

Considerations When Looking for Good Land

There are many things to consider when you set out to buy a farm, or lease one, for that matter. In this section, we'll look at what I think are the most important. The more farms you visit, the better you'll know what you're looking for. Give serious thought to the subject and make your own list of criteria. Here's my checklist to get you started.

A Place Where You'll Be Happy

First and foremost, look for a place that appeals to you, a place where you'll be happy to wake up in the morning. One of the main goals of this book is to help you make a decent living as a farmer. That said, not many people take up organic farming with the sole objective of making money. There are easier ways to do that. Most of us farm because it's a life we believe in and want to live. But where we choose to put down roots and live out our dream does make a difference. If you like flat land and wide expanses of open space, that's what you should look for. If you feel good when you're on top of a hill with a good view, look for high ground (but be prepared for wind). If you prefer privacy, you'll probably be happier with a place that is surrounded by trees and set back off the road.

Also consider the broader area in which you are concentrating your search for land. Is the climate suitable for the type of farming you plan to do? How long is the growing season? Does the area contain a community you'd be glad to be part of? What about schools, cultural facilities, and opportunities for outdoor activity? How far would you have to travel to visit important family members? How close would you like to be to a thriving urban center so that you can spend a night on the town once in a while? Or do you prefer the deep boondocks and the quiet life?

No doubt you'll have more than a few criteria of your own, but be willing to compromise a little. It's unlikely you'll find everything you want on one piece of land, in one part of the country. Besides, perfection is overrated.

Good Soil

You can create a decent garden almost anywhere by importing topsoil and compost and perhaps building raised beds to overcome a base soil that is too shallow or has poor drainage. But that's a garden. If you plan to produce crops for market on more than a very small scale, you need to have productive soil in place and at least a few acres of it.

Local Cooperative Extension agents or Soil and Water Conservation District offices can provide soil maps of a particular property and a description of the soils contained therein and, more importantly, their suitability for different uses. These days, you can get all this stuff online as well, by using the USDA's Web Soil Survey. Class I and Class II soils are usually the best for growing crops, but it wouldn't hurt to have some less-ideal soils that are suitable for woods and pastureland.

Prior or current land use is often an indicator of a soil's potential. Fields that have been used for grazing livestock may not be suitable for growing crops. Neither is forested land likely to be. But open fields that are already growing hay, corn, or other crops should have good potential. These are important things to find out.

If you prefer not to rely only on soil maps and descriptions, you might try visiting a few of the nearby neighbors and asking them what the land you're considering has been used for over as many years as they can remember. Longtime neighbors may be able to give an on-the-ground sense of the land's potential or alert you to serious constraints. Before we put down a deposit on our farm, I walked around the fields with a shovel and dug down a foot or two in several spots to get a good look at the soil I was thinking about buying.

The Lay of the Land

After you've checked out the soil maps and are satisfied that the land you're looking at is suitable for growing vegetables, it would be wise to consider three other land-related variables: slope, aspect, and exposure to wind.

Slope

The slope of a field is expressed as a percentage. It is determined by dividing the vertical rise by the horizontal distance and then multiplying by 100. Thus, a field with a vertical rise of 12 feet over a horizontal distance of 150 feet has a slope of 8 percent.

$(12 \div 150) \times 100 = 8\%$

The USDA uses the letters A, B, C, D, and so on following the name of a soil to denote slope, as follows:

A = 0 to 3 percent (nearly level)
B = 3 to 8 percent (gentle slope)
C = 8 to 15 percent (moderate slope)
D = 15 to 25 percent (steep slope)

Thus, the soil symbol MdB describes a Mardin soil (Md) (see box on page 197) with a slope from 3 to 8 percent (B). If no capital letter follows the abbreviation of the soil name, the land being mapped is level. See chapter 9 to find out how to identify your soil type, degree of slope, and other useful descriptive information.

Nearly level ground (0- to 3-percent slope) is generally easier to farm and less prone to erosion than sloping land.

Gently sloping fields (3- to 8-percent slope) can have good agricultural potential but need to be managed conservatively to minimize soil erosion. To manage such land, you can reduce tillage, employ permanent planting beds, and liberally use cover crops.

Moderately sloping fields (9- to 15-percent slope) need to be treated with much more care. Using tractors on these slopes can be hazardous. Both tractor and tractor implements are likely to drift in the direction of the slope, making it a challenge to maintain parallel rows. The potential for erosion is higher on moderate slopes, especially when the soil is freshly tilled or left without cover during winter. None of this is to say that moderately sloping fields cannot be put to good use. They might lend themselves to strip cropping (alternating strips of annual and perennial crops) or terracing — two time-tested practices for controlling erosion. They might also be suitable for an orchard, berry bushes, Christmas trees, or grazing livestock.

Steep slopes (more than 15-percent slope) are positively dangerous if you're using any sort of heavy equipment or machinery, and, if tilled, are highly erodible. They should be left with permanent cover whenever possible. Suitable uses for such land include fruit and nut trees, berry bushes, Christmas trees, and livestock grazing.

Aspect

"Aspect" refers to the direction in which a slope faces. In the Northern Hemisphere, south-facing slopes receive earlier sun and more of it than north-facing slopes. They warm up faster and generally get spring-planted crops off to a better start. Heat-loving crops tend to thrive on south-facing slopes; crops that are better suited to cooler temperatures will probably be happier on a north-facing slope. Most growers like to have at least some level or south-facing land.

Wind

Excessive wind is stressful to plants. It slows growth and dries them out. Plants exposed to a lot of wind require more irrigation. Watch out for valleys and other landforms that might act as natural wind funnels. Trees planted as windbreaks can help a lot.

Peregrine Farm

Owners: Alex and Betsy Hitt
Location: Graham, North Carolina
Year Started: 1982
Farm Size: 26 acres, with 5 acres in production
Crops: Mixed vegetables, small fruits, cut flowers, pastured turkeys

A LEX HITT STUDIED SOILS AND HORTICULTURE in college and might have gone to work for the Soil Conservation Service (now renamed the Natural Resources Conservation Service [NRCS]) but decided to become a farmer instead. He and his wife Betsy, who has been farming with him since 1982, have no regrets. They've had the freedom to chart their own course and build a viable business while working outdoors and with plants, which is what they both love to do.

The 5 acres under cultivation at Peregrine Farm are divided more or less evenly between mixed vegetables and cut flowers, with some blueberries and other perennials on the side. Eight years ago, the Hitts began adding about 100 pastured turkeys annually to their farm — more for the birds' bug-eating and manure-spreading skills than their talent for gracing the Thanksgiving table.

Alex and Betsy produce 250,000 transplants each year in their passive solar greenhouse. On cold nights, they use heat mats to keep germinating flats warm. The Hitts also do a fair amount of in-ground growing under plastic. In 2004, they purchased two ¼-acre Haygrove tunnels, each of which has four bays, that are 24 feet wide and 100 feet long. These are used primarily for flowers and tomatoes, rather than season extension. The farm also has six 16- by 48-foot movable tunnels that slide on wooden rails. Each year, these smaller tunnels are moved with either a tractor or the muscle power of four or five sturdy humans. The moving tunnels fit nicely into the farm's rotation plan and make it easy to plant cover crops.

There is a minimal supply of equipment at Peregrine Farm. The main items are a 30-horsepower two-wheel-drive tractor, a tractor-mounted rototiller, a spring-tooth cultivator, a disc, and a flail mower.

Alex and Betsy sell most of what they grow at two nearby farmers' markets. Their season runs from mid-March to early October, and they return a few times just before Thanksgiving and Christmas. The turkeys, all of which are sold in advance, go out two days before Thanksgiving.

The Hitts hire two part-time workers (32 hours/week) during their growing season. They let their organic certification lapse some years ago, after they stopped selling wholesale, but they still follow organic methods.

Land Acquisition

Fresh out of college, Alex and Betsy wanted to farm but didn't have enough money to buy land. They came up with a creative solution to their problem: They became incorporated, wrote up a prospectus, and went out looking for investors to fund their farm purchase and start-up costs. Their plan worked. Within a year, they were able to assemble 17 shareholders, each of whom was willing to invest between $3,000 and $10,000 in a future farm. Some of the shareholders were friends and family. Others were simply interested in what Alex and Betsy were trying to do. They raised $80,000, which was enough to buy land and get started.

From the shareholders' point of view, the investment was relatively safe. The land itself was held as collateral. The contract they signed allowed them to ask for their money back after 5 years. Were the farming venture to fail, the land and house would be sold and shareholders would be able to retrieve most or all of their investment. The farm did not fail, though it didn't become profitable for several years. After 12 years, all 17 shareholders were paid back, with a modest gain on their investment — the contract contained a buy-back clause, enabling Alex and Betsy to purchase any investor's share at any time. During the early years of unprofitability, shareholders were also able to write off portions of the money they had invested on their tax returns.

In describing their uncommon land acquisition strategy, Alex stressed the importance of a clear contractual agreement that protected the interests of all parties, and a good lawyer to draw it up. He thinks that none of the 17 investors who helped him and Betsy were dissatisfied with the deal.

Plenty of Water

To grow well, most plants need water and a steady supply of it. Unless you plan to grow just hay, or perhaps grains to feed cows or horses, you don't want to leave yourself at the mercy of rainfall. Almost all farms are dependent on wells for household water. If the well on your farm starts running low after one member of the family has taken a long shower, then it's unlikely you'll have enough water to irrigate crops. A good well that can provide at least 10 to 20 gallons of water per minute, all day long, is an important asset on a farm.

If you don't have a generous well, you will need some other source of uncontaminated water. A small river, or perennial stream, running through your property might fit the bill. Or a pond. If there's no pond, you can always dig one, but first do some research into the underlying geology — you need to be sure the pond you dig can hold water. The best ponds are fed by springs or a creek. A pond that simply holds rainwater is less desirable, unless it is located at the bottom of a long slope where it receives plenty of runoff. (These days, it is recommended that all sur-

The Value of Land

Your planning and production capacity will depend to a large extent on the amount of land you have available. More tillable acres will give you the option to grow a wider variety of crops and more of each variety. More land will make it easier to leave perennial cover crops in place and to leave land fallow for a season or more. It will allow you to separate crops more effectively in your rotation plan. Greater separation of the same crop (or crops in the same family) from one year to the next will reduce pest and disease problems.

Here's an example of why separation pays off: In many parts of North America, adult Colorado potato beetles (a major pest of potatoes) can overwinter in the soil. If you plant potatoes adjacent to where you had them the year before but only 10 feet away, the beetles won't have far to go to find something they like to eat. If, on the other hand, you can separate the two sequential plantings by several hundred feet, your second-year potatoes are likely to be spared early damage.

face water be treated before being applied to the aboveground portion of growing plants. More on this in chapter 8.)

Watch out for floodplains. Even land that floods only once every 10 or 20 years can cause big trouble, especially now that our planet is experiencing more frequent and intense extremes of weather. Bottomland or alluvial soil along the banks of streams and rivers can be very fertile but is often susceptible to seasonal or periodic flooding. That's not to say this type of land should be avoided — traditionally, it has been prized by vegetable growers — but it would be nice to have some high ground as well. You can order a floodplain map for the property you are considering through the Federal Emergency Management Agency (FEMA) Map Service Center website. You might also want to hear what local residents have to say on the subject of flooding.

Enough Land

When my wife and I started looking for a farm, we thought that 5 or 10 acres would be enough, but along the way I got a taste for land and decided that more would be better. We started looking farther afield, where property values were not so high. We ended up with an old dairy farm, 88 acres, with 25 acres of tillable ground and the balance in pasture and woods. I've never felt it was too big.

The more land you have, the easier it is to rotate crops, create buffer zones and wind breaks, dig ponds, and leave land fallow. For most small farmers, 10 to 20 acres of fertile land should be enough, but it wouldn't hurt to have more. More land than you need in the beginning will allow room for expansion in the future. You may have a predetermined scale of operation in mind, but its economic viability has not been verified. After several years of planting 5 or 10 acres and just scraping by financially, many growers have found they could increase their bottom lines dramatically by adding another 5 acres or so. Once you have honed your growing skills, have the right equipment in place, and are confident in your markets, modest expansion can be a natural step.

If you want an ecologically diverse farm, plan on its having some pasture for livestock and forested areas for wildlife. A well-managed woodlot can provide you with an ongoing supply of firewood and lumber, as well as a nice buffer between you and your neighbors.

It should also be noted that more than a few growers these days are producing impressive amounts of food and surviving financially on just

2 or 3 acres. At this scale, expenses can be kept to a minimum. Paid help, tractors, and other large equipment are probably not necessary. By selecting high-value crops that require less space and by using intensive techniques, these new growers are able to make do on relatively small plots of land.

Access to Markets

A farmer needs to be a good businessperson. Get used to that idea. After all, a farm is a small business — probably, in fact, the original model of a small business, with perhaps the one exception of the oldest profession known to humankind. Every successful businessperson knows you need a ready market for what you produce or the services you provide. With the growing popularity of local food, small farmers can benefit greatly by being close to urban centers with multitudes of potential customers ready to gobble up what they have to offer.

Unfortunately, arable land that's close to urban centers is scarce these days and tends to be very expensive. Suburban sprawl has taken a heavy toll on farmland over the last 30 years. If you can afford a farm within 50 or 75 miles of a major city, you're doing well. I know a lot of farmers who travel greater distances to market their wares. Others focus on smaller markets in small or midsize cities, often going to three or four a week.

Many diversified farms market through CSAs, selling shares of their produce to local residents. Others sell to upscale restaurants that are eager to buy fresh, local food and are willing to pay what a farmer needs to earn. More recently, local growers' cooperatives that take care of marketing and distribution are proving fruitful. There are many different models, and you don't need to be tied to any single one. When buying a farm, however, consider who your potential customers will be and how far away they are. Cheap land in a remote region, far from a good marketing opportunity, may not end up being such a bargain. You could be forced to wholesale what you produce, and selling wholesale means you will receive a lot less money.

Existing Buildings and Infrastructure

It's almost always cheaper to buy an existing farm with a house, barn, and other outbuildings than to start from scratch on a raw piece of land. An existing farm will have a well, a septic system, a driveway, and electric and phone hookup. All of these can cost big bucks if you have to build,

The Case of the Small Dairy Farmer in the United States

Dairy farmers are being paid so little for their milk these days (often less than it costs them to produce it) that many are going out of business. It's a tragic and almost criminal situation, since everyone else along the milk chain, from the truckers, to the processing plants, to the cheese makers, to the supermarkets, are making their share. Only the farmer gets the short end of the stick, or no stick at all. The hard fact is that many dairy farmers are being forced to sell their farms at bargain prices. This is especially true in the post-subprime mortgage era, now that developers are no longer knocking on their doors.

develop, or install them yourself, and doing so can use up a lot of time. It's a trade-off. Raw land may enable you to create an infrastructure that better suits your needs, but you will have to pay a lot more to do this, unless you are extraordinarily multitalented.

In the Northeast and in other parts of the country, dairy farms that are up for sale are worth considering. They usually have all the buildings and infrastructure you will ever need, with the exception of such things as greenhouses and packing sheds.

Your House

A well-built house in a good state of repair is, of course, highly desirable but not as critical to your success as the criteria already discussed. You can always restore, renovate, or enlarge an old house as time and resources become available. It's a lot more difficult to transform poor soil into good soil or find enough water where little exists.

That said, there are specific questions to ask when looking at your future home:

• Is it well insulated?

• Is the foundation solid?

• Is the roof in good shape?

- Are the wiring and plumbing safe and up to code?

- Is the septic system adequate?

- Are there signs of termites or wood rot?

- Do you like the place?

It's a good idea to hire an engineer or professional home inspector to take a close look and give you a written assessment of the condition of your possible future home. Many mortgage providers will require an engineer's report, and it may help you to negotiate a better price.

Affordability

This is a tough one these days. Land that's close to urban centers, where you can sell what you grow, is usually priced for development, not for farmers. The recent tumble in the housing market has brought land prices down a bit, but they are still too high for many would-be farmers. Sometimes family members or friends are willing to put up the capital

Surprises

> My Story

A MAJOR SURPRISE we encountered on our farm was the location of the water tank, which functions as a sort of pressurized halfway house for water coming up from the well. The water tank was in the barn instead of the house basement, which is where water tanks normally are. Just before the onset of our first winter, we were warned by a neighboring farmer that without the body heat generated by cows in the barn, our water tank would certainly freeze — cutting off our supply of water to the house and splitting pipes in the barn. To correct this situation, in late November, as the ground began to freeze, we had to hire a backhoe operator and two plumbers and spend a couple of thousand dollars. It would have helped to know this in advance.

to get a young farmer started, with the understanding that they will be repaid as the business prospers. In such cases the farm can be used as collateral against loans.

Look for land on which the development rights have been sold or donated to the state, or to a land trust or conservation organization. If this land is for sale, it should cost less because it can never be developed. If it's not for sale, it may be available to a farmer for long-term lease. Seek out retiring farmers and landowners, especially those with environmental leanings, who own property that is not in use or is used only lightly (for a crop such as hay). These folks may be willing to enter into a long-term lease arrangement with a young farmer who commits to using organic or sustainable practices. Such an arrangement will often reduce the landowner's property tax bill.

There are a growing number of organizations that will help match up future and retiring farmers in lots of creative ways. The International Farm Transition Network (see Resources) holds seminars and publishes materials with the goal of fostering the next generation of farmers and assisting those who are making the transition away from farming. See Resources for a list of Land Link and Farm Link programs that participate in their network.

Previous Land Use

If you plan to farm organically, you might want to avoid land that has received chemical fertilizer, been sprayed with pesticides, or been subjected to other polluting practices. Current organic certification standards require three chemical-free years before food coming off the land can be called organic, but this doesn't necessarily have to be a deal breaker. Many future organic farmers go through a transition period, during which they introduce organic methods.

Watch out for old orchards, many of which have been sprayed with arsenic over many years. Arsenic buildup in a soil can reach toxic levels. If you're looking at an old orchard, have the soil tested for toxicity.

Neighbors

It's hard to have much control over who your neighbors are, or will be, but there are certain types of neighbors you would be wise to steer clear of, such as those who operate landfills or gravel pits or polluting industries. Then, there are the regular suburban folks who might object to the smell of manure or a rooster crowing at 6:00 AM. It's also a good idea to avoid properties that underlie existing or proposed transmission lines or areas where natural gas drilling may occur, and you may prefer not to be next to a large conventional farm where pesticide runoff or spray drift could be a problem. Local zoning codes, tax maps, and ownership records can provide useful information of this sort. Google Earth will give you a close look at any prospective property and its surroundings.

There's a lot to think about when buying a farm, and it's not something to rush into. The more farms you see, the better you'll understand what you're looking for. At least wanting to be a farmer these days is not as wacky an idea as it used to be. The public's appetite for fresh, local food has never been stronger. Small, sustainable farms, with a diversity of crops and possibly animals, are more likely to succeed today than in a long time. For those who are drawn to the agrarian life, buying or leasing a farm and starting a business can be a wonderful adventure. It'll keep you busy, that's for sure. But the work is varied and satisfying, and it will put you in touch with timeless forces that can breathe vitality into your life.

CHAPTER 2 RECAP

Eleven Criteria to Consider When Looking for Land

1. A place where you'll be happy
2. Good soil
3. The lay of the land
4. Plenty of water
5. Enough land
6. Access to markets
7. Existing buildings and infrastructure
8. Your house
9. Affordability
10. Previous land use
11. Neighbors

Chapter 3

....

The Farm Plan

S MALL FARMERS ARE OPTIMISTS AT HEART. The coming year is going to be better for us than the one we've just had — at least that's what we'd like to believe. Of course, this does not always turn out to be the case. Just wanting something to happen doesn't necessarily make it so, and most seasoned farmers know that reality carries a big stick.

Missteps and miscalculations will inevitably be made. Equipment will break down at critical junctures. Nature and ill chance will deal us a bad hand now and then. What's important is that we learn from our mistakes rather than repeat them, have backup plans in case of equipment failure, and devise systems for dodging Mother Nature's curveballs when they come our way.

To my mind, the challenges and variability of farming are what make it eternally interesting. Setbacks and adversity, though seldom enjoyed, are not things to be mortally afraid of. In the long run, they make us wiser and better farmers, as long as we are able to absorb the lessons they teach.

For most new farmers, there's going to be a fairly steep learning curve, which is why it's better not to count on major success in your first couple of years. No one becomes a doctor, dentist, mechanic, or carpenter overnight. Acquiring the skills you need to run a commercial vegetable operation can be just as daunting. If you're new to farming, give yourself a little time to learn the trade. And while you're learning, it wouldn't hurt to have a part-time job or some other source of income to keep you solvent and reduce stress. But once you've jumped in and are on the learning curve, the idea is to move along it as efficiently as possible, assuming it's where you still want to be.

Details, details, details. On any diversified, organic farm growing multiple crops for daily or weekly markets, there will be a lot of details to remember. Things such as:

• Varieties of each vegetable or herb planted

- Amounts planted, along with locations and dates
- Spacing between plants and rows
- Amendments applied
- Special actions taken
- Days to harvest for each crop
- Successes, failures, and problems encountered
- Dates and amounts of harvest
- Amounts sold and prices received

If your brain is anything like mine, you won't be able to keep all this information neatly filed inside your head over the course of a long and eventful season. This is why you need to develop a comprehensive farm plan and good systems for recording relevant data. Keeping track of what happens will enable you to forge ahead, maximize your successes, and avoid past mistakes.

True, most farmers would prefer to spend their time in the field getting tangible work done. Dealing with numbers and records is rarely what attracted us to the agrarian life. Nonetheless, to succeed, we have to buckle down and sharpen our planning, recording, and accounting skills. These are just a few of the many hats a farmer has to wear.

Planning and record keeping. The remainder of this chapter will present a planning and record-keeping system that has evolved over the course of more than 20 years and become the backbone of our farm operation. It is not a static system. Every year it gets tweaked and improved a little bit. Because it is tailored to our farm, our resources, and our markets, it should be viewed as one farmer's approach, rather than as a single blueprint for success.

Not all parts of the system presented below will be pertinent to all growers, especially those with different approaches to marketing. But there are some big things all of us have in common. We plant crops, we harvest them, and we sell them. And if we're smart, we'll keep track of the details along the way. Doing so will speed up the learning process and give any farmer more control over his or her future.

Maps of Farm and Fields

They say a picture is worth a thousand words, and this is very often the case, but when you're dealing with land and physical resources, a map may be worth even more. Here, we'll consider some of the maps that are essential to a farm plan:

- Farm map (for the big picture)

- Soils map (if you have more than one soil type)

- Field map(s)

- Planting beds map(s)

Farm Map

Every farmer should have a basic farm map showing property boundaries and major features such as wooded areas, pasture, cropland, streams, ponds, driveway, physical structures (house, barn, outbuildings), and adjacent land uses (see example on next page). If some of your land is not level, you should note this. Slopes are more likely to erode if left with no plant cover, especially during periods of heavy rain and spring thaw. The orientation of sloping fields is also relevant. As already noted, in the Northern Hemisphere, south-facing slopes receive more sun and warm up faster than north-facing slopes.

An existing survey, shrunk to a manageable size, or even an expanded tax map, could serve as a starting point for a farm map. Better yet, go to the Internet and find your farm and surrounding land on the Web Soil Survey created by the USDA's NRCS. This site will provide an aerial-photo map of your land and acreage calculations for any area you designate. It will identify soil types and enable you to pick out structures such as houses and barns. If printed and perhaps enlarged, it could serve as a base for your farm map, or you might prefer to create your own personalized map from scratch.

Whatever base map you choose to work with, it will be up to you to decide on a suitable scale and to sketch in relevant details. Strive for reasonable accuracy, but there's no reason to go overboard with detail. A good farm map should give you the big picture of your land.

Soils Map

If the land you are farming has more than one soil type, it makes sense to obtain a soils map that can be superimposed on your farm map. As already noted, a soils map, as well as descriptive data for each soil type, can be found on the USDA's Web Soil Survey site. On our farm, because of the erratic movement of the last glacier some 10,000 or 12,000 years ago, we have six different soil types and varying degrees of slope.

For vegetables we mostly use our Mardin gravelly silt loam (symbol, MdB), which is a reasonably well-drained and fertile soil. The poorly

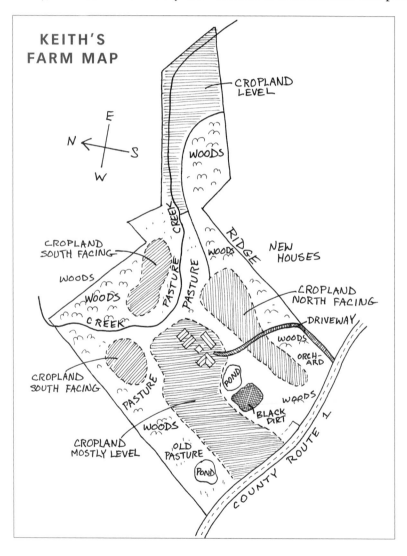

drained Erie gravelly silt loam (symbol, ErB) has served as pasture in years past, and the well-drained Bath-Nassau shaly silt loam (symbol, BnB) is used for Christmas trees and a small orchard.

KEITH'S FARM SOILS MAP

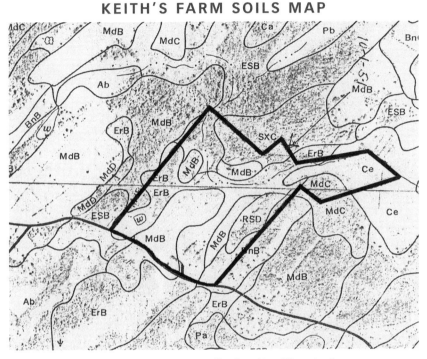

This USDA soils map shows our property (outlined) and its different soils.

Field Map

The third map you need is a more detailed representation of your actual growing area. This might be one large field or several smaller ones divided by hedgerows, access routes, buffer zones, or other natural boundaries. Note all significant features on this field map, especially those that pertain to your farm operation. Measure and record the dimensions of each field, then calculate acreage or square footage, so you can determine exactly how much planting area is available.

In case you're not too good with numbers, an acre contains 43,560 square feet. If you have a field that is 355 feet long and 180 feet wide, you have a total of 63,900 (355 × 180) square feet, or 1.47 acres, which most farmers would be comfortable calling 1.5 acres. Each field should receive an identifying name or number.

KEITH'S FARM FIELD MAP

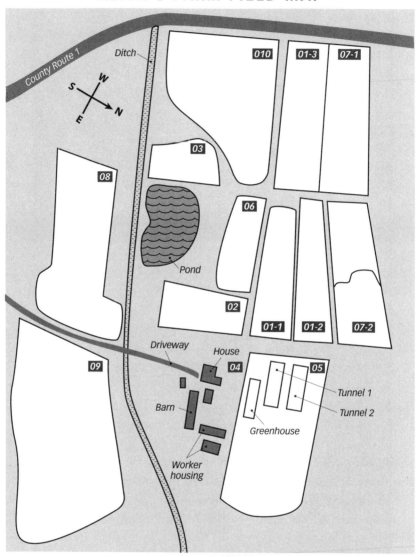

Note: Field 04 includes orchard around house and other buildings.

FIELD SIZES					
Field Number	Acreage	Field Number	Acreage	Field Number	Acreage
01-1	0.6	04	0.5	08	1.5
01-2	0.7	05	1	09	2.0
01-3	1.2	06	0.5	010	1.8
02	0.4	07-1	1		
03	0.5	07-2	1		

Separate maps for each field. Instead of, or in addition to, a single field map, you might decide to create a separate map for each field. Depending on the size and number of fields, this will allow you to accommodate more detail if you feel it is necessary.

The field map, whether it be a single map or a series of individual field maps, will assist you in creating a planting schedule, which I'll discuss later in this chapter. In chapter 11, Crop Rotation, we'll take a further step and consider crop rotation units and the maps used to describe them.

Planting Beds Map

Most small farmers use a bed system for planting some or all of their crops. Beds can be temporary and serve for one season only or they might be semipermanent, staying right where they are for several years or longer. The width of each bed will usually be defined by the size of your tillage equipment, while the bed length will correspond to the width or length of the field. Typically, beds are between 45 and 55 inches wide.

Bed Width

The width of the bed is the critical dimension because it will allow you to accommodate a specific number of rows of vegetables. Individual crops have different space requirements. For example, a 52-inch-wide bed, which is what we use, allows us to accommodate any of the following crop spacings, as illustrated on the following page:

- Four rows of lettuces, with 12-inch spacing between rows

- Three rows of broccoli, with 18 inches between rows

- Two rows of beans, with 36 inches between rows

- One row of winter squash

In our system, the outer rows of each bed are always positioned 8 inches in from the edge of the bed to allow room for the plants to grow. A strip about 24 inches wide is left open between the beds. These strips or pathways allow foot access for activities such as planting, hoeing, trellising, and harvesting. They should be wide enough to enable a tractor's wheels to straddle the beds without touching the plants so that future tractor cultivation is possible. By concentrating all traffic in the strips or pathways between the beds, we avoid compacting soil in the critical growing area.

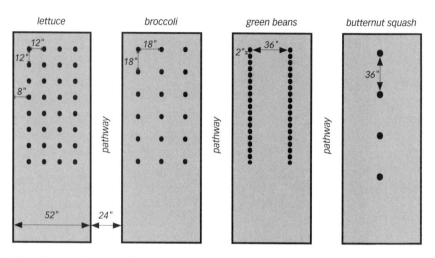

52"-wide beds showing different spacing for lettuces, broccoli, green beans, and butternut squash

Calculate Number of Plants per Bed

The important point here is that the bed is the basic planting unit. If you know its width and its length, you can calculate, with a high degree of accuracy, how much of a given crop will fit into it. This will enable you to estimate yield. In the case of a crop such as lettuces, this is straightforward. A 200-foot-long bed with four rows of lettuces, with 1-foot spacing between rows and 1 foot between plants within each row, will yield a maximum of 800 lettuces:

$(200 \div 1) \times 4 = 800$

The same bed with three rows of cabbages or broccoli plants, with 18 inches between rows and 18 inches between plants within each row, will yield a maximum of 400 cabbages or broccolis:

$(200 \div 1.5) \times 3 = 400$

But we know that things don't always work out quite as planned. For a wide variety of reasons, some plants might be lost. To be on the safe side, let's reduce these numbers by 15 percent to arrive at more realistic yield estimates. Now we can expect to harvest 680 lettuces and 340 cabbages or broccolis.

KEITH'S FARM PLANTING BEDS MAP

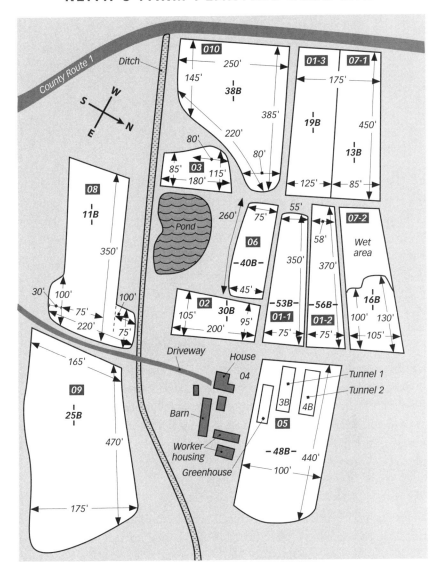

Note: Field 04 includes orchard around house and other buildings.

Field map showing dimensions of fields, approximate number of beds per field, and direction of beds (indicated by dashes on either side of the bed number).

On Center

Growers often use the term "on center" when describing the width of their beds. To use the example given above, 52-inch-wide beds with 24-inch-wide tractor and access paths between them would be referred to as 76 inches on center — that is, 76 inches from the center of one bed to the center of the next.

The next step is to calculate how many beds can fit into each field and enter this data onto a copy of the field map. Once you know the dimensions of your fields and the width of your beds, this should be relatively easy. Remember to include the width of the access path along with the width of the bed when making your calculation. Depending on how precise you are with your tractor and tillage equipment, the number of beds per field might have a margin of error of plus or minus 1.

Deciding How Much to Grow

In the preceding section, we looked at the capacity of the land at your disposal — how much acreage you have, how many beds will fit into a field, and how much of any individual crop each bed can contain. Now we will consider the importance of growing the right amount of produce to meet anticipated demand and of estimating what you can reasonably produce given your resources.

How Much Can You Sell?

Before planting crops, even before buying seed, every farmer should have a fair idea of how much he or she can expect to sell — not only over the course of a season but also over the course of a month, a week, and even a single market. There's no valid reason to grow more than you can sell, unless you plan to give it away.

Growing more than you can sell means you've squandered resources and time, while growing less means you've missed out on potential sales. Farmer graveyards are filled with once innocent and enthusiastic souls who failed to research their market before going into the field.

Experienced growers should be able to estimate with some accuracy how much of each crop they can move in a week or a month, assuming they don't make radical changes in their marketing. For the beginning farmer without much of a frame of reference, figuring out market demand will be a far more difficult task, which is a major reason it's wise to start out with modest expectations. There are steps, however, that beginning and less-practiced growers can take.

CSAs

Those who are planning to sell through a CSA should know how many shares they will be providing each week, what the shares will cost, and how much of each vegetable or herb they plan to put in every share. The shares should be sold and paid for before planting begins.

Some CSA growers calculate their total expenses for the year, including salaries for themselves and their workers, and divide this figure by the number of shares they plan to offer (and expect to sell) to arrive at a per-share price. There's a tendency to underestimate expenses, some of which are hard to predict (e.g., a major tractor repair job or a replacement fan for your greenhouse), so don't lowball this figure.

Share prices should also correlate with the quantity and value of produce that you expect to provide each week, and the number of weeks the share will last. Many growers offer shares of different sizes and different time durations and price them accordingly. Larger and longer running shares obviously will cost more. Thirty to forty shares per acre of land farmed is fairly typical for the average fresh-produce CSA that runs through summer and fall. For more on CSAs, see chapter 17.

Restaurants

If you will be selling to restaurants, you should establish accounts in advance. Give them a price list. Find out how much they will be willing to order. Try to get some form of commitment, then make sure you can provide what you have promised. In the beginning, a nicely presented package of free samples might be a good way to get your foot in the door and assist chefs in planning future menus.

Farm Stands and Farmers' Markets

If you're going to sell at your own farm stand or at a farmers' market, you need to get an idea of how much business you can expect in a day and how demand for your produce may change through the course of a season. Some markets are better attended in summer; others in fall when kids are back in school. Markets that occur just before big food holidays, such as Thanksgiving, can be very strong. Getting a handle on all this in your first couple of years will not be easy, but any advance research should pay off.

Estimate traffic flow. You might estimate the flow of vehicular traffic that passes your farm stand or the flow of foot traffic at the market you will be attending and try to gauge what percentage of potential customers actually stop to buy. Talk to farmers already selling at that market, or ask the market manager how much business a typical new farmer can expect in an average day — in spring, summer, fall, and possibly even winter.

What seems to be selling? Take note of which vegetables and herbs are available at your future market and the rate at which they sell. Fancy and exotic items (such as Asian greens or unusual herbs) can be reasonable sellers, especially in big cities with more food-savvy consumers, but even so, many people will pass them by. Don't overlook the less glamorous mainstay vegetables. They are what most of us eat most of the time.

Ethnic tastes. You might also do a little research into the ethnic background and food preferences of the community surrounding the market you plan to attend. Planting the right vegetables and varieties could be a key to your success.

Just visiting the market several times and casting about an observant eye, in advance of setting up your own stand, will give you a sense of your prospects. But don't expect customers to flock to your stand at the first market, unless you have something very unusual to offer. Give them a little time to get to know you. If you keep showing up, week after week, with good-looking vegetables and a smile on your face, they will come. Persistence pays off.

The Wholesale Market

Wholesaling to a supermarket such as Whole Foods can be an option for beginning farmers, but quality standards are very high, and sometimes verbal agreements don't pan out. A local co-op, though smaller, might pay off in community connection as well as sales. Also, be sure that you can actually make a profit at the wholesale price, which is often about half the retail price.

How Much Can You Grow?

Arriving at a good estimate of how much you can sell is crucial to your farm plan. Equally important is figuring out how much you can grow. This will depend on the amount of land that is available to you, your level of knowledge and experience, and the equipment and labor at your disposal. Here, we'll focus on the last two: equipment and labor.

Equipment

Preparing 5 acres of land for planting with a walk-behind rototiller is going to take a lot more time and physical energy than doing the same job with a larger, tractor-mounted tiller. Hand-hilling a half acre of potato plants with a hoe, or a shovel and rake, could consume the better part of a day, whereas a tractor with a tool bar and two sets of hilling discs could rip through the job in half an hour. A midmount cultivator on a small tractor will gobble up weeds faster than several individuals working with hand hoes. I could give many more examples. The point is, when you've got the right equipment for the job, more work gets done.

Labor

Tractors, rototillers, hilling discs, and cultivators are great when you need them, but there will be many times when what you need is a ready supply of human labor. Weeding newly seeded carrots or beans is a task that requires human hands, and often knees. Tractor-mounted cultivators may take care of weeds between rows of vegetables, but they are less likely to touch the weeds that are growing between plants within rows. Workers with hoes are needed for this.

On small, diversified farms, most harvesting is done by hand. I don't know of any machine that can pick tomatoes, or cut and bunch herbs, or load a truck for market. More labor will translate into more tomatoes and more herbs, and probably more of everything else.

So at the risk of stating the obvious, let me emphasize once again that the type of equipment you have and the size and skill of your labor force will have a very large influence on how much produce you are able to plant, tend, and bring to fruition. These two variables should be considered at least as seriously as anticipated market demand.

Having more work than you (and your crew) can handle will mean some tasks are done poorly or not at all. Losing crops you've already invested in, because you don't have time to tend them properly, is a painful experience. It will cause you stress and will seriously affect your bottom line. Again, start out modestly. The answers to the questions "How much can you sell?" and "How much can you grow?" will become evident as you gain experience.

The Weekly Planting Schedule

Once you've defined your growing area and estimated how much you can sell at different times of the season, you're ready to draw up a planting plan that will enable you to grow enough vegetables to satisfy the demand that you anticipate. On our farm, we call this the Weekly Planting Schedule because it tracks our planting by the week rather than by the day. Each week brings its own set of challenges in the form of weather (sometimes too wet or too cold), field conditions (soil may be too damp or too dry for planting a specific crop), and random unexpected events (a disabled vehicle or a sick worker). For these reasons, we'd rather not be pinned down to the exact day on which we must do a particular task.

The main purpose of the Weekly Planting Schedule is to tell us what to plant in a given week and exactly how much. Since we prefer to use transplants whenever possible, much of our planting begins in a greenhouse, though some crops — such as peas, beans, and carrots — are seeded directly into the field. Regardless of where they start their journey, all crops are listed in the Weekly Planting Schedule.

Other useful information. Over the years, our Weekly Planting Schedule has been expanded to include other useful pieces of information, advice, and reminders — such as the ideal soil temperature for germination of a specific crop, either in the greenhouse or in the field. Lettuces, for example, germinate well when the soil temperature is around 70°F (21°C); when the temperature is much above 80°F (27°C), germination can drop drastically.

Spinach will germinate most quickly at 70°F (21°C) or thereabouts. Tomatoes and peppers, on the other hand, like it hot. Temperatures around 80°F (27°C) are good for them. Because we start most of our seeds in trays or flats that sit on thermostatically controlled heat mats in our greenhouse, it's easy to adjust the soil temperature for a particular crop. We'll look into this in more detail when discussing greenhouse growing, chapter 6.

Onions

AT OUR FARMERS' MARKET, we usually sell out of onions by midfall, which means there's a couple of months left in our marketing season when we could be selling onions if we had them. Frequently, customers ask me why we don't grow more. They know that onions are a storage crop (at least most cooking onions are), and they know they would buy them from us in November and December if we had them. What they usually don't know is that organic onions are a very time-consuming proposition for a farmer. They grow slowly, and their slender leaves do not compete well with weeds.

If we planted a lot more onions, we would very likely not have enough time to keep them well weeded and irrigated. There are many other crops to be planted and tended throughout the season. Too many onions would mean that other crops might get shortchanged, and onions are not our only allium: We also grow shallots and garlic — lots of garlic, in fact. All three, being sensitive to changes in day length, are ready for harvest within the same 3- or 4-week period. Throw in the weather, and you might have to dig or pull a lot of alliums in a short stretch of time. There is also the challenge of curing and keeping them dry — both of which take time and space.

The moral of the story: We could grow more onions, and we could sell more onions. But when the bigger picture is taken into account, we don't have the time to do it.

Worden Farm

Owners: Eva and Chris Worden

Location: Punta Gorda, Florida

Year Started: 2003

Acreage: 85 acres, with 40 acres in production

Crops: Mixed vegetables and herbs

EVA AND CHRIS WORDEN STARTED GROWING VEGETABLES on a leased land-trust property in Connecticut while still in grad school (both were earning PhDs in agriculture-related fields). In 2003, they took the big step of purchasing 56 acres of level ground nestled among pine flatwoods forest and orange groves in southwestern Florida. In their first year, they farmed just 5 acres and had no trouble selling what they produced. In 2010, Eva and Chris purchased an adjacent property, expanding their farm to a total of 85 acres.

Each year, the Wordens have increased the size of their planting, and today they grow more than 50 different vegetables and herbs on 40 acres. Root crops, especially carrots, and assorted greens are among their more popular items. Most of what they produce goes to four farmers' markets and a 500-member CSA. Their season runs from August to June. July, when it is just too hot and rainy in Florida to grow most vegetables, is their well-earned month off. The Wordens use a crew of about 10 local workers and 4 interns. Their farm is certified organic by Quality Certification Services.

Eva describes her husband, Chris, as an "equipment aficionado." It's easy to see why. Worden Farm has nearly 20 tractors, among them several Allis Chalmers Gs with cultivators, an International 274 high-crop cultivation tractor (20 hp), a Case IH JX95 high-crop tractor (90 hp), and several other machines equipped with forklifts and front-end loaders. Implements for all these tractors include rototillers, bed shapers, a manure spreader, and vegetable crop seeders.

Worden Farm is divided into 16 fields that range from 175 to 450 feet long and average 200 feet wide — wide enough to accommodate about 30 beds. Between the fields are bio-strips planted to noninvasive eucalyptus and *Corymbia* trees, sunflowers, and species *Ammi*, such as white dill (also known as false Queen Anne's lace). The trees in the bio-strips stabilize the soil and serve as windbreaks. The white dill and sunflowers attract beneficial insects and are sometimes harvested for cut flowers. Chris and Eva grow all their crops on 6-inch-high raised beds on 6-foot bed centers. To optimize drainage in crop-growing areas, they regrade fields and reshape beds each season.

Extensive subsurface irrigation lines form the framework of Worden Farm's bed system. Main lines carry water to each field, while sublines and risers distribute water every few beds within the fields. Water is sourced from two ponds and two shallow wells. The growing crops can receive the precise amount of moisture they need through either overhead sprinklers or drip irrigation. The Wordens developed this efficient irrigation system with input from the USDA NRCS.

For summer cover crops, Eva and Chris prefer a combination of sorghum sudangrass (for biomass) and cowpeas (for nitrogen). The Wordens make some of their own compost and use composted poultry litter from off-farm sources.

Worden Farm has two 30- by 96-foot tunnels or hoop houses — one for transplant production and one for inground growing of heat-loving plants such as tomatoes and peppers. The main function of these hoop houses is to protect transplants and inground crops from rain and wind. The hoop houses have no fans, louvers, or dedicated heaters. Roll-up sides provide ventilation, and on rare chilly nights portable propane heaters are called into service.

Now in its tenth year, Worden Farm is a good deal more than a highly successful organic vegetable operation. Educational outreach and involvement with the local community are important to Chris and Eva. They believe that to have a healthier food system and satisfied customers, information about the food we eat and how it is produced must be widely disseminated. To this end, they offer an average of 10 on-farm workshops each year on subjects ranging from gardening techniques to culinary matters. The Wordens also host larger events in conjunction with other organizations that promote sustainable agriculture and healthy eating. At all of these gatherings, they provide activities for children (Eva and Chris have two of their own). Finally, as if all this were not enough, the Wordens offer farmer training workshops and serve as consultants in organic vegetable production and marketing. This is one busy farm.

In addition to soil temperatures, notes reminding us to trim the tops off our greenhouse onions to create stockier plants and stimulate root growth, or to soak parsley seed for a couple of days to hasten germination, or to drip-irrigate newly seeded carrots in the field, have all found their way into our Weekly Planting Schedule. Most experienced growers know that these are good things to do, but whether they always remember to do them at just the right time is another matter. I'm happy to have the reminders.

The notes in the Weekly Planting Schedule are also a good resource for our workforce. They provide growing tips everyone can learn from.

Soil Temperature and Potatoes

Soil temperature is also a factor to consider when seeding crops directly into the field. Newly planted potatoes, for example, can rot if the soil remains cold and wet for a couple of weeks. More than once, I've made the mistake of planting my spuds too early, in hopes of reaping the rewards of an early harvest, only to get hit with a cold, wet spell and find out a few weeks later that half of our planting stock never made it above ground.

This is why our Weekly Planting Schedule now has an entry around potato planting time reminding us to be sure the soil temperature in the field is above 55°F (13°C) day and night before we set out to plant. A reliable soil thermometer is an easy way to check soil temperature. Or you might simply wait until the dandelions on your lawn have bloomed — that's another time-tested way to tell when the ground is getting ready to receive potatoes. But to be on the safe side, I'd check the long-range weather forecast for the week ahead, even after the dandelions bloom, just in case there's a cold snap coming your way.

Sample Page of Weekly Planning Schedule

Date	Item	Number of Plants or Row Feet	Soil Temp Range/ Optimum Temp (°F)	Greenhouse Temp (°F)	Transplant Date
3/8	Dandelion (Clio)	400	65–75/70	55–75	4/15–4/25
	Peppers		75–85/80	60–80	5/5–5/15 in high tunnel 5/25–6/5 in field
	Peppers: Peacework	150			
	Peppers: Hot Portugal	150			
	Peppers: Carmen	75			
	Peppers: Lipstick	75			
	Peppers: Apple Sweet Pimento	25			
	Tomatoes: Group 1		75–85/80	60–80	5/1–5/5 in high tunnel
	Tomatoes: Jetstar	150			
	Tomatoes: Cobra	75			
	Tomatoes: Apero	50			
	Tomatoes: Arbason	75			
	Lettuce		65–75/70	55–75	4/15–4/25
	Lettuce: Red Salad Bowl	200			
	Lettuce: Jericho	200			
	Lettuce: Magenta	200			
	Kale: Red Russian	400	65–75/70	55–75	4/20–4/30
	Swiss Chard: Green	600	70–80/75	55–75	4/20–4/30
	Swiss Chard: Golden	300	70–80/75	55–75	4/20–4/30
	Spinach: Perpetual	300	70–80/75	55–75	4/20–4/30
	Scallions	800	60–70/65	55–75	4/20–4/30
3/15	Basil: Genovese	400	75–85/80	60–80	5/5–5/15 in high tunnel
	Peas (Direct-seed in field): Sugarsnap	200 row feet	55–85/75		
	Peas: Mammoth Melting Snow	200 row feet			

Onion Notes:
Trim tops off onion seedlings — to about 3" tall. Remove debris. Fertilize with fish emulsion and kelp, or Fertrell 2-4-2. Repeat in 2 weeks, and again in 4 weeks, trimming a little higher.

Allow for growth. Our Weekly Planting Schedule started out as a 4- or 5-page document many years ago. It now runs about 20 pages. Some of the same crops show up repeatedly, and there's a good reason for this. Whether you're growing for your own farm stand, a farmers' market, a CSA, or a restaurant, you want to have a steady and continuous supply of fresh, perishable crops. Lettuces and other salad greens are in demand all season at most markets, but that doesn't mean you can sell an unlimited quantity in a single day. If your market can absorb only four hundred heads of lettuce in a week, having two thousand ready for harvest at the same time won't do you any good, unless you're willing to wholesale and take a big price cut — assuming you can find a buyer on short notice.

The goal is to have four hundred lettuces ready for the week and another four hundred ready for the next week, and so on. This is where the Weekly Planting Schedule comes in very handy, especially when used in conjunction with two other forms: The Pick List and the Harvest and Sales Record. We'll get to these after we consider the crop rotation map.

A Cautionary Note Regarding Lettuces

Having 400 heads of lettuce ready for harvest each week sounds like a good plan. And it is, if your market can handle them. But good plans don't always come to fruition. Sometimes the weather gets in the way. If you have 2 or 3 weeks of rain, you probably won't be able to get your lettuces transplanted into the field. This means that their maturity and harvest date will be postponed. But that's not all that can go wrong.

Lettuces that have waited too long to be planted often do not perform well once they get to the field. They tend to bolt before reaching a marketable size. We've learned the hard way that it's usually wiser to throw away lettuce seedlings that have been sitting in their flats for more than 6 weeks. It's better to plant younger ones. The same cautionary note can be applied to many other plants in the greenhouse, though most give you a longer grace period than do lettuces.

Items to Include in a Weekly Planting Schedule

Date the week starts. For us, this is always a Monday.

Vegetable or herb to be planted and variety.

Amount. This means the number of cells in a flat in the greenhouse, or, for direct-seeded crops, row feet in the field.

Optimum soil temperature for germination. This is especially relevant in the early and late parts of the season, when temperatures are on the cold side (we use heat mats to speed up germination and therefore have some control over soil temperature in the greenhouse).

Optimum ambient greenhouse temperature. This is relevant after germination, when seedlings are removed from heat mats.

Expected transplant date (for plants started in greenhouse only).

Assorted advice and reminders.

Crop Rotation Maps

The crop rotation map or plan is similar to the field map, but it has a lot more data on it. It shows which crops will be planted in which fields and exactly where. The idea is to have available just the right amount of field space in the right location for each crop that is listed in the Weekly Planting Schedule.

Looking back at the sample page from the Weekly Planting Schedule (page 55), you'll see that in the week starting Monday, March 8, we seeded two hundred cells each of three different types of lettuce:

- 'Red Salad Bowl' (a red oak leaf variety)

- 'Jericho' (a romaine)

- 'Magenta' (a very tasty French crisp variety)

KEITH'S FARM 2010
SPRING/SUMMER CROP ROTATION

Note: Field 04 includes orchard around house and other buildings.

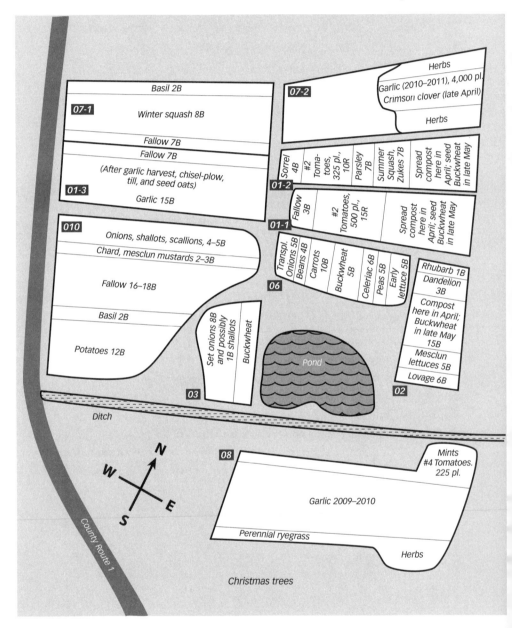

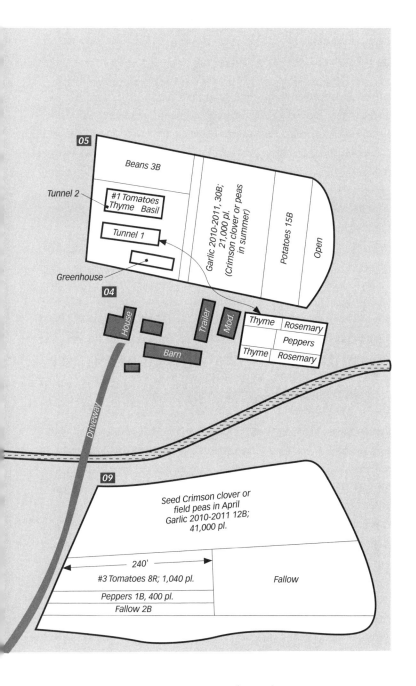

If all goes well in the greenhouse, we will end up with six hundred lettuce seedlings ready to be transplanted 5 to 6 weeks later, at which point we will need a place to put them. As noted earlier, our normal spacing for lettuces is four rows per bed, with 12 inches between rows and 12 inches between plants within each row. This means we will need 600 row feet or 150 bed feet for this particular planting.

Over the course of a season, we normally start around 15,000 lettuces in our greenhouse, but we have to take into account the likelihood that setbacks will occur. So let's assume we will lose 10 percent, or 1,500 plants, while they are in the greenhouse, from such factors as poor germination or excessive or inadequate watering. That would leave us with 13,500 lettuces that will require a total of 13,500 row feet or 3,375 bed feet of real estate in the field. Since our beds are 76 inches (6.33 feet) on center, after allowing for pathways, we can calculate the acreage needed as follows:

3,375 × 6.33 ft = 21,364 sq ft (or a little under ½ acre)

This, then, is the amount of space we must allocate to lettuces on our crop rotation plan. They don't all have to be next to each other. We could end up with lettuces in a few different fields over the course of a season, as long as there were no lettuces in those locations in the previous 4 years. Not having successive plantings of lettuce, or any other crop for that matter, next to each other in the same field is a good thing. Having different crops between them will interrupt the movement of pests and diseases.

Two crop rotation maps. You might have noticed that this section is titled "Crop Rotation Maps" rather than "Crop Rotation Map." That's because we have two of them for each year — one for spring/summer crops and another for summer/fall crops. Fields or portions of fields that are fallow or in cover crop on the spring/summer rotation map are usually planted with such crops as broccoli, kale, carrots, or turnips in midsummer.

Make extra copies. Rotation being what it is, annual vegetables, herbs, and cover crops will move around on the crop rotation map from one year to the next. For this reason it's important to make copies of your rotation maps and file them away carefully, so you can always look back at previous years to see precisely what was planted, how much, and where. If you're handy enough with a computer to draw rotation maps, that might work even better. Just don't count on memory alone. It's an unreliable faculty.

Crop rotation is one of the cornerstones of organic farming. It is a large and complex topic. I'll zoom in on it in chapter 11.

Pick List and Harvest and Sales Record

Most diversified organic growers want to have a steady supply of certain perishable crops throughout as much of the growing season as possible — a point I've already made more than once. Sudden, large surges in production are usually not desirable. It doesn't do any good to have 200 pounds of mesclun salad greens ready for harvest at the same time if your farmers' market can absorb just 40 pounds of mesclun a week and if the excess plants will get too big or go to seed if left in the field much longer.

Neither will it be very satisfying to have only 20 pounds of mesclun to take to market and see it all gone by midday when you know that you could have sold twice that amount. In both instances you've missed the mark. By closely tracking how much your CSA needs or how much you can move at each market or sell to each restaurant from week to week, you can get a good idea of future demand. This is where the Pick List and Harvest and Sales Record forms come in very handy.

The Pick List

The Pick List we use on our farm is a one-page sheet listing the approximately 80 different crops we grow, though not necessarily every variety. We almost always harvest perishable crops on the day before market, which we refer to as a "pick day." Storage crops like potatoes, onions, and garlic are harvested when they are mature and in much larger quantities.

The Pick List is usually filled out the night before the pick day. It tells us the amount of each crop to harvest or take from storage. Quantities are determined by how much we have available in the field, or in storage, and anticipated market demand, which can be reasonably estimated by looking at recent sales figures from previous markets.

Summary of Data on Pick List

- Harvest Date
- Market Day and Date
- Item
- Weather Forecast
- Amount to Be Harvested (in pounds, bunches, heads, lugs, coolers, pints, pieces)

Keith's Farm Pick List

Weather Forecast: Partly Cloudy, High 80°F/Low 55°F (27°C/13°C)
Harvest Date: 9/24
Market Day and Date: Sat. 9/25

VEGETABLES

Item	Amount	Item	Amount
Beans	50 #	Shallots	30 Pt
Peas: Sugar Snap	n/a	Celeriac	30 #
Peas: Snow	n/a	Peppers: Sweet	30 #
Broccoli	n/a	Peppers: Hot	15 #
Broccoli Leaves	n/a	Potatoes: White	110 #
Collards	40 Bu	Potatoes: Yellow	75 #
Kale: Winterbor	70 Bu	Potatoes: Red	75 #
Kale: Red Russian	60 Bu	Potatoes: Blue	45 #
Kale: White Russian	n/a	Carrots	70 #
Kale: Lacinato	100 Bu	Radishes	n/a
Kale: Rainbow Lacinato	60 Bu	Tomatoes:	
Kale: Siberian	n/a	Cherry: Red/Orange	50/80 Pt
Mustard: Green	40 Bu	Heirloom	200 #
Mustard: Red	30 Bu	Red: Large/Small	300 #
Turnips: White	120 Bu	Plum	50 #
Turnips: Gold	n/a	Green	50 #
Turnips: Purple	30 Bu	Mesclun (coolers)	5 coolers
Tatsoi	35 Hd	Arugula (coolers)	4 coolers
Mizuna	50 Bu	Swiss Chard	70 Bu
Sorrel	1.5 Lg	Ppt Spinach	35 Bu
Dandelion	110 Bu	Summer Squash	n/a
Endive	n/a	Zucchini	n/a
Escarole	30 Hd	Winter Squash: Acorn	50 #
Lettuce: Fr Crisp	120 Hd	Winter Squash: Butternut	75 #
Lettuce: Red Oak	95 Hd	Winter Squash: Delicata	n/a
Lettuce: Romaine	75 Hd	Winter Squash: Sweet Dumpling	n/a
Lettuce: Other	60 Hd	Winter Squash: Other	n/a
Garlic (Jumbo)	700 Pc	Pumpkins: Lge/Sm	n/a
Onions: Cooking	50 #	Pumpkins: Mini	n/a
Onions: Salad	50 #		
Scallions	60 Bu		

HERBS

Item	Amount	Item	Amount
Anise Hyssop	12 Bu	Rosemary	65 Bu
Basil	140 Bu	Sage	40 Bu
Catnip	10 Bu	Savory S/W	15/10 Bu
Chives	n/a	Tarragon	20 Bu
Cilantro	40 Bu	Thyme	70 Bu
Lavender	15 Bu	Mint: Choc.	40 Bu
Lovage	55 Bu	Mint: KC	50 Bu
Marjoram	30 Bu	Other: Apples	50 #
Oregano	30 Bu		
Parsley: Italian	150 Bu		
Parsley: Curly	45 Bu		

KEY
# = pound
Bu = bunch
Lg = lug
Hd = head
Pc = piece
Pt = pint

At the beginning of the season, the number of crops tagged for harvest on the Pick List is small because many vegetables have not yet ripened, matured, or reached marketable size. As the season progresses, more get added. By late summer and early fall, we are at peak production, and the Pick List has a lot more items checked on it.

Weather forecast. The weather forecast will also affect how much of each item we harvest. On wet, cold, or oppressively hot days, fewer people show up at outdoor markets. If the forecast is bad, we'll reduce the amount of perishable crops, such as lettuces, kale, and mesclun, by about 20 percent. Storage crops, such as garlic, winter squash, and onions, are in a different category — the amounts of these brought to market don't fluctuate much in response to weather forecasts. If we don't sell what we bring to market, we can always take them home and bring them back another day.

Pounds, bunches, heads, lugs, coolers, pints, pieces. Quantities on the Pick List are expressed in different ways depending on the vegetable in question. Potatoes, onions, squash, and tomatoes, for example, are usually recorded and sold by the pound (#); herbs and certain greens, such as kale, collards, or Swiss chard, go by the bunch (Bu); lettuces, escarole, and bok choy go by the head (Hd); loose (unbunched) greens, like sorrel, are usually counted by the lug (Lg); washed mesclun and arugula go by the cooler; cherry tomatoes and shallots go by the pint (Pt); and garlic goes by the piece (Pc).

No scale necessary. In case you're wondering, we rarely take a scale into the field. Vegetables such as carrots or potatoes that are sold by the pound are harvested in black plastic lugs or containers with a capacity of about 1.8 cubic feet, or 1.5 bushels. We know, from experience, that approximately 50 pounds of carrots or potatoes will fill one lug. In other words, we could just as easily record these crops by the lug as by the pound, and in fact, sometimes we do.

The Harvest and Sales Record

Like the Pick List, the Harvest and Sales Record is a one-page sheet that lists all the crops we grow throughout the season. It follows roughly the same format as the Pick List, differing from it in a few respects: First, the Harvest and Sales Record records exactly how much of each item we did harvest or take from storage for a particular market. The Pick List is more of a wish list and a reflection of what we think is available and what we estimate we can sell. In the field, conditions are constantly changing, and the current reality may not always accommodate everything the Pick List asks for.

For example, whoever is assigned to harvest lettuces might notice that many of the heads of variety A are still a little small and would be better left in the field for another week, while the heads of variety B are large and a few of them are showing signs of bolting. The picker might, therefore, reduce the number of heads of variety A and increase the number of variety B and make note of these changes on an index card or scrap of paper in the field. These altered numbers will be entered on the Harvest and Sales Record at the end of the day, as we are loading the truck.

Summary of Data on Harvest and Sales Record

- Harvest Date
- Market Date
- Item
- Amount Harvested (may differ from amount on Pick List)
- Field Number
- Amount Sold
- Estimate of How Much More Could Be Sold
- Market Weather
- Initials of Recorder

Don't Skip Markets

Bad weather may put a damper on sales, but it is no reason to skip a market. You want your loyal customers to know that you'll always be there, come rain or shine. Once they understand that you're willing to endure difficult conditions to bring fresh food to them, a surprising number will return the favor and come out in bad weather to support you. And of course, predicted bad weather doesn't always come to pass. A slow-moving front may dissipate before arrival, or move in a different direction, leaving you with a perfectly fine day. Weather forecasts are valuable, but don't put too much weight on them.

The Harvest and Sales Record has other important functions, such as noting the field from which each individual crop is harvested. This is required by NOFA, our certification organization — the rationale being that a crop can be traced back to a particular field, rather than the whole farm, should questions ever arise concerning its organic status.

Sales. More importantly, the Harvest and Sales Record has a column for keeping track of sales. At the end of the market day or on the drive home, we fill out the "Amount Sold" column to the best of our abilities. These numbers are of immense value when the time comes to create a Pick List for the following week's market.

In my early years of farming, I mostly relied on memory when developing pick lists. It did not always serve me well. Now, with so many more crops and, I suspect, fewer functioning brain cells, I could get very seriously lost. The "Amount Sold" column of the Harvest and Sales Record is my savior. With all the information I need from the previous week on one sheet of paper, I am able to make a well-targeted Pick List in short order.

Sometimes I even go back to a previous year's Harvest and Sales Record to see what quantity of a certain item sold at a particular market. How else would I know how many pumpkins to bring to a pre-Halloween market, or how many bunches of sage, thyme, and rosemary we might sell for turkey stuffing on the Wednesday before Thanksgiving? Believe me, it could easily be five times as many as we sold at a market 2 or 3 weeks earlier.

Keith's Farm Harvest and Sales Record

Harvest Date: 9/24

Market Date: Sat. 9/25

Market Weather: Sunny 80°F (27°C)

VEGETABLES			
Item	Field	Amount	Amount Sold
Beans	02	55 #	55 + 5
Collards	05	40 Bu	36
Kale: Winterbor	01	70 Bu	65
Kale: Red Russian	01	60 Bu	60
Kale: Lacinato	01	100 Bu	100 + 20
Kale: Rainbow Lacinato	01	60 Bu	60 + 10
Kale: Siberian	n/a	n/a	n/a
Mustard: Green	02	40 Bu	35
Mustard: Red	02	30 Bu	28
Turnips: White	09	120 Bu	116
Turnips: Purple	09	30 Bu	29
Tatsoi	05	35 Hd	35 + 5
Mizuna	05	50 Bu	40
Sorrel	02	1.5 Lg	1.5 + 0.5
Dandelion	02	110 Bu	105
Endive	n/a	n/a	n/a
Escarole	01	30 Hd	30 + 10
Lettuce: French Crisp	01	75 Hd	75 + 30
Lettuce: Red Oak	01	95 Hd	90
Lettuce: Romaine	01	100 Hd	90
Lettuce: Other	01	60 Hd	60 + 10
Garlic (Jumbo)	09	700 Pc	660
Onions: Cook	06	50 #	48
Onions: Salad	06	50 #	50 + 20
Scallions	10	60 Bu	55
Shallots	10	30 Pt	25
Celeriac	06	30 #	25
Peppers: Sweet	09	25 #	25 + 10
Peppers: Hot	09	14 #	12
Potatoes: White	05	110 #	90
Potatoes: Yellow	05	75 #	65
Potatoes: Red	05	65 #	60

Key	
# = pound	
Bu = bunch	
Lg = lug	
Hd = head	
Pc = piece	
Pt = pint	

VEGETABLES *(CONTINUED)*			
Item	Field	Amount	Amount Sold
Potatoes: Blue	05	35 #	35 + 15
Carrots	06	75 #	75 + 15
Tomatoes: Cherry, Red/Orange	01	45/90 Pt	45 + 10/85
Tomatoes: Heirloom	01	245 #	190
Tomatoes: Red, Lg/Sm	01	330 #	275
Tomatoes: Plum	01	60 #	50
Tomatoes: Green	01	55 #	45
Mesclun (coolers)	08	5 coolers	5
Arugula (coolers)	08	4 coolers	4 + 2
Swiss Chard	01	70 Bu	65
Spinach: Perp	06	30 Bu	30 + 10
Squash: Summer	n/a	n/a	n/a
Zucchini	n/a	n/a	n/a
Squash: Acorn	07	50 #	50
Squash: Butternut	07	80 #	75

HERBS			
Anise Hyssop	09	12 Bu	11
Basil	07	70 Bu	70 + 40
Catnip	08	10 Bu	7
Chives	n/a	n/a	n/a
Cilantro	07	40 Bu	35
Lavender	09	15 Bu	15
Lovage	02	55 Bu	45
Marjoram	09	30 Bu	25
Oregano	08	30 Bu	30 + 5
Parsley: Italian	01	150 Bu	145
Parsley: Curly	01	45 Bu	35
Rosemary	Tunnel 1	65 Bu	65 + 5
Sage	08	40 Bu	40 + 5
Savory: Summer/Winter	09	15/10 Bu	15/8
Tarragon	08	15 Bu	15 + 10
Thyme	09	70 Bu	70
Mint: Chocolate	08	40 Bu	35
Mint: Kentucky Colonel	08	50 Bu	50 + 5
Other: Apples	04	50 #	45

Initials of Recorder: KDS

How much could we have sold? You may have noticed on the sample Harvest and Sales Record (see pages 66–67) that some of the items listed have a "+" symbol followed by a number (e.g., + 10) in the "Amount Sold" column. This number is entered at the end of the market day for crops that were sold out. It is a round-number estimate of how many more bunches, heads, pounds, or pints we think we could have sold on that day. These estimates will be taken into account when I create a Pick List for the next market.

Weather. On the Harvest and Sales Record, we also note the weather conditions that prevailed while we were at market. A partly sunny day with temperatures in the low 80s (high 20s °C) will almost always result in better sales than a day with a hot and humid morning and an afternoon peppered with violent thunderstorms. No surprise there. Those of us who sell in open-air markets are, to a large extent, at the mercy of the weather, even though we have tents for shelter. A hurricane could shut the market down altogether. This is why both past and predicted weather are highly significant when preparing a Pick List.

Estimates will do. The numbers and weather information entered on the Pick List and the Harvest and Sales Record are all pretty basic, and sometimes they are approximations. Whether you sold exactly 73 or 78 bunches of thyme is not that relevant. A guesstimate of 75 bunches will do just fine. Estimated numbers are a lot better than no numbers. The point is that these details have a way of getting blurred and misremembered. A system for recording relevant (if not always precisely accurate) data that is suited to your farm operation will serve you well. It'll simplify decision making, save time, and keep you on target.

To-Do Lists

For me, mental clarity is an elusive faculty — not something I can reach for at the drop of a hat. First thing in the morning, my brain is usually functioning as well as it can, which isn't saying much. Sometimes later in the evening, after a short nap and with no distractions, I can assemble a few clear thoughts. Sunday afternoons, too, can offer fertile ground. I've learned to take advantage of these times to get as much mental work done as possible and, in particular, to prepare tailored and distinct to-do lists to guide me into the days ahead. If you're anything like me, I recommend this strategy. Here are some of the lists I use.

Notebook versus iPad

My weekly to-do lists are handwritten in a simple exercise book that usually lives on our kitchen table, where I have access to it during breakfast and lunch. Most people these days probably prefer to use an iPad or some other digital device with instant recall and vast storage capacity, and I can understand that — I've got an iPad myself. But perhaps because I grew up in a nondigital time, I still like the old-fashioned paper. I can scribble notes in the margin while eating my cereal, and it's easy to leaf back to the previous week to see what tasks did not get finished, or perhaps even started. Missed tasks are likely to get to the head of the line for the new week. Of course, all of this could be done on the iPad.

The Weekly To-Do List

As the growing season gets underway, each week there is usually a mountain of work to get done and not quite enough time in which to do it. Some tasks are more pressing than others. Before the week begins, I make a list of everything I would like to accomplish and then prioritize it so that the most critical jobs don't get lost in the shuffle.

Step 1: Survey Your Crops

On Sunday afternoon, I take an hour to walk the fields and observe how the existing crops are progressing. I bring with me a small digital recorder that fits in the palm of my hand. I note on the recorder which vegetables are ready for harvest and which need weeding, irrigation, mulching, staking, or any other form of attention. I keep an eye out for early signs of plant disease, insect damage, and other symptoms of stress. I check those crops that are most palatable to herbivores such as woodchucks, rabbits, and deer, to see if any have been eaten.

I tour the greenhouse and inspect the hardening-off structures to determine which crops are ready to be transplanted into the field. I look inside the high tunnels to see how the protected, early-season tomatoes and basil are doing. I check to see that all tools have been put away in their proper places and that the lugs and coolers we will need for our next market are clean and ready to use.

I look at the Weekly Planting Schedule and note which crops are due to be seeded in both the greenhouse and field. Everything that needs to be done the following week and every issue that needs to be addressed gets noted on the recorder. Well, "everything" may be an exaggeration here. I probably miss a few things more often than not. But with any luck, I'll catch them later in the week, or my crew will fill me in.

Step 2: Create the Lists

Back inside, I sit down with a notebook, hit the replay button on the recorder, and create a to-do list for the coming week. The list is divided into two parts: The first and major part itemizes all the general farm work that needs to get done. Tasks are organized into major categories, such as planting, weeding, mowing, irrigation, and other chores. Doing this helps me visualize the week ahead and get a sense of how much can be realistically accomplished.

The second part of the weekly to-do list focuses on tractor work (usually done by me or a more experienced crew member) and business matters that need to be attended to (usually done by me). The part 2 list might include ordering supplies, renewing truck registration, paying the latest insurance bill, and sundry other things any small-business person has to contend with. To overlook these irksome and often time-consuming chores is to invite major headaches down the road.

Reality check. Sometimes there will be too much on the weekly to-do list for one week, in which case less vital tasks may need to get bumped. Of course, I would prefer that everything get done, but if that's not possible, the next best thing is to know this in advance. This is much better than to arrive at the end of the week and be shocked to discover that we didn't have enough time to weed the new carrots or plant the first round of arugula for fall—failing to weed carrots soon after they emerge means tackling a tougher weeding job the next week, and not having arugula for fall markets means missing out on easy sales. The more I know at the beginning of the week, the better I can prioritize tasks and allocate our labor.

The 3-day week. Most of the work on the weekly to-do list has to be done in just 3 days: Monday, Wednesday, and Thursday. Some of us (usually three or four people) go to market on Wednesdays and Saturdays. This means that on Tuesdays and Fridays most of our energies go to harvesting and preparing for market. There may be time for a few other chores on these "pick days," but getting ready for market takes priority. On Wednesdays we are short a few people on the farm but still manage to get plenty of work done.

Sample Weekly To-Do List, Part 1

Monday, 6/7

Transplanting:
2 beds (800 RF) lettuces in Field 06
1 bed (600 RF) Swiss chard in Field 010
8 rows (3500 RF) winter squash in Field 07.1
2 beds mesclun greens in Field 02
200 RF cilantro in Field 07.2
Hand-plant remaining herbs (lovage, anise hyssop, marjoram)
    in Field 08

Seeding:
Seed 200 RF beans in Field 06
Seed buckwheat cover crop in fallow portions of Fields 01.1 and 02
Seed in greenhouse, per Weekly Planting Schedule

Weeding:
Wheel-hoe (WH) and hand-hoe (HH) 010 onions and shallots
WH/HH last 4 beds of lettuces and 2 beds of Italian dandelion in 06
WH 07 and 08 garlic
Hand-weed (HW) direct-seeded parsley in 06
HW herbs (sage, thyme, oregano, marjoram) in 08

Other:
Drip-irrigate per schedule
Hill 05 potatoes
Set Florida weave stakes in 02 tomatoes
Florida-weave 01.1 tomatoes
Seed NZ white clover in 09 pepper and tomato pathways
Cut back seedy sorrel in 02 and seedy oregano in 09
Mow garlic and onion pathways in 01.2, 08, 010
Start removing garlic scapes   *(List continued on next page)*

Harden off more lettuces and basil
Thin newly seeded lettuces and brassicas in greenhouse
Prune and trellis T2 tomatoes

Sample Weekly To-Do List, Part 2

(Tractor Work and Business Matters)
Monday, 6/7

Payroll
Payroll tax deposit
Deposit cash
Pay Workers' Comp insurance
Rototill: 02 for lettuces and mesclun; 06 for chard
Disc 01.1 and 02 before and after seeding buckwheat
Turn compost at top of 09
Transplant with water-wheel transplanter: lettuces, chard, winter
    squash, mesclun in fields 06, 010, 07.1, and 02, respectively
Hill 05 potatoes
Brush-hog: weeds on left side of 03; winter rye in 010; all field
    perimeters
Inspection and oil change for Ford F150
Order more tomato trellis clips
Order work gloves: 1 dozen small; 2 dozen medium (have plenty
    of larges)
Call John's Mower regarding chain saw repair
Put away manure spreader
Buy propane for flame weeder

Again, weather: the wild card. There is also the weather to consider. If
you've come to farming from an indoor line of work — say, teaching or
sitting at a computer terminal — you'll discover in a hurry that nature
and its elements will now be playing a major role in your life.

Enter the 3-day forecast. Admittedly, weather forecasting is no more
than just that — forecasting — but it's a lot better than the alternative,
which is hoping for the best. Knowing that an ominous front is heading
in your direction but is still 2 or 3 days away will greatly affect how you
plan your week.

A high probability of rain in midweek tells me to get any needed tractor work or planting done in the first couple of days. Driving a tractor around in a wet field damages soil structure and leads to compaction. Better to schedule lighter tasks, such as mulching or hand-weeding herbs or seeding in the greenhouse, for the days when the probability of rain is high. (Note: 5-day and 7-day forecasts are readily available, but at least in the Northeast, they are notoriously unreliable.)

When basing decisions on a local weather forecast, consider your farm's exact location and topography, especially when nighttime temperatures begin to drop. Remember: Warm air is lighter and therefore rises; cold air is heavier and tends to sink to a lower elevation. If you're living on the floor of a valley, you're likely to get hit with a frost before locations higher up on either side. Ask other growers in your area what they use for weather forecasts. Some states have special weather forecast systems designed to help farmers.

More on Weather

Weather forecasts can be had from many sources. I've bookmarked the town nearest to our farm on the website I use, so that if I'm within reach of my computer, the latest weather conditions and forecast for our area are never more than a few keystrokes away.

Another good site to keep with your bookmarks or favorites is the National Weather Service's Enhanced Radar Mosaic Loop. This site has maps with moving images that show something called "base reflectivity," which is used to measure and track precipitation. You can tell at a glance if rain is headed your way, how intense it will be, and approximately when it will arrive.

Base reflectivity maps are continuously updated and very easy to read. Early in the season I take some pleasure in announcing to my newer crew members, after they return from lunch, that they can expect rain on the farm within a half hour. When the announcement proves correct, which it usually does, they are suitably impressed and may even attribute to me godlike divining powers — that is, until they encounter their first base reflectivity map.

The Daily To-Do List

However daunting the week ahead may seem, the weekly to-do list reassures me that there is still order in the universe. At least I have a plan and that plan should enable me to enjoy a night of untroubled sleep. But come Monday morning, unless I was especially diligent on Sunday night, there is yet another list to make. This one will enumerate all the tasks to be tackled on that particular day.

On rising, I check the latest weather forecast to make sure there are no surprises in store. If there's a chance of precipitation, I go to our region's base reflectivity map to see when rain might reach us, then I sit down with the weekly to-do list and draw up a shorter list for the day. The most urgent work will most likely be at the top of the list. But there are several other factors to consider when scheduling a day's work.

Scheduling Considerations

A carefully considered daily to-do list will do more than enumerate all the tasks you hope to complete, or possibly just work on, over the course of a day. It will also schedule and prioritize tasks so that each is performed at the optimum time. Depending on the jobs in question, there are several variables to consider. Here are some that I routinely take into account.

Weeding. Whether you're using a hand hoe, a wheel hoe, or a tractor and cultivator, weeds are best dealt with when the sun is shining and still has a few hours of shining to do. Generally speaking, the goal is to uproot the offending plants and have them dry out as quickly as possible so there's minimal chance they'll be able to reestablish themselves. Weeding in damp, overcast weather is less effective.

Transplanting. Transplanting is better relegated to the latter part of the day when the sun is low in the sky and seedlings will not be unduly stressed as they make their transition to the open field. Of course, on a damp overcast day, one might safely transplant at any time.

Wet foliage. Some plants, such as tomatoes, are best handled when their foliage is dry — fungal and bacterial diseases can easily be spread by hand when they are wet. So damp conditions would rule out pruning or trellising tomatoes.

Workers. It's also wise to consider your workforce, even if it is just yourself. Most people prefer not to do the same tedious job all day long, especially if it keeps them in a physically demanding position. For example, weeding onions for several hours, on hands and knees, is going to

Sample Daily To-Do List

Monday, June 7

PS 78/52; 40COR after 3:00 PM

(Partly sunny in morning; high 78; low 52; 40% chance of rain after 3:00 PM)

TASK	Suggested No. Persons	Suggested No. Hours
Wheel-hoe garlic in Field 01.2 (AM)	4	2
Wheel-hoe/hand-hoe shallots and onions in 010 (AM)	4	2
Wheel-hoe and hand-hoe lettuces and dandelion in 06 (AM)	3	1
Hand-weed direct-seeded parsley in 02 (late AM)	4	2
Flame-weed carrots seeded last week in 06, but only if they have not yet emerged (late AM)	1	0.5
Seed two rows of green beans in 06 (late AM)	1	0.5
Mulch celeriac in 06 with compost and wood shavings (late AM or early PM)	2	2
Set Florida weave stakes in 02 tomatoes (PM)	3	3
Seed white clover in 09 pepper and tomato pathways (PM)	1	1
Clean empty flats outside greenhouse (PM)	1	1
Plant leftover lovage, marjoram, anise hyssop in 08 (PM)	2	2
If no rain, transplant winter squash in Field 07 and Swiss chard in Field 09 (mid-PM)	3	3
Prune and trellis tomatoes in Tunnel 2 (late PM)	3	2
If rain, seed in greenhouse, as per Weekly Planting Schedule and thin seedlings (late PM)	6	'til end of day

take a toll on even the most stalwart and motivated among us. It is far better to spend 2 or 3 hours on such a tiresome job, then move on to something more physically engaging that calls for an upright stance.

Variety. I try to structure each day so there is a variety of work for each of us and so our bodies will be used in different ways. To this end, I might suggest that a finite amount of time (say, 3 hours) be dedicated to a large and irksome job and that every able-bodied person go to work on it. With more people, the work gets done faster, and we can take advantage of the dynamic created by a larger group making good progress. If the job doesn't get finished in the allotted time, we can always come back to it the next day and be glad about the progress we've already made.

Most years, there are one or two members of my crew who like to work alone in the field, at least some of the time, and they can be very productive doing so. It's smart to take their preferences into account, too.

Push the envelope. Typically, my daily to-do list will have a little more on it than can be accomplished easily in a single day. This serves to motivate us and limit the amount of time spent standing around wondering what to do next. On days when we are able to check off every job on the list, we get to feel especially pleased with ourselves.

The Morning Meeting

Once the daily to-do list is completed, I make copies of it — one for my crew leader, one for myself, and one to be posted in a central location (usually the greenhouse) for everyone to see. At the beginning of the workday, we meet as a group and go over the list together. Some tasks may be clear cut and self-explanatory; others may need clarification. There are often questions regarding the best tool to use for a particular job or where certain supplies — such as stakes and trellises for supporting peas — can be found.

For 15 or 20 minutes, everyone has the opportunity to ask questions and provide input. The idea is to have the whole crew (usually there are seven or eight of us) get a sense of what the day ahead will hold and the work we intend to accomplish. The daily to-do list helps a lot in this respect, but we view it as a plan, not an edict.

Suggestions, Not Imperatives

Small, diversified farms are busy, dynamic places, and farmers need to be adaptable. It helps to stay nimble on your feet and ready to respond to new contingencies when they arise. Well-considered plans are important and necessary. They help you allocate time and energy, and they create the impression among your crew that you know what you're doing — even when you're not totally confident that this is the case.

But sticking rigidly to a plan just because it can be viewed on a computer screen, or is written down on paper, is not always the wisest choice. Unexpected problems sometimes come up and need to be dealt with. Priorities can shift. Conditions in the field might call for a different approach.

In the same vein, the number of people assigned to each task on the daily to-do list, and the hours allocated and time of day (AM or PM), are suggestions only. Often, they are not closely adhered to. Some jobs will get done more quickly than expected, while others will take longer, but these numbers do serve a purpose: They imply priorities, guide our energies, and give everyone a sense of what they can expect from the day. The tasks are listed in rough order of priority.

Planting Records Book

The Planting Records Book bears the same relationship to the Weekly Planting Schedule that the Harvest and Sales Record bears to the Pick List. It is a record of what is planted in each field over the course of a season, as opposed to what is called for in the Weekly Planting Schedule. Remember, the Weekly Planting Schedule is a plan, and plans don't always come to pass exactly as written, especially when you're working in the realm of nature. The Scottish poet Robert Burns put it rather nicely when he wrote:

> *The best laid schemes o' mice an' men*
> *Gang aft agley* (go oft awry)

We hope our plans don't go awry too often, and usually, they don't. Either way, the Planting Records Book is the final word. We can go back to it with confidence and find out exactly what did happen, when, and where.

Our Planting Records Book is organized by the field. Each field is allocated several pages, and each page has several columns on it. Many people may find it easier to use a spreadsheet program for this type of information. Regardless of your preferences, the type of information recorded should be the same. Below is an example of the data entered in our Planting Records Book.

Selective data entry. If you're hoping for some relief from lists and records, you'll be pleased to know that not all the data listed on the next page needs to be entered for every planting. We almost always plant kale at the spacing given in the table, so a symbol such as "SAU" — same as usual — could be used. If we deviate from this spacing, we would note this down. You might think the harvest period, which is defined by the first and last harvest dates, would also fall into this category. It doesn't. The reason is day length. Let me explain.

Information Recorded in Planting Records Book

Column Heads	Example
Item Planted and Variety	Kale: Red Russian
Date Seeded in Greenhouse or Direct-Seeded in Field	3/26/10
Date Transplanted	4/30/10
Amount in Row Feet	300
Date of First Harvest	6/5/10
Date of Last Harvest	6/25/10
Spacing Between Plants & Rows	12" betw plants; 18" betw rows
Amendments Applied	None
Notes	Floating row cover for flea beetles Bolted toward end of June

Changing day length. Lettuces seeded in the greenhouse in early May and transplanted in early or mid-June will grow faster and form salable heads more quickly than lettuces seeded in mid-July and planted in mid-August. The main reason is day length. A secondary reason is lower temperatures in fall. The shorter days in August and September mean less growing time, so the lettuces take longer to size up. By October and November this phenomenon is a lot more exaggerated. Late-season lettuces may need an extra month or more over their spring counterparts to reach salable size.

There is, however, an upside to later plantings. In the cooler fall weather, lettuces will remain in good condition in the field for several weeks, extending the harvest period. In midsummer heat, lettuces can quickly bolt or go to seed, which causes them to taste bitter. You might have a window of only 2 to 3 weeks in which to take them to market.

I learned the importance of day length the hard way with my broccoli crop. In my early days of farming, it took me a few years to figure out why I couldn't get decent heads of broccoli in fall. Invariably, they would be only 2 or 3 inches in diameter by November, when damaging frosts are likely to occur. I gave the plants more fertilizer and more generous spacing, but these measures didn't help much.

My mistake, it turned out, was to assume that the "days to harvest" numbers given in the seed catalogs were reliable throughout the growing season. Indeed, they were fairly accurate if we seeded broccoli in spring and allowed it to enjoy the longer growing days. But eventually, it became clear to me that to get decent heads of broccoli in fall, it was necessary to give the plants at least another month of growing time.

Growing Degree Days

Aside from a simple visual inspection, one way to gauge when a specific crop should be ready for harvest is to keep track of growing degree days (GDDs), which are a cumulative measure of warmth over time. In most parts of North America, GDDs (they are really units rather than actual days) occur when the average temperature for the day is above 50°F (10°C). For example, if the high temperature on May 15 was 78°F and the low 42°F, the average for the day would be 60°F.

$$(78 + 42) \div 2 = 60$$

Since 60°F is 10 degrees above the baseline of 50°F, we have an accumulation of 10 GDDs. Average daily temperatures below 50°F are ignored, rather than subtracted from the running total. (Depending on their location and subject of study, scientists might use a GDD baseline that is above or below 50°F.)

Because plant growth and development are strongly influenced by the amount of warmth that is accumulated over time, the total GDDs accrued during a given period can be used to determine when a specific crop should be mature and ready for harvest. GDDs are an aspect of the very interesting science of phenology — the study of how climatic and cyclic phenomena influence plant and animal growth and behavior.

A farmer might use GDDs to predict other phenological cycles, such as when certain insect pests will emerge from dormancy and become active. This method of keeping track of accumulated warmth is a more reliable indicator of plant growth and other cyclic events in nature than the calendar method, which provides dates based only on the historical record for a given area. Various websites keep track of GDDs for different North American regions and locales. Just type "growing degree days" in your web browser to learn more.

This is critical information, and it's the reason we track first and last harvest dates for any crops that are seeded at different times during the growing season. Ultimately, it enables us to create a more fine-tuned Weekly Planting Schedule, which, in turn, enables us to have the right amount of seasonally available produce to sell from the first market we attend in May to the last in December, all of which will be reflected in our Pick Lists and Harvest and Sales Records.

FIELD SCHEMATIC FOR SUMMER/FALL CROPS

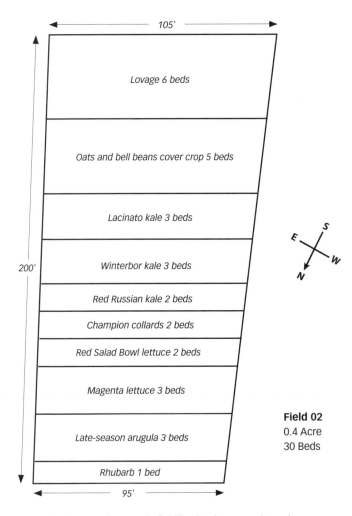

Schematic diagram of a sample field in Planting Records Book

Schematic for each field. In addition to the data shown on page 78, the Planting Records Book records the location of each crop within a field. This is done using a simple schematic diagram. If everything has gone according to plan, this schematic should be identical to or closely resemble the layout of crops on the crop rotation maps. But as already noted more than once, on a small diversified farm, things don't always go according to plan, so it might be a little different. Either way, the Planting Records Book provides a quick and easy way to look back at a previous year and see exactly what was planted, when, and where.

More Planning and Paperwork

In the event your appetite for lists and plans is not yet satisfied, you'll be glad to know that still others will be needed. Or you might already have had enough and want to stop reading this book right now. I would understand.

Weekly irrigation schedule. This is one thing every vegetable grower should have. We'll address this in chapter 8, where we'll consider the larger subject of irrigation and plant water needs.

Business tools. There will also be various accounting and business matters to take care of. You'll need a Farm Account Book or computerized system for keeping track of all inputs and expenses.

You'll need systems for filing receipts, maintaining employment records, and making sure you're on time with payroll taxes, disability, and workers' compensation insurance and other state and federal requirements that bedevil many small employers. These matters will be touched on in chapter 23. If you want to be certified organic, you'll need to develop an Organic Farm Plan that satisfies the requirements of the National Organic Program (NOP).

For a more exhaustive and well-organized look at the business end of farming, see Richard Wiswall's *The Organic Farmer's Business Handbook* (Chelsea Green, 2009).

Profitability assessment. Some farmers track and compare the profitability of individual crops to determine which to grow more of and which less, and which to drop altogether. This entails keeping a record of the cost of all inputs (seeds, fertilizer, compost, irrigation supplies, protective measures against pests and diseases, etc.) and an Hours-Worked Log, in which you record all the time you and your crew (if you have one) spend planting, weeding, irrigating, harvesting, transporting, and selling the crop in question.

After you've assigned a dollar value to your labor and added that number to your other expenses, you should be able to arrive at a cost-of-production figure. Compare this figure with harvest and sales numbers, and you'll find out whether an individual crop is a good bet or a bad one.

At first glance, the planning and recording systems outlined in this chapter may be a little overwhelming to those who are new to farming. My advice: Give them a try. They really are quite easy to put in place, especially if you tackle them one at a time.

These systems, or versions of them, should help any grower — neophyte or old hand — make informed decisions and maintain some control over the endless flow of details and data that come with the farming life. They are concerned with practical, on-the-ground realities. Their ultimate purpose is to help you get organized, maintain your sanity, and run a profitable and enjoyable business. Not all of the maps and lists we've covered need to be in place and in final form on day one. Many will, and should, evolve with your farm operation.

It goes without saying that learning how to look after your land and grow healthy food in a sustainable manner is of the highest importance. But the truth is, it's not enough. At the end of the season, you need to see a healthy profit as well. Otherwise, you'll have trouble paying your bills, you'll feel dispirited, and your farming career might come to an end. Think about the big picture, and you'll see that the various plans, maps, lists, and other assorted data described in this chapter are interlocking parts of the overall farm plan. One often depends on or leads to another. Leave out one or two vital parts and your otherwise best-laid plans may indeed go awry. It does pay to be organized.

PART

TOOLS OF THE TRADE

Chapter 4

Tractors and Tractor
Implements

L
ONG AGO, mechanization revolutionized agriculture. Today, it's
hard to get by without it. You can garden with a shovel, a fork,
a rake, and a few other simple tools, but to farm anything larger
than a small plot, you need more than this.

Tractors and the equipment they operate are designed to do a multitude of tasks. Both come in a dizzying array of shapes, sizes, and models. The subject is so vast it could fill the pages of many a hefty tome. Some folks are great tractor and farm machinery buffs. Not me, I'm afraid. Having never excelled in the mechanical domain, the broad subject of farm equipment is not my strong suit. This personal deficiency puts me at a disadvantage when interacting with farmers and others who are well versed on these manly matters. Fortunately, though, I've pieced together enough knowledge to get by, and my ability to run a small farm does not seem to have suffered greatly.

Without a doubt, I am heavily reliant on the three tractors I own and the several implements they operate. Over the years, I've gained a fair understanding of what they can and cannot do. I've taught myself to look after each of them to the best of my ability. I've become reasonably competent with a grease gun, an oil filter wrench, and a few other basic maintenance tools. I've learned to be on the lookout for telltale signs (and sounds) that a tractor or piece of equipment needs attention. I've developed good relations with neighboring farmers and mechanics who are willing to share their knowledge with me and tackle repairs when necessary. Their support is vital. If an important piece of equipment fails — be it the water-wheel transplanter, the manure spreader, or the truck that goes to market — I want it back in running order as quickly as possible.

I've grown wary of being totally dependent on one tractor or one piece of equipment to perform an important, time-sensitive task, knowing that sooner or later, Murphy's Law will take hold. The best defense

(84)

against Murphy, I've learned, is having a backup plan and some built-in redundancy — in other words, more than one way to achieve the same end, and more than one piece of equipment to get me there.

It's been a bit of a struggle, but the truth is, despite my mechanical limitations, I've grown surprisingly fond of my tractors and other farm machinery. When well maintained and accorded proper respect, they do the heavy work that is asked of them year after year. If they were taken away, I'd sorely miss them, and more to the point, I'd have to give up farming for a living.

Start with a Tractor

To the uninitiated, a tractor might appear to be not much more than a slow-moving vehicle with large rear tires and good traction. Indeed, by itself, there's not a whole lot you can do with such a machine. It's only when you pair your tractor with an appropriately sized farm implement that its utility becomes evident. Expand the range of implements, and a whole world of possibilities opens up.

Tractor Essentials

There are four basic features that make modern tractors (i.e., those manufactured since the mid-1900s) so useful; however, be sure that any tractor you buy, however old, has all of them, in working order. These four features are:

- Draw bar

- Hydraulic hose lines

- Three-point hitch

- Power take-off shaft

Draw Bar

The draw bar is a simple heavy-duty tow bar that enables you to pull objects with wheels, such as carts and wagons. It will also allow you to pull drag implements that don't have wheels, such as discs, harrows, and some old cultivators. Usually, it's just a matter of lining up the tow bar with the implement you want to pull and conjoining the two with a hitch pin. Almost all tractors have draw bars.

Hydraulic Valve Coupling

A remote hydraulic valve allows you to lift any drag implement that has a hydraulic cylinder and is attached to the tractor's draw bar, so you can transport it to the field without tearing up your driveway or lawn. Once you get to the field, you release the hydraulic pressure, which allows the implement to drop into position, whereupon you can go to work. Hydraulic couplings and hose lines can be used to do other things as well, such as open and close the gate of a manure spreader or adjust the position of an implement while on a slope. Make sure that any tractor you buy has at least one remote hydraulic breakaway coupling that will accept a hydraulic hose.

Three-Point Hitch

A tractor's three-point hitch comprises two sturdy arms that can be raised and lowered hydraulically and a third point known as a top link. Think of the two arms as the bottom corners of a roughly equilateral triangle and the top link as the central apex. Unlike the two arms, the top link has no hydraulic capability, though it has a telescoping capacity and can be adjusted manually.

Many useful implements are three-point-hitch compatible, meaning you can easily attach them to your tractor, transport them to the field, then lower them to the appropriate working depth. For example, you might want a chisel plow to penetrate 10 or 12 inches into the soil, but a rototiller to go only 4 or 5 inches deep. A brush hog, on the other hand, might be set to mow 6 inches above the soil surface. The three-point hitch lever, which is accessible from the driver's seat, enables you to adjust the implement to the depth that you want.

If you're considering a larger tractor (40 hp or above), make sure that the lower arms of the three-point hitch have a telescoping capability, so you can back up to within a few inches of an implement you wish to attach and use the telescoping arms to make the connection. This feature will make it a lot easier to attach heavy implements.

Power Take-Off Shaft

This splined (or grooved) shaft provides serious rotary power to implements with moving parts, like mowers, rototillers, spaders, and balers. It is usually located in the rear of the tractor and should be treated with great caution. Never make adjustments to a rear-mounted implement

Hydraulic valve
coupling

Arms of
3-point-hitch

Top link

PTO shaft

Drawbar

Rear view of tractor

while the PTO shaft is engaged and the engine running. It is quite capable of ripping your arm off at the socket. The PTO shaft functions like a second engine on your tractor and can provide power to many different implements. It can even run a stationary piece of equipment such as a log splitter, wood chipper, post-hole digger, or PTO-driven generator.

What Size Tractor?

Tractors and the implements they operate come in a range of different sizes. You want a tractor that suits the scale and nature of your operation. How big it should be will depend on the answers to the following questions.

- How big is your farm?

- How many acres of vegetables are you planning to grow?

- What is the terrain like? Is it level, hilly, undulating?

- What type of soil do you have? Is it a light, well-drained sandy loam or a heavier clay soil? Does it have lots of rocks?

- What implements are you planning to use with your tractor?

- What do you want your standard bed width to be?

- Do you want a four-wheel-drive tractor?

- How mechanized do you wish to be?

After having posed these questions, let me make a few qualifying and very general remarks about each of them. View my remarks as a way of thinking about the question of tractor size, not as ironclad advice. Just about everything I say could be disputed — even by me!

Farm size. As you might expect, bigger farms call for bigger tractors, even when the acres in vegetable production are few. Sooner or later, you'll find yourself needing a tractor to help with clearing land, harvesting timber, mowing fields, fixing fences, and transporting heavy objects or equipment.

Acres in production. Again, the answer is fairly obvious. If you're growing just a few acres of vegetables, a small tractor — say, under 40 horsepower — should be adequate. Now, I'll go out on a limb and suggest, somewhat arbitrarily, a ratio of horsepower to acres under cultivation:

2 acres or less: A 15- to 25-horsepower garden tractor, or possibly a heavy-duty walk-behind rototiller with a few different attachments

2 to 5 acres: A 25- to 40-horsepower tractor

5 to 10 acres: A 40- to 60-horsepower tractor

10 acres and above: Preferably two (or more) tractors — one in the 40- to 50-horsepower range, the others 60 horsepower or above

Terrain. Uneven and hilly terrain calls for a somewhat larger tractor. It takes more power to pull a heavy implement, such as a chisel plow, uphill than it does to pull the same implement over level ground.

Soil. A light, easy-to-work, rock-free soil requires less tractor power than a heavier soil with lots of rocks in it. Well-drained soils are easier to work than poorly drained soils.

Implements. This is an important one that you absolutely have to get right. Your tractor must be heavy enough and powerful enough to handle the implements you already have or intend to purchase. For example, it's unlikely you'll be able to run a 72-inch-wide rototiller with a 40-horsepower tractor. The tractor's three-point hitch probably couldn't even pick up a tiller of that width. In other words, you have to match implement size with the tractor you have, or vice versa. Anyone who deals in farm machinery should be able to advise you on these matters.

Bed width. This is another important consideration. The distance between the rear wheels of a tractor usually determines bed width. Typically, but by no means universally, beds on diversified vegetable farms are between 45 and 55 inches wide, with an additional 18 to 24 inches on either side for the tractor tires. See chapter 3 for more on this subject.

If you're planning on a 50-inch-wide bed, you'll need a tractor, or tractors, that have at least 50 inches between the rear tractor wheels, measured on the inside. The front wheels also need to be taken into account — the center-to-center distance between them and the two rear wheels should be roughly equal. This categorically rules out the old tricycle-style tractors that have a dual wheel right in the middle. It should also be noted that on most tractors, the distance between wheels can be adjusted, but it's not something you'll want to do too often.

Four-wheel drive. Tractors with four-wheel drive cost more, but they also have better traction and more pulling power than two-wheel drive machines of the same horsepower. This makes them more versatile and means that a lighter four-wheel-drive machine can often compete well with a heavier two-wheel-drive model. Four-wheel drive can also provide insurance against getting stuck. If you have to venture onto soft ground, do so in two-wheel drive. If the rear wheels start to spin, switch into four-wheel drive and exit the scene.

Level of mechanization. The more mechanized you are, the less human labor you will need on the farm. On the other hand, if you have a large labor force and a small piece of land, you can probably get by with less tractor power. But in general, mechanization will get certain jobs done a lot faster.

The Case for Big

In my mind, it makes sense to have more tractor power — say, an additional 15 or 20 horsepower — than you think you need, rather than less. A bigger tractor is built to do heavier work with relative ease. It will pull a large implement, like a chisel plow, more easily and at greater speed than a smaller tractor. This will get the job done faster and save you time.

Plan for that big job. Though you may anticipate doing only light tractor work, a heavy, demanding job is bound to come along at some point. When it does, you'll be glad to have a machine that can meet the challenge. Bigger tractors can handle both large and small implements, whereas smaller tractors are limited to smaller implements. If you run a small tractor too hard and ask it to do more than it was designed to do, you'll wear it out before its time.

Smaller tractors do have a few advantages, though. They can be ideal for light work such as shallow, in-row cultivation. They generally require less turning area, which can be a plus when you're working in cramped quarters, and they don't require as much storage space. A small tractor with narrow wheels will enable you to leave less space between beds.

Compaction. Smaller tractors weigh less and therefore are less likely to compact your soil, though this is not always the case — narrow wheels on a small tractor may concentrate as much weight per square inch as wider wheels on a heavier machine. This is because the big wheels spread the weight out over a larger area.

If all other things are equal, choose a tractor with large, reasonably wide rear wheels (about 15 inches), because it will reduce compaction and give you much better traction than the same machine with smaller, narrow wheels. Wider wheels will prove themselves when pulling an implement like a chisel plow that creates a lot of drag.

Cost. Another reason to opt for a modestly larger tractor is that they often don't cost much more than smaller ones. In some parts of the country, prices of small tractors have increased due to a surplus of weekend farmers with oversized backyards. The bigger machines, of which there are many, are more than these folks can handle.

Other Considerations

Unless you already have experience with tractors and farm equipment, my advice would be to take some time to educate yourself before making any decisions. Better to spend a season or two figuring out what will work best for you, rather than rush in and make large purchases that you'll regret later on. Find out what other farmers are using and what works well for them and what doesn't work so well. Here are more things to keep in mind.

Buy Used and Buy Old

My preference is for older, used equipment, especially when it comes to tractors. Older tractors cost a lot less than new ones — $5,000 to $10,000 should get you a respectable machine, versus $30,000 to $40,000 for a new one. Older tractors were built to last, and they are relatively easy to work on. They don't have sensitive and expensive electronics for which you need special instruments and training to understand. A 50-year-old tractor that has been well maintained can still have a lot of life in it, and if you stick with well-known brands (Ford, Case, International, Farmall, John Deere, Massey Ferguson, Allis Chalmers), you should have little trouble finding parts.

The Hours Meter

One way to assess how much life is still left in a tractor is to see how many recorded hours it has. Unlike cars that record miles driven on their odometers, tractors record hours of use; however, not all tractors, especially older models, do this in the same way. For example, some tractors count an hour of idling or light use (such as pulling a cart) as less than an hour — perhaps just half an hour. On the other hand, an hour of heavy use (such as pulling a chisel plow through heavy soil), with the engine running at 2,000 rpm or above, would normally be counted as a full hour. In other words, such a tractor is more accurately recording engine wear and tear, rather than just the hours that it is running.

Generally speaking, a tractor with fewer hours will cost more money. It's difficult to say how many hours are too many because not all tractors are the same, not all of them record hours in the same way, and there are many other factors to consider, such as:

- Make and size of the tractor

- Physical condition

- Ownership history

- Maintenance record (if you can obtain one)

- Type of use

- Sound of the engine

- Appearance of the engine and transmission oil

Personally, if all other factors were equal, I would be wary of smaller gasoline-powered tractors (say, under 40 hp) with more than 3,000 hours of use, and larger diesel tractors (say, over 50 hp) with more than 6,000 or 7,000 hours. That said, my largest tractor, a 90-horsepower John Deere 4040, has nearly 9,000 hours, and it's running just fine.

Maintenance

The best way to keep a tractor running is to maintain it well, preferably doing the basic work yourself. The operator's manual (always acquire one of these from the seller, from a dealer, or online) will tell you when to undertake various maintenance work, usually based on meter hours.

We change the engine oil and filters on our tractors once a year, even when the hours of use are under what the manual recommends. We also apply grease to all grease fittings, or nipples, at least annually. At less frequent intervals, we change transmission oil and replace air, fuel, and transmission filters. We keep a maintenance log in which we record each maintenance event, noting the amount and type of oil used and the identifying numbers of any filters replaced.

I strongly recommend a maintenance log for each tractor you own, even if it is just a sheet of card stock in the back of the operator's manual. It'll serve as a reminder of what maintenance work to do and when you should do it, and it will save you from having to research the types of oil and stock numbers of filters that you need.

I should mention that I have a good farmer friend and neighbor, Ted Stephens, who knows a lot about tractors and farm machinery. On all my trips to look at used tractors, Ted has come along to conduct a thorough inspection and test run. His knowledge and advice have guided me well. When Ted nods his head in approval, I know it's time to reach for my checkbook, or at least make an offer. When he looks down and moves his head almost imperceptibly to the side, it's a signal that we should look elsewhere. My point, of course, is that if you don't know much about the tractor or other equipment you're about to buy, it sure helps to have someone come along who does.

Diesel or Gasoline

Most older and larger tractors have diesel engines, and most serious farmers (i.e., those who farm for a living) are happy about this. Diesel engines don't have a carburetor or spark plugs and distributor points. There's less that can go wrong with them. They run cooler than gasoline engines and have more power at lower rpm. A well-maintained diesel tractor should last significantly longer than its gasoline counterpart and have fewer problems along the way.

One minor drawback with diesel engines is that they are hard to start in cold weather. This can be overcome by installing a block heater, if the

Sample John Deere 2350 Maintenance Log

Date	Meter Hours	Type of Work	Details	Amount
7/13/10	6387	Changed engine oil and filter	Rotella 15W40 JD filter T19044	9 qts
7/13/10	6387	Changed transmission/ hydraulic fluid and filter	JD Hy-Guard Trans/ Hydraulic Fluid JD Filter AR69444	7.5 gals
7/13/10	6387	Changed fuel filter Greased all nipples	JDAR50041	
6/18/11	6476	Changed engine oil and filter Greased all nipples	Rotella 15W40 Napa Gold 1243	9 qts
6/18/11	6476	Changed air filters	Inner element JDAE31724 Outer element Napa Gold 2631	
6/5/12	6564	Changed engine oil and filter Greased all nipples Cleaned air filter Cleaned radiator and screen	Rotella 15W40 Napa Gold 1243	9 qts

Here's a case where it's handy to have two tractors. The one on the right is coupled to a manure spreader; the other, which has a bucket, is loading the spreader. With only one tractor, the spreader would have to be uncoupled for each new load.

tractor doesn't already have one. At temperatures below freezing, you simply have to remember to plug the block heater into an electrical outlet about half an hour before starting the engine.

Many smaller tractors have gasoline engines, and they will function just fine, though they will be less capable of prolonged, heavy work, and their lifespan is likely to be shorter. On the positive side, a gas-powered engine should start in cold weather. Also, if you have some mechanical skills, a gasoline engine will be somewhat easier to work on. When things go seriously wrong with a diesel engine, you'll probably need the services of a qualified diesel mechanic.

Bucket

If you're going to own just one tractor, and if you can afford it, go for one with a bucket or front loader. They come in very handy. With a bucket you can pick up heavy objects and transport them around. You can turn compost or scoop it up and load it into a spreader. You can level out soil, remove large rocks, push snow out of the way, knock over small tree stumps, and do any number of other fun and useful things. Once you've owned a tractor with a bucket, you'll find that life without one is less agreeable.

Tractor Implements

I've already mentioned a few of the more common tractor implements that are used on diversified vegetable farms. Now, we'll take a closer look at some of them and the types of work they are designed to do. Let's start by dividing tractor work into five major categories, then consider the implements suited to each category:

- Tillage

- Planting

- Cultivation

- Composting and compost spreading

- Mowing

Note, this is by no means an exhaustive list. These five categories are the most relevant to organic vegetable growers. There are many other implements available and tasks they can perform. I'll list some of them at the end of this section.

Tillage

It's tough, if not impossible, to get vegetables to grow on ground that has other plants, such as weeds or a cover crop, already in residence. Tillage usually involves incorporating such plants into the soil so that they won't compete with the crop that you're planning to grow. If you have a backyard garden, tillage can be undertaken with a shovel, fork, mattock, or hoe. On a farm you're more likely to use a moldboard plow, disc, chisel plow, rototiller, or spader — all implements that would be attached to a tractor.

Primary and secondary tillage. Farmers often undertake tillage in two stages, which are referred to as primary and secondary tillage. Primary tillage may involve breaking up sod or incorporating a green manure, cover crop, or heavy stand of weeds. Secondary tillage is more often undertaken just before planting. Its main purpose is to prepare a satisfactory seedbed, or a bed for receiving transplants.

The Moldboard Plow

The moldboard plow is the traditional primary tillage implement. These days, almost all moldboard plows are three-point-hitch compatible. The plowshares cut 6 to 10 inches into the soil and turn it over completely, burying any existing plants in the process. Moldboard plows do a good job of turning over sod, weeds, and cover crops, but they leave an uneven surface, with multiple ridges, that is difficult to plant into.

The case against the moldboard plow. The majority of organic growers are not big fans of the moldboard plow because they feel that inverting the soil is a bad idea. Much of a soil's biological activity occurs in its top few inches, which is literally teeming with organisms, from the tiniest

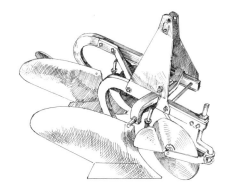

Two-bottom moldboard plow

microbes all the way up to larger critters like beetles and earthworms. Taking the residents of the penthouse and consigning them to the basement, as one author has put it, is seriously disruptive.

Repeatedly using a moldboard plow can also create a compacted layer, about 9 inches down, where the plow bottoms out. This is known as a plowpan, and it is not a good thing for your soil to have. A plowpan is likely to impede drainage and present a barrier to plant roots. Using a moldboard plow excessively can also expose a soil to increased erosion. Still, one can make an argument for the occasional use of this implement. When it comes to turning in a heavy cover crop or sod that has been in place for a long time, a moldboard plow is hard to beat.

Disc Harrow

After plowing a field, many farmers use a disc harrow to level out the furrows and ridges created by the plow. This implement leaves ground that is more receptive to seeds or transplants. Depending on the model you select, a disc harrow can be either three-point-hitch compatible or raised and lowered hydraulically. It will have a series of slightly concave steel discs that are usually arranged in two (sometimes four) sections and offset in such a way as to do a reasonable job of loosening and chopping up compacted soil. When used after a moldboard plow, a disc harrow would constitute secondary tillage.

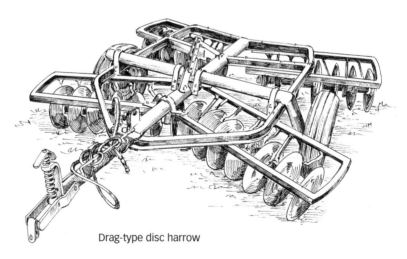

Drag-type disc harrow

My Tractors

My Story

OUR FARM STARTED OUT as a one-man operation (me), with 1½ acres of planted land and an assortment of garden tools, most of them acquired at yard sales. In the first year, I paid a neighboring farmer to plow a portion of a hay field. I then rented, from another neighbor, an old tractor and disc to level out the plowed land and create a seedbed. After that, it was pretty much rakes, shovels, forks, and a simple seeder.

Toward the end of my first year, I bought a used garden tractor from a retiring gardener with a half-acre plot. The machine came with a small one-bottom plow, a harrow, a rototiller, a cart, and a mower deck. The price for everything was $750. It was a good assortment of pieces to further my education, but before long, I could see the need for equipment with a lot more muscle.

At the beginning of year three, I bought my first tractor — a 1960s-vintage 38-horsepower Allis Chalmers D15 with a primitive bucket. I also picked up a used two-bottom plow and a disc. Together, the three pieces cost under $5,000. They are probably worth at least that much today. The Allis Chalmers is still in good running order and gets used frequently for lighter work such as cultivating and mowing. It has served me very well for more than 20 years. The disc still gets pulled into service now and then, but long ago I abandoned the two-bottom plow in favor of a chisel plow and rear-mount rototiller, both of which can be used on my two larger tractors, a John Deere 2350 and a John Deere 4040.

Spring-Tooth Harrow

A spring-tooth harrow, sometimes called a drag harrow, is another secondary tillage implement that might be used to smooth out the furrows created by a moldboard plow. This harrow relies on a series of flexible iron teeth to loosen and aerate the soil and to create a more level seedbed. Because they are strictly drag implements, with no hydraulic capability, spring-tooth harrows are somewhat outdated and not in high demand. This means that used models that still work quite well can be purchased inexpensively.

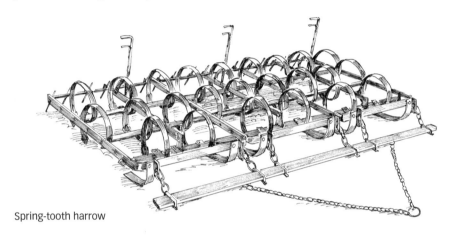

Spring-tooth harrow

Chisel Plow

A chisel plow is a primary, deep-tillage implement favored by many organic farmers. It opens up, loosens, and aerates the soil without inverting it. Crop residues and organic matter stay on or near the soil surface. Chisel plows attach to a tractor's three-point-hitch mechanism. They are usually set to penetrate to a depth of 8 to 12 inches, but some models can go deeper than that, if you have enough tractor power to pull them. At greater depths, chisel plows can do a good job of breaking up hardpans and plowpans, thus improving drainage and opening up to plant roots the subsoil and its storehouse of nutrients.

Typically, chisel plows have three, five, or seven shanks, usually set 10 to 12 inches apart. They are relatively inexpensive, but don't buy a chisel plow that is too big for your tractor. If you do, you'll be limited to using it at a shallow depth. Figure on needing 10 to 15 horsepower in your tractor for each shank on a chisel plow.

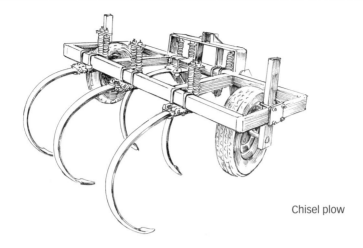

Chisel plow

A downside to the chisel plow. This implement doesn't do such a good job of killing weeds and grasses in long-standing sod or perennial covers such as clover and vetch.

Rototillers

A chisel plow is a good piece of equipment to own. It is among the least destructive of soil implements, but usually it will not leave you with a satisfactory seedbed. For this, you might use a rototiller (also known as a rotary tiller, rotary hoe, or rotavator), which works the soil with L-shaped, rotating blades or tines (usually four to six per flange). It connects to the tractor's three-point hitch and takes its power from the PTO shaft.

A rototiller can be used as either a primary or a secondary tillage implement. In some cases it may be the sole implement that is needed to prepare the ground for planting, in which case it is a primary and secondary tillage implement rolled into one. When a rototiller is used after a chisel plow has already loosened the soil, it would be considered a secondary tillage implement.

The downside to rototillers. Rototillers do a good job of incorporating surface plant material into the soil without burying it too deeply, which is why they leave behind a good seedbed; however, they are somewhat harmful to the soil's aggregate structure. Overuse can result in a pulverized and too finely ground soil that is less able to hold itself together and therefore is more prone to erosion. This problem can, to some extent, be ameliorated by using a rototiller with an adjustable gearbox that allows you to slow down the rate at which the tines turn.

A tractor's forward speed will also affect the turning speed of the rototiller tines. In general, it is wise to rototill as slowly as you can and no more deeply than is necessary to incorporate surface material. Some older tractors don't go slowly enough to rototill at an appropriate speed.

Rototiller size. Tractor-mounted rototillers come in different sizes, ranging from about 3 to 10 feet wide. The bigger the tiller, the larger the tractor you will need to operate it. For most small- and medium-size vegetable farms, tillers in the 3- to 6-foot range should be adequate. Most growers let the size of their tiller define the width of their planting beds, sometimes allowing some of that bed width to be taken up by the tractor tires. Figure on needing a tractor with at least 10 horsepower for each foot of rototiller width.

Depth adjustment. If you're in the market for a rototiller, look for one that allows you to adjust the till depth easily. To incorporate a cover crop after it is mowed, you'll need to till to a depth of about 6 inches. On the other hand, when tilling in annual weeds less than 6 inches tall, or preparing a bed for planting, you probably don't want to (or need to) disturb more than the top 2 or 3 inches of soil. You can adjust the depth on many tillers by loosening nuts and bolts and repositioning skids on either side of the implement. If you find this chore irksome and time-consuming, you're likely to let a one-size-fits-all depth prevail, which is counterproductive. Our rototiller has something called a rear-cage roller, which enables us to change the till depth in less than a minute by reaching back from the driver's seat and turning a crank handle.

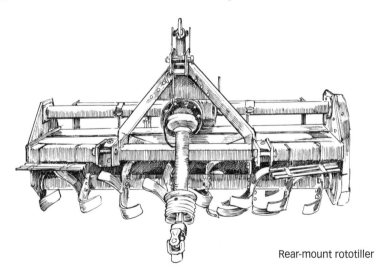

Rear-mount rototiller

Walk-Behind Rototillers

Sometimes referred to as walk-behind or two-wheel tractors, these machines are a definite option for beginning farmers who do not want to invest in a full-size tractor and accompanying implements. They also may make sense for growers who are working less than 2 acres of land.

The best walk-behind machines come with multiple attachments, such as rotary and sickle bar mowers, snowblowers, chipper/shredders, spaders, minidozer blades, hiller/furrowers, and, more to the point, rototillers. But when purchased new, these high-quality machines and their implements are expensive. And there is not an abundance of used models on the market, especially ones in good working order. BCS rototillers and walk-behind tractors are widely recognized as among the best and most versatile available.

Because you have to walk behind and physically guide a two-wheel tractor, the action and vibration of these machines can take a toll on your body, and they are slow going. They are best suited to light, easy-to-work soils, tight spots, and situations that call for limited use. Many growers who start out with a walk-behind tractor find themselves, a year or two later, in the market for a regular, four-wheel, sit-down machine.

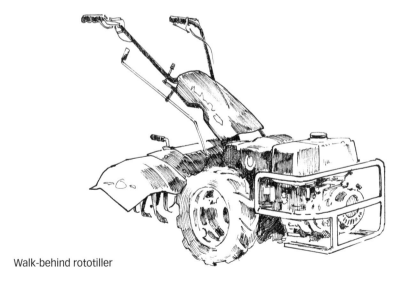

Walk-behind rototiller

Spaders

A spader is an interesting alternative to a rototiller. Instead of L-shaped, rotating blades, most spaders have a series of reciprocating or articulating shovels. These shovels have an up-and-down motion and penetrate the soil to different depths, usually 6 to 12 inches. They work with a digging action that loosens and opens the soil but does not turn it over. A point in favor of reciprocating spaders is that they don't require a lot of horsepower to operate, so they can be used on relatively small tractors. The spading action actually propels the tractor forward. Another type of spader — the rotary spader — requires a lot more tractor power. Its action is more circular and closer to that of a rototiller.

Proponents of spaders claim that they are easier on the soil than rototillers because they are less likely to break up the crumb structure. Detractors say they don't create as level a seedbed or do as good a job of incorporating plant material. Of course, not all spaders are created equal. Some probably do a better job of preparing a seedbed than others.

Spaders are more expensive than rototillers, and they require more maintenance and sometimes costly repair. Because they have a lot of connecting arms and moving parts, there is plenty of opportunity for something to break. In sandy, rock-free soils, they can perform well; however, if your soil has its fair share of rocks, including some big ones, you should exercise due diligence before buying a spader. If you do decide on this implement, be sure to get one with a PTO slip clutch, which will offer some protection in the event you encounter a large rock.

Reciprocating spader (rear view)

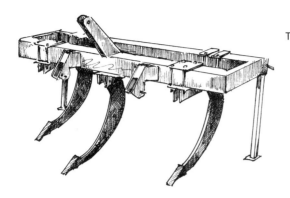

Three-shank subsoiler

Subsoilers

Though not, strictly speaking, a tillage implement, and not often found on small vegetable farms, the subsoiler can play a useful role. It is a three-point-hitch implement that resembles a chisel plow but has just one, two, or three heavy-duty shanks. Subsoilers can cut as deeply as 18 inches into the soil. They are good for breaking up plowpans, fragipans, and other impermeable layers in the subsoil. They improve drainage and aeration and allow plant roots greater access to subterranean minerals. Because they go so deep, you need substantial tractor power to pull a subsoiler — at least 25 horsepower per shank, along with good tire traction. If this doesn't do the job, try using a bulldozer.

Planting

It is not uncommon for small, commercial vegetable growers to set transplants in the ground by hand. We did this on our farm for more than 10 years. The process went as follows: We used a rototiller to prepare a planting bed, then straddled the bed with a tractor-mounted cultivator and cut narrow furrows a few inches deep. When planting lettuces we set up four tines on the cultivator, 12 inches apart, and cut four parallel furrows. For broccoli or kale, we set up three tines 18 inches apart. For winter squash, it was one row per bed.

Next, we came in with trays of transplants and pressed the individual seedlings into the furrows at whatever in-row spacing we chose. The final step was to water-in each seedling. It was an effective but somewhat slow process that required a lot of bending. Now we use a tractor and water-wheel transplanter, and the job gets done a lot faster — and with less stress on our backs.

(103)

The Water-Wheel Transplanter

The water-wheel transplanter is not the only type of vegetable transplanter on the market, but it is the first choice of many growers. This implement speeds up the planting process, promotes uniformity, and eliminates the need to hand-water after planting — a chore that can use up a lot of time.

The turning wheels of a water-wheel transplanter punch holes in the soil a few inches deep. They will also punch clean holes into plastic mulch, which most other transplanters cannot do. Water from two 80-gallon tanks mounted above the transplanter flows into the holes. Workers sitting on the back of the transplanter press seedlings into the wet holes as the tractor moves down the bed.

Bed size and spacing. The water-wheel transplanter we use is designed to straddle a 42-inch-wide bed. It can accommodate up to four parallel rows at a time. The spacing of the spikes on the wheels determines the spacing of plants within the rows. Standard options include 6 inches, 12 inches, 18 inches, 24 inches, and 36 inches. Custom spacing is also available.

Plastic mulch and bare soil. The water-wheel transplanter is primarily designed for planting into plastic mulch, but on our farm, we've found that it can also do a good job of planting into bare ground. In bare ground, wet soil can cake around the spikes. This is unlikely to be a bothersome issue in sandy soils, but it might be a serious impediment in soils with a high proportion of clay. You can reduce caking and mud buildup by wrapping strips of old inner tubing around the spiked wheels, and making holes for the spikes to protrude through. Even so, you might want to do a trial run with a water-wheel transplanter in your soil before investing in one of these potentially valuable implements.

Weight. Because water is so heavy (two 80-gallon tanks weigh over 1,300 pounds), you need a fairly good-size tractor (60 hp or above) to pick up a water-wheel transplanter using a three-point hitch. For smaller tractors, pull-type transplanters are available. These are attached to the tractor draw bar, rather than the three-point hitch.

Ground speed. Another factor to consider before buying a water-wheel transplanter, or any other transplanter for that matter, is the minimum ground speed of your tractor. At their slowest speed, most tractors will move too fast for a human to set a plant in the ground every 6 inches. Even at a 12-inch spacing, workers sitting on the back of the transplanter

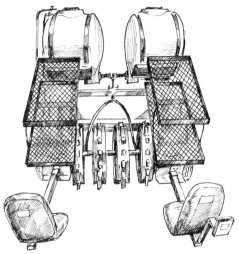

Water-wheel transplanter

This water-wheel transplanter is being used to transplant four rows of lettuce, with 12 inches between the rows and 12 inches between plants within the rows.

might have a hard time keeping up in a four-row bed system. This will not be a problem if you can find a tractor with a creeper gear — a supplemental transmission that permits very slow ground speeds.

We spent a couple of years looking for a used tractor with a creeper gear and enough horsepower to pick up a three-point-hitch water-wheel transplanter. Now we have one — a John Deere 4040 — and it can move as slowly as half a mile per hour. We use it for almost all transplants, including bare-root onions, which we space four rows to a bed, 6 inches between plants.

Cultivation

The word "cultivation" can be used in a few different ways, with somewhat different meanings. In the context of farming, one meaning is controlling and destroying weeds using tractors and implements known as cultivators. This is known as mechanical weed control, as opposed to chemical weed control, which relies on the use of herbicides. The subject of mechanical weed control, or cultivation, is a large one, and there are numerous options available to a farmer — far more than there is space for in this book. Here, we'll take a look at just some of the different tractor-operated cultivators that are available. There will be more on weed control in chapter 20.

Oldies but Goodies

Two old-time but still popular midmount cultivating tractors are the Allis Chalmers G and the Farmall Cub. Both are lightweight machines with narrow tires. They are easy to maneuver and give the operator an unobstructed view of the plants being cultivated, which allows for precise weeding. The G's engine is behind the driver's seat, so you don't breathe any exhaust fumes. Though the last Gs were made in the late 1950s, many have been refurbished and are in use today. They are still sought after by small- and medium-size organic growers.

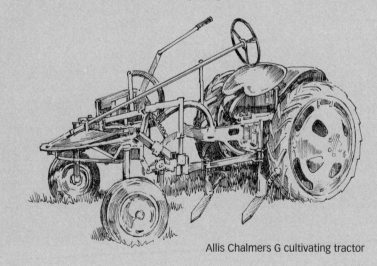

Allis Chalmers G cultivating tractor

Determining the best cultivator for the job will depend on such variables as:

• The vegetables you are growing and their size

• The type and size of the weeds

• The type of soil in which both vegetables and weeds are growing

Before investing in cultivation equipment, it would be wise to consult with your local Cooperative Extension agent and organic association, or talk to neighboring farmers who are growing similar crops under similar soil conditions. Find out what they are using and what they like best.

Cultivators are usually picked up and lowered by the three-point hitch at the rear of the tractor, although a few are attached to the front of the tractor, where a bucket might otherwise be. And some are mounted in the middle, right under the driver's seat — these are known as midmount cultivators. Compared with the standard three-point-hitch cultivator, front- and midmount cultivators give the operator a better view of what he or she is doing and therefore allow for closer cultivation. The problem is that you need specialized and specially outfitted tractors to accommodate these mid- and front-mount implements. But, if you're planning to do a lot of tractor cultivation, they are good things to have.

To cultivate a crop plant completely, you must remove weeds in three different places: Between the rows of plants, between the plants within each row, and in the area between the beds where the tractor tires go. The second place — between the plants within each row — is the most difficult to achieve of these three when using a tractor and cultivator. If you don't have a cultivator that can do this, you can always use hand hoes to dispatch these in-row weeds.

Blind Cultivators

Also called spring-tine or flex-tine weeders, these cultivators have a large number (50, 60, or more, depending on the width of the implement) of narrow, L-shaped, rake-like tines that drag across the entire planting area, hitting weeds and crop plants alike. They work quite well when the weeds are tiny and easily uprooted and the crop plants are big enough to withstand the moderate dragging pressure that the spring tines exert. The tines can be adjusted for tension. Gauge wheels, which are usually a worthwhile add-on, allow you to easily adjust the implement's depth.

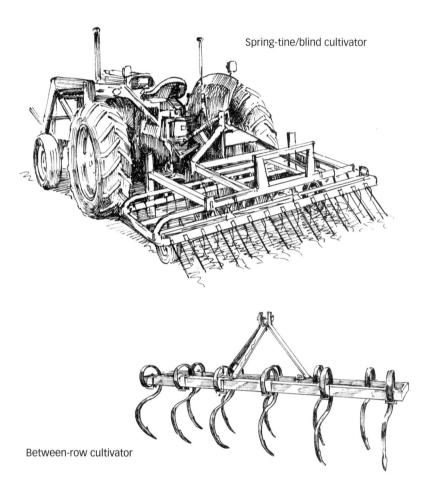

Spring-tine/blind cultivator

Between-row cultivator

We use this type of cultivator extensively on our garlic in spring and early summer. We've uprooted literally millions of small weeds but never a single garlic plant. We also use it on direct-seeded crops such as carrots, turnips, and radishes when the plants are at least a few inches tall and in little danger of being dislodged. We don't use this implement on leafy greens such as lettuces and kale because it has a tendency to tear the leaves.

Blind cultivators are useful for covering green manure and cover crop seed when these are sowed with a broadcast seeder rather than a seed drill. They can also be used to break up a crusty seedbed and eliminate small weeds immediately before direct-seeding or transplanting a crop plant. They do very little damage to the soil structure.

Between-Row Cultivators

These are the most common cultivators, and most of them are three-point-hitch compatible. They often use sweeps or S tines mounted on a tool bar. The sweeps or tines, which are available in different sizes, can be adjusted to fit adequately between rows of vegetables. The idea is to uproot as many weeds as possible without disturbing the crop plant. Some between-row cultivators come with gauge wheels that allow one to better control the depth of the sweeps or S tines.

In-Row Cultivators

These belong to a more complex group of cultivators that usually incorporates more than one function. Some farmers use sweeps or S tines to cultivate between rows, then rely on discs, spring hoes, or side knives to hill the soil around the crop plants, with the intent of burying small weeds. Others use starfishlike rotating fingers to disturb the soil as close to the crop plant as possible. Mid- or belly-mounted in-row cultivators will provide the greatest accuracy.

Another type of in-row cultivator is the basket weeder. This device uses metal cages to uproot shallow weeds without hilling the soil on either side, which makes it useful for cultivating small plants at an early stage of growth. Basket weeders work best in friable, rock-free soils.

Disc harrows and spring-tooth harrows, mentioned above, can be used to control weeds on land before planting without causing much damage to the soil. *Steel in the Field: A Farmer's Guide to Weed Management Tools* (see Resources) is a good book on weed control implements.

Composting and Compost-Spreading Equipment

Small compost piles can be easily turned by hand with a fork or shovel, but if you want to make enough compost to spread across an acre or more of land, you'll find that a tractor with a bucket will come in very handy. The bucket will enable you to turn, mix, and aerate your pile. This will speed up the composting process and create a more uniform finished product.

Manure Spreader

Once compost is made, it can be spread across a field before planting. A typical application rate is 25 tons per acre. This will amount to roughly ¼ inch of compost across the entire acre. On a garden plot, this can easily

Rear-throw manure
(or compost) spreader

be done with a wheelbarrow or a pickup truck and a shovel. For larger areas — say 0.5 acre or more — a manure spreader, and a tractor to pull it, will do the job very nicely and save you a lot of work.

Most manure spreaders are hitched to the tractor's tow bar and rely on the PTO shaft to throw the compost (or manure, or whatever you are spreading) out to the side or rear of the spreader. Those that throw the material to the side work best with dairy manure, which is usually in a somewhat liquid state. For compost, a rear ejection spreader is better.

New manure spreaders are expensive, and old ones, because they have gearboxes and moving parts, are often a little worse for wear. Before buying a used spreader, make sure it is in working order. In particular, check the gearbox, chains, and sweeps.

Mowing

Lawn mowing is not the only type of mowing that needs to be done on an organic vegetable farm. Often, cover crops are mowed before being incorporated into the soil. It's also wise to mow weeds before they go to seed. Field perimeters and access routes for tractors and pickup trucks need periodic mowing, unless you're prepared to wade through grasses and weeds several feet tall. The three most commonly used mowers on vegetable farms are:

- Brush hogs
- Sickle bar mowers
- Flail mowers

Brush Hogs

The brush hog is a relatively inexpensive rotary-type mower with two heavy-duty blades that, unlike lawnmower blades, are not rigidly attached to an axle. Rather than cut through vegetation, the free-swinging blades whack it down. If they encounter a rock or stump, the blades recoil on themselves, then bounce back without causing damage to the implement, though the blades themselves might be dented or nicked.

Brush hogs do a reasonable, though somewhat rough, job of felling cover crops and high grasses, as well as heavier brush, scrub, and small saplings. They have two significant drawbacks, however: They throw the cut material out to one side, rather than spreading it evenly, and because they are mounted to the three-point-hitch in the rear of the tractor, they miss a lot of the material that the tires ride over and flatten.

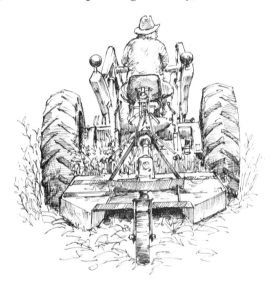

Brush hogs run off the tractor's PTO shaft and can be dangerous. Don't use them in the vicinity of people or animals. They can cast a rock or other object out to the side at great velocity and over a distance of a few hundred feet. The original brush hogs were made by the Bush Hog Company and carry that name.

Brush hog

Sickle Bar Mowers

Sickle bar mowers operate off to the side of a tractor, rather than behind. They use a series of reciprocating teeth mounted on a bar that is usually 6 or 8 feet long. Sickle bar mowers run off the PTO shaft and usually attach to the three-point hitch in the rear of the tractor, though there are some mid mount models.

Sickle bar mowers are good for cutting grasses and most cover crops but are not suitable for cutting heavy brush or scrub and do not appreciate encountering rocks or other solid objects. Their major advantage is that they cut freestanding vegetation off to the side before it is laid flat by the tractor wheels — something a brush hog is not able to do.

Flail Mowers

Flail mowers use small blades mounted on the ends of chains to cut through grasses, cover crops, and scrubby vegetation. They do a cleaner cut than a brush hog and chop up the material they encounter, leaving it on the ground where it can act as a mulch. Flail mowers can operate on rough and uneven ground without getting fouled up, and they can deal with small rocks. Like the sickle bar mower and brush hog, they attach to the tractor's three-point hitch and rely on the PTO shaft for power. Because they are rear-mounted, they miss much of the vegetation in the wheel tracks, just like brush hogs.

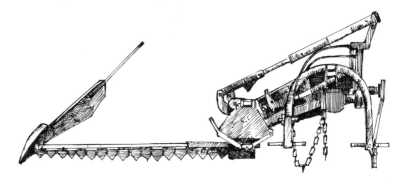

Sickle bar mower

Flail mower

Walk-Behind Brush Mower

These machines are like walk-behind lawn mowers, but they have just one heavy-duty, rigid blade, and they usually cut at a higher level than lawn mowers, meaning that they are less likely to hit rocks or other obstructions. They are useful for cutting tall grasses and weeds in the access paths or strips between beds of vegetables. The access paths are where the tractor tires have gone. These paths are often somewhat compacted, and therefore not easy to cultivate.

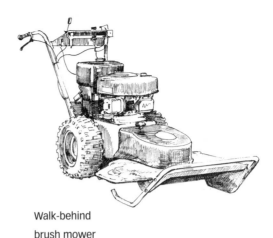

Walk-behind
brush mower

Other Tractor Implements

As I already noted, the subject of farm equipment, like many of the other matters discussed in this book, is a big one. One could spend a lifetime and still not have a mastery of all the material. There are many different paths to the same end — a healthy, productive, and profitable farm. Over time you'll discover what works best for you.

The four categories of implements just discussed are those most commonly found on small and mid-size vegetable farms, both organic and otherwise. But the list is not exhaustive. There are other entire categories of tractor-compatible implements that might, sooner or later, find a rightful place on your farm, depending on your land, your crops, your markets, and your resources. Here are some of them:

- Plastic and paper mulch layers

- Plastic mulch lifters and winders

- Fork lifts (some attach to a three-point hitch)

- Raised-bed shapers

- Hilling discs

- Potato diggers

- Boom sprayers

- Air-blast sprayers

- Fertilizer spreaders

- Rock pickers

- Broadcast and drill seeders

- Assorted harvesting equipment

In case you're wondering about the inclusion of sprayers and fertilizer spreaders on an organic farm — yes, there can be a place for them. Every year, as demand for organic produce increases, there are more organically approved pesticides and natural fertilizers on the market. But that is not to say you will necessarily need to use them.

Animal Traction

Before the invention of tractors and the implements they operate, nearly all grain and vegetable farmers relied on animals — usually oxen or horses — to do much of the heavy lifting and pulling. For the longest time, man and his beasts of burden were codependents, working side by side. All mowing, tilling, planting, and harvesting equipment was designed to be used with animals. The equipment was not PTO or three-point hitch compatible, neither was it receptive to hydraulic hose lines, but it got the job done in its own way, on a smaller, more intimate scale than modern industrial agriculture tolerates.

Times have changed, but not altogether. Though no longer common, the practice and art of farming with animals has not disappeared and, in fact, appears to be on the rise. Draft animals are making a modest comeback, and for some good reasons. Of the approximately 160 young, and sometimes not-so-young, people who have worked on our farm, at least half a dozen have gone on to pursue farming careers that employ large animals, though not always to the total exclusion of tractors. Here are some of the reasons they give for taking this perhaps more challenging path:

Animal Traction: PROS

- Draft animals don't require increasingly expensive petrochemical oil and fuel.
- They are truly solar-powered engines, running on water, grass, and grain.
- They don't pollute our atmosphere and exacerbate global warming.
- They compact the soil less than tractors.
- They seldom malfunction or break down in the middle of a job.
- They don't require costly repairs.
- They can be used individually or in combination, depending on the task and the pulling power needed.
- They breed and, in so doing, perpetuate themselves, at little cost to a farmer.
- They make very little noise.
- They can work safely on steep slopes.
- They can turn and maneuver under tight conditions.
- They provide manure that can be used to make compost and enrich the soil.
- They are less expensive than tractors.
- They are a pleasure to bond and work with.
- Implements used with animals tend to be less expensive and relatively easy to maintain.

These do seem to be compelling reasons to opt for horses or oxen over tractors, but before you make that decision, let me mention some of the drawbacks:

Animal Traction: CONS

- Draft animals need to be provided with food and water every day of the year.
- They might require the services of a vet, which can be expensive.
- They need a fair amount of attention — feeding, grooming, and moving to fresh pasture.
- They might need shelter in winter, and if so, this shelter will need to be kept clean.
- In most instances, they will be slower than tractors, often significantly.
- They require much training and skillful handling.

- Working with draft animals can be physically demanding.
- They need to be worked frequently to remain in good fettle and remember their commands.
- They are large, powerful animals and potentially dangerous, especially to unskilled handlers and children.
- Animal traction implements are specialized and not always easy to obtain, and the range of applications is limited.

Working with draft animals will call for considerable time and commitment on your part. If it's the direction you want to take, I strongly recommend that you spend a year, at the very least, working with a farmer who has ample experience in this field. Even if you do decide to use draft animals, you might find that a tractor with a bucket will come in handy, too.

CHAPTER 4 RECAP

Tractor Buying Tips

▶ Bigger is generally better. Be sure to buy a tractor that is compatible with, and big enough to handle, the implements you already have or plan to purchase.

▶ All things being equal, choose a tractor with about 15-inch-wide rear wheels to reduce compaction and give you good traction.

▶ Older, used tractors are cheaper and easier to repair than new ones.

▶ Consider other factors — such as make and size, physical condition, and maintenance record — when assessing meter hours.

▶ If you can afford it, buy a tractor with a bucket or front loader.

▶ If you have a friend who knows tractors well, take him or her with you on shopping trips.

Chapter 5

....

Small Equipment and Tools

U NLESS YOU HAVE AN UNLIMITED SUPPLY of willing workers, it's hard to grow more than a couple of acres of vegetables
without the aid of a tractor (or possibly a horse) and a few basic
implements. This is especially true when it comes to field preparations
such as chisel plowing and rototilling. But it doesn't mean that hand tools
have no place on a small diversified farm — far from it. Many organic
growers, including this one, rely on an assortment of wheel hoes, hand
hoes, rakes, forks, shovels, wheelbarrows, and other miscellaneous tools
on a daily basis.

Crop diversity and numerous sequential plantings of the same crop
are the norm on small organic farms. Under these conditions, the use of
hand tools — especially for weed control — often makes the most sense.
Let's use Swiss chard as an example.

In a typical year, we seed Swiss chard in our greenhouse on seven
separate occasions — once in early March, once in April, once in May,
three times in June, and one last time in the first half of July. We aim for
anywhere between 300 and 1,200 seedlings per sowing, depending on
the month, with the larger numbers occurring in June so we have a good
supply of chard to take us through fall. The total for the season is usually about 4,500. In aggregate, this is a fair number of seedlings, but the
amount of land required for each individual planting is quite small. Let's
take a closer look at the numbers.

In the field, we plant Swiss chard three rows to a bed, with 18 inches
between rows and 12 inches between plants within each row; therefore,
a 200-foot row would contain two hundred plants, and a 200-foot bed
would contain six hundred. Remember, our beds are 76 inches (6.33 feet)
on center, so a 200-foot bed will require 1,266 square feet (200 × 6.33),
which is only 0.03 acre (1,266 ÷ 43,560). Not a lot of land.

In my view, at this scale, tractor cultivation would be overkill — both
unnecessary and impractical. Moreover, it would increase the likelihood
of soil compaction. Instead, we rely on workers who are proficient with

wheel hoes and hand hoes to keep weeds at bay during the few weeks that the plants are most vulnerable to competition. One or two persons weeding for a half hour, every other week, is usually enough to take care of business, as long as we don't let the weeds get too far ahead of us.

Of course, one might argue that if there were numerous plantings of similar crops at the same row spacing, a small cultivating tractor might do the work more expeditiously. And this is probably true.

The Tool Shed

Near the corner of our closest field, we have a small (12' × 12') shed sitting on a concrete pad. This shed is stocked with hand tools that are brought into service almost every day. Let's take a look at some of the standbys and what they are used for:

Wheel Hoes

A wheel hoe is a human-powered, walk-behind cultivator with either one or two wheels or, in some rare instances, four. They are used primarily for controlling weeds between parallel rows of vegetables. The one-wheeled models work between rows. The two-wheeled models can straddle a single row of vegetables and take out weeds on either side — while the vegetables are still small. As the crop grows larger, they can be used between rows, just as the one-wheeled models.

Some wheel hoes use small cultivating tines or teeth to scratch their way through the soil up to a few inches deep. These tines do a fair job of uprooting both small and larger weeds but, because of their hooked shape and the depth to which they penetrate, they cause significant soil disturbance. This often brings dormant weed seeds closer to the surface, where they are likely to take advantage of newfound light and promptly germinate. In other words, the tine-type hoes can solve your immediate weed problem but, in the process, create a future one.

Other wheel hoes use blades, sweeps, or oscillating stirrups to undercut weeds, rather than uproot them. These hoes are used for shallow cultivation and work best when weeds are small — less than a few inches high. They disturb the soil surface less than the tine hoes and are therefore less likely to bring on another flush of weeds. They also require less physical effort. Most growers prefer this type of hoe. We have at least half a dozen wheel hoes in our shed, and a few different models. The ones we reach for most often are the oscillating stirrup type. Depending on the spacing between our rows, we use interchangeable stirrups that are 4, 6, or 8 inches wide.

Height adjustment. If you're in the market for a wheel hoe, look for one that has a height adjustment feature that allows the handles to be set easily to match the height of the user. Taller people will want the handles set higher, and shorter folks will want them set lower. Most people lean forward slightly when using a wheel hoe. The idea is to let the legs do most of the pushing, rather than the arms and upper body. A reasonably fit individual, with a correctly adjusted wheel hoe, should have no trouble working for a couple of hours at a stretch.

Offset adjustment. This is another good feature in a wheel hoe. The offset adjustment enables you to cultivate off to one side and avoid walking on the area the hoe has just passed over and loosened.

Attachments. Some wheel hoe manufacturers offer a few different attachments. A good one to get is a V-shaped furrower. This can come in handy when you need a planting furrow and don't want to use a tractor to make one. It can also be used for hilling plants such as carrots and leeks.

Prices. The better-known wheel hoes on the market are made by three different companies — Valley Oak, Glaser, and Hoss. You can buy them online or through various seed and garden supply catalogs. Prices range from $250 to $400.

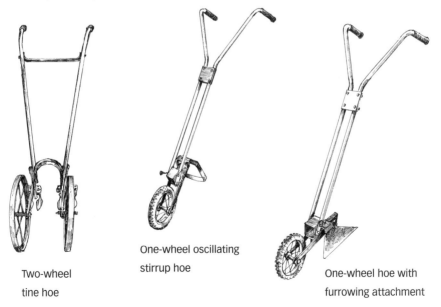

Two-wheel
tine hoe

One-wheel oscillating
stirrup hoe

One-wheel hoe with
furrowing attachment

Long-Handled Hoes

Wheel hoes control weeds between rows of vegetables. Hand hoes can do this, too, but they are especially good for working between plants within a row, which is something that a wheel hoe is not designed to do. There are plenty of different hand hoes to choose from — they come in a variety of shapes and designs, both long- and short-handled, each touting its own virtues and applications.

Some long-handled hoes are suited to shallow cultivation and small weeds; others, especially the tine-style hoes, tackle weeds both large and small, but like tine-style wheel hoes, these more versatile tools tend to rough up the soil surface and expose weed seeds to light. Some hoes have small blades and more precise action — they are ideal for getting between closely spaced plants, such as onions or mesclun salad greens; other hoes are heavier and sport broader blades that can dispatch a larger number of weeds in one pass.

If you're anything like us, you'll end up owning several different hoes and will take some satisfaction in selecting the most suitable one for the job at hand. Here's a rundown on some of the better hoes that are available today.

Stirrup hoe. Popular and easy to use, most stirrup hoes have oscillating stirrups with thin blades that cut through small weeds just below the soil surface. They are used with a back-and-forth or push-and-pull motion. The user has full visibility of the two sides of the hoe at all times, which reduces the chance of nicking crop plants. The stirrups come in different widths, from 3 to 8 inches. Blades are replaceable, which is a good thing, because they wear down after substantial use in rocky soil.

Trapezoid hoe. This sturdy hoe has a cutting blade with sharp corners and a beveled shape. It is good for larger weeds, especially those with taproots. The blades are replaceable.

Collinear hoe. Designed by Eliot Coleman, the collinear hoe slices through the soil just below the surface. It works well in tight spaces and when handled skillfully can get under the leaves of heading crops such as cabbages, escarole, and lettuces without causing damage.

Toothed cultivator hoe. This hoe is designed to penetrate more deeply into the soil. It is used to uproot larger weeds and loosen a crusty soil surface. It can also be used to incorporate compost or organic fertilizer.

The CobraHead hoe. This is a versatile, all-purpose hoe that can work the soil surface or dig more deeply to uproot larger weeds. The manufacturer compares it to a "steel fingernail," and with some justification. The CobraHead can also be used to cut shallow furrows for planting. Its sturdy but small blade allows for precision weeding.

The Hooke 'n Crooke hoe. This extremely sturdy undercutting hoe has become one of our favorites. The flat, rigid blade is used in a back-and-forth motion to slice through weeds both large and small, with minimal disturbance to the soil surface. The grooved tip can be used to isolate and uproot individual weeds. If flipped over, the Hooke 'n Crooke can be used like a small rake to remove cut weeds or smooth out the soil surface. This hoe is available in two sizes: the larger Heron model with a 6-inch blade and the smaller, somewhat lighter Honey Bee, which has a 4-inch blade.

Prices. Well-designed and well-made hand hoes are not inexpensive. You can purchase replacement blades for most of them, and the best ones will last for many years. Over the course of a farming career, you're likely to spend countless hours working with hoes. Quality tools are a pleasure to use; they prove their worth in the long run, and often in the short run, as well. If a friend or family member is having a hard time decid-

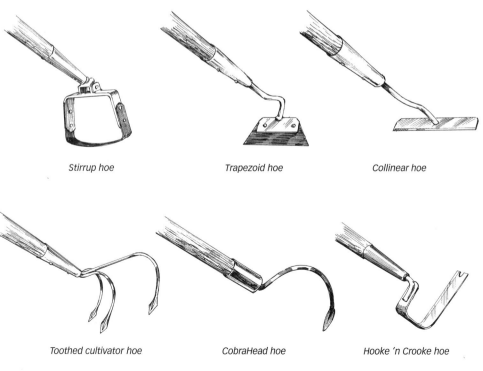

Stirrup hoe   Trapezoid hoe   Collinear hoe

Toothed cultivator hoe   CobraHead hoe   Hooke 'n Crooke hoe

ing what to give you for your birthday, tell them you'd be happy with a good hand hoe.

Short-Handled Hoes and Cultivators

These hoes are held in one hand and are normally used when kneeling or sitting. They are good for very delicate weeding around direct-seeded crops such as carrots or onions soon after germination, when a long-handled hoe might be too blunt an instrument. There are several models on the market, and as is the case with long-handled hoes, it makes sense to try at least a few of them to find out which ones work the best for you.

Both Cobra and Hooke 'n Crooke have one-handed versions of their larger hoes. Both are well made and good for their intended purposes: The Cobra is more of a precision scalping and digging tool, while the Hooke 'n Crooke Hummingbird model does a nice job of undercutting small- and medium-size weeds. Check out garden supply and seed catalogs to view the different hand hoes available.

Other Useful Field Tools

Shovels. Every farm needs a few good shovels. You'll use them for planting trees, turning over small patches of ground, cleaning out a manure spreader, burying the edges of plastic mulch or row cover, and just plain shoveling. Go for quality.

Forks. A good digging fork is hard to find but worth looking for. We use forks for digging up carrots, turnips, potatoes, celeriac, and sometimes garlic. The problem with cheap forks is that they tend to break under heavy use. After they are sunk all the way into the ground, many forks cannot withstand the full weight of a human body bearing down on them — at least not repeatedly.

Even with a high-quality, high-priced fork, it is better to ease off a little and withdraw the tines a few inches when the soil is dry and offering a lot of resistance. Don't test the tool's limits. Hay and manure forks, or pitchforks, have longer handles and thinner tines. These are no good for digging but work well for pitching around straw, hay, manure, or piles of weeds.

Right-angle trowel. This is a useful tool for setting plants in the ground. With an easy motion, it enables you to create a wedge-shaped hole big enough to accommodate the plug or root-ball of your transplant. It is good for herbs and other limited plantings.

Other Field Tools

Here's a list of other field tools that come in handy on our farm:

- Rakes
- Wheelbarrows
- Garden carts
- Hedge clippers
- Lopping shears
- Trowels
- Dandelion weeders
- Flame weeder
- Sickles
- Scythes
- Chain saw and protective gear
- Weed whacker

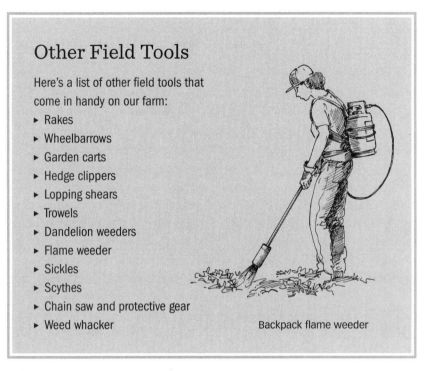

Backpack flame weeder

Stake pounders. If you grow tomatoes, peas, or other crops that will benefit from staking or some form of trellising, a stake pounder is a useful tool to have. It will work better than a sledgehammer, which is likely to chip, split, or mushroom your stakes. It will also make it easier to set your stakes vertically, rather than at different angles. A good stake pounder should be heavy enough to drive a stake into the ground without your having to exert too much downward pressure. Instead, focus your energy on the lifting. But even then, stake pounding can be vigorous work, especially in dry or rocky soil. Wait for some moisture, if you can.

Stake pounder
and stake

The Workshop

To my mind, a good workshop will be spacious and have good lighting, a concrete floor, a large and very solid bench with a vise grip on it, plenty of electrical outlets (including at least one 220 outlet), ample shelving, and a heavy-duty pegboard. Finally, it would be nice to have a large sliding door that would permit you to bring in a vehicle or tractor. But these are things to aspire to. A lesser workshop will take you a long way.

Workshop Tools

Depending on your needs and do-it-yourself skills, this can be a long list. We started out with a few hammers, screwdrivers, pliers, wrenches, socket sets, and the like, many of them picked up for pennies on the dollar at auctions or yard sales. Next came a number of power tools: A drill, skill saw, chop saw, table saw, grinder, and sander. Within a few years, we had acquired an air compressor and a battery charger. Later, an arc welder and then an oxyacetylene welder found their way onto the farm, but I have to admit that I'm far from proficient with either of these — they are generally used by others on the farm who have proper training. In New York State, the Boards of Cooperative Educational Services (BOCES) offers adult education courses in practical subjects ranging from welding to small engine repair. Many other states offer similar training and adult education programs.

The longer you farm, the more you'll understand which tools will make your life a little easier and enable you to get work done more expeditiously. You don't need a fully equipped workshop on day one. A few simple items are enough to get started. In good time, your supply of tools will increase. Each new addition will come because you have a need for it.

Most farmers will keep acquiring tools and equipment for a long time, perhaps right up to retirement and even beyond. In a way, it's part of the job description, part of the learning process. There are a lot of useful tools out there, many of them not specifically designed for farmers and gardeners; nonetheless, it's good to have them on a farmstead and become familiar with them.

Suggested Tools

Here are some of the tools I've used to stock my workshop. The list is by no means exhaustive.

BASIC HAND TOOLS
- Hammers
- Screwdrivers
- Wrenches (English and metric)
- Socket sets (English and metric)

POWER TOOLS
- Drill
- Skill saw
- Chop saw
- Table saw
- Grinder
- Sander

LATER ACQUISITIONS
- Air compressor
- Battery charger
- Welding equipment

MORE USEFUL ITEMS
- Solid workshop bench
- Vise to go on one corner of the bench
- Stepladder and an extension ladder
- Hacksaw, jigsaw, and bow saw
- Fencing pliers
- Wire cutter
- Bolt cutter
- Pipe wrenches of different sizes
- Assorted nuts, bolts, screws, and nails
- Assortment of spring clamps
- Crowbar and pry bar
- Grease gun
- Hydraulic jack
- Staple gun
- Pocketknife (I'm partial to the useful Leatherman multitool; however, it takes a few seconds to access the blade.)

Root Mass Farm

Proprietors: Landon Jefferies and Lindsey Shapiro
Location: Oley, Pennsylvania
Year Started: 2011
Acreage: 2 acres
Crops: Mixed vegetables, herbs, small fruits

LANDON JEFFERIES HAS WORKED AROUND FARMS since he was 14 years old. He started out working summers at a farm stand near his hometown of Westford, Massachusetts. A few years later, while attending Vassar College, he signed on for summer work at Butterbrook Farm — a diversified organic farm not far from his home. Then, when he was just 20 and still a little shy, he came to work at our place in 2005. He put in two seasons with us, the second in the role of crew manager — our youngest ever. He helped assign tasks, worked alongside the rest of the crew, and took charge of our Wednesday farmers' market in New York City.

After graduating with a degree in political science (the major, he informed me, with the fewest requirements), Landon moved to Philadelphia to be with his girlfriend, Lindsey Shapiro. For a few years, he managed farmers' markets around Philadelphia, worked at farm stands, and managed a small urban farm on a historical property. But what he really wanted to do was start a farm of his own. The problem was finding land to do it.

Land for Vegetables

In 2010, Landon found what he was looking for in the rural town of Oley, Pennsylvania, an hour northwest of Philadelphia. A large family with farming aspirations, some idle land, and plenty of mouths to feed, agreed to lease a few acres and a small house to him and Lindsey, in exchange for a regular supply of organically grown vegetables.

In summer 2010, Landon and Lindsey set about creating Root Mass Farm. They were able to use the landowner's tractor and moldboard plow to work up an acre of clay loam soil that had been in pasture for several years. They seeded buckwheat as a smother crop at the end of June and followed it with a cover of alfalfa in early September. The following spring, they planted some 30 different vegetables (including two thousand heads of garlic — most of it from our Italian 'Rocambole' stock but also some of a sharper hardneck variety called 'Georgian Fire'), along with assorted herbs, melons, and berries. They signed up with two farmers' markets in Philadelphia and established what they call a "stakeholder account" — their term for a debit-style, farmers' market CSA — to bring in some cash up front.

When I spoke with Landon, he told me he was satisfied with Root Mass Farm's first season and was planning to bring another acre into production in 2012 and, in all likelihood, a third acre in 2013. And there are possibilities for expanding beyond that. Landon and Lindsey have the use of the landowners' equipment, which, in addition to the tractor and plow, includes a set of discs, a bed shaper, and a brush hog mower. They have access to a steady supply of cow manure and a shallow well, which they use for drip irrigation. They use only organic methods but are not certified.

Except for when they got hit with torrential rains late in the season, the first-year crops at Root Mass Farm grew nicely. One of the farmers' markets in Philadelphia worked out well, the other less so. They'll probably drop the weaker market and look for a better one. Landon and Lindsey are excited about their prospects. They like being in business for themselves and making their own decisions. They are working hard, and both of them believe in and enjoy what they do. But Landon cautions: "We'll see where it takes us."

Chapter 6

The Greenhouse

M Y FIRST THREE YEARS OF FARMING, I made do without a greenhouse for the simple reason that I didn't have one. I managed, with mixed success, to start some trays of seedlings on a table in our front room. The seeds germinated well enough, but the plants themselves grew tall and thin as they competed for the limited amount of light that was available. Outside, I seeded many crops directly in the field and, as a consequence of this, spent a lot of time on my hands and knees pulling weeds that were engulfing the tiny seedlings I was trying to rescue.

In late spring and early summer of those first years, I was able to purchase some nonorganic tomato and pepper transplants from a local nursery. (At that time organic growers were not required to start with organically grown transplants, as they are now.) The nursery's selection of plants was limited and geared to backyard gardeners rather than to commercial growers. There wasn't much for me to choose from, and often the plants I came home with didn't adapt very well to the open field.

Looking back, it's hard to believe I lasted without a greenhouse as long as I did. Today the greenhouse is an integral part of our operation; running the farm without it would be almost unthinkable.

Seven Reasons You Need a Greenhouse

A greenhouse enables you to captain your own ship to a far greater degree than is possible without one. It gives you flexibility and reduces your dependence on outside sources. It will allow you to work with a schedule of your own design and more readily adapt to changing weather. Without a greenhouse, you'll find yourself faced with unwelcome constraints. In this section, we'll look at these and other reasons a greenhouse is such a necessary component of a diversified, organic vegetable farm.

1. Organic Transplants Are Often Not Easy to Find

NOP rules now require that organic growers use transplants grown according to organic standards, in certified organic greenhouses; that is, grown with organic seed (when this is available), in approved potting mixes, and without the use of synthetic fertilizer or pesticides. This fact alone makes having your own greenhouse almost a necessity.

Today, there are more commercial greenhouses and nurseries producing organic transplants than there were 25 years ago, but the number is still not high. Many of those that do exist are mail-order operations or businesses that cater more to home gardeners than to farmers. Relying on others to grow your transplants will limit your options and significantly increase your production costs.

2. Stay One Step Ahead of the Weeds

In the struggle against weeds, a greenhouse is an important ally. For a moment, imagine that you had the perverse desire to plant the seed of a known weed (say, a lamb's quarters or a redroot pigweed) next to the seed of a vegetable or herb. Invariably, the weed seed will germinate sooner and grow more vigorously. It will hungrily consume nutrients, water, and light that might otherwise have gone to your crop plant, causing the latter to become a paltry version of what it might have been. This is exactly what happens to many crops that are seeded directly into the field, unless the grower is vigilant about protecting his or her germinating plants from the onslaught of weeds. And that vigilance, of course, demands a significant input of labor.

If you take a month-old, 3- or 4-inch-tall transplant from your greenhouse and set it into freshly tilled ground, with no weeds in sight, the transplant will have several days to adapt to its new environment before any fast-growing annual weeds emerge. When weeds do appear, they will be smaller than your crop plant and should be relatively easy to dispatch with a hoe or cultivator. This is why most organic growers who do not have herbicides at their disposal prefer to use transplants whenever they can.

Some Vegetables Don't Like to Be Transplanted

Some crops — most notably beans, peas, carrots, radishes, and turnips — don't like being transplanted. For them, direct seeding in the field is the only option. Fortunately, most of these crops germinate and grow fairly quickly. They will give the weeds a run for their money and give you time to go to work with a hoe without disrupting your crop plants.

Carrots, though, can be difficult. If carrot seeds are kept moist and warm, they can germinate in a week, but they get off to a slow start and can soon be overrun by fast-growing annual weeds. For the first few weeks after a carrot germinates, you will usually need to undertake some painstaking weeding. See page 308 for strategies to reduce weeding time in carrots.

3. Choose the Varieties You Want

When you have your own greenhouse, you can select exactly what you want to grow. There are thousands of varieties of vegetables and herbs to choose from. Commercial greenhouses, whether they cater to an organic clientele or not, usually grow only a few of the better-known varieties of any given vegetable — they simply don't have the space to grow everything that is available.

Once you have a little experience under your belt, you'll want to try multiple varieties to see which perform best on your farm and please your customers the most. If you don't have a greenhouse, you might approach commercial operations and ask if they will custom-grow for you.

4. Have Transplants Ready When You Need Them

Most commercial nurseries (and remember, you would need one that conforms to organic standards if you plan to be certified organic) sell the lion's share of their vegetable transplants in late spring and early summer. At other times, the selection may be very limited or lacking entirely.

Take a crop such as lettuce: Lettuces are in demand by the general public throughout the growing season, and most small farmers who directly market their produce want to have a steady supply of them for as long as possible. For those of us in southern New York, this means planting lettuces almost weekly from early March through late August. The only way to ensure we can do this (besides direct seeding lettuces in the field, which, for reason number 2 given above, is not practical) is to start the varieties of lettuces we want in our own greenhouse. This way, we can have transplants available exactly when we want them and in the right quantities so we can keep our customers happy.

5. Get a Jump on the Season

If you live in any part of North America that has freezing temperatures in winter, a heated greenhouse enables you to start seeds in a warm environment long before you can venture out into the field. This gives you a big jump on the season. For example, we start onions from seed in late February and set them out at the end of April.

By early March, when there's often still plenty of snow on the ground, you can find us in our greenhouse, seeding perennial herbs, lettuces, kale, chard, and, before long, peppers and tomatoes. As soon as soil conditions are suitable (i.e., not too wet and not too cold, which for us is usually not until late April), we set out sturdy transplants of multiple crops in the field.

A greenhouse can also be of good service at the end of the season. We seed mesclun greens in our greenhouse in September and transplant the seedlings into one of our high tunnels a month later, for late November and December harvest.

6. A More Secure Environment

Crops in the field live at the pleasure or displeasure of Mother Nature. If she drops 6 inches of rain over a couple of days, young outdoor plants may be harshly pelted and, if they happen to be on low-lying ground, may find themselves in standing water, unable to survive. Hail, gale-force winds, or a late-spring frost can damage or slow the growth of plants in the field.

Plants in a greenhouse are protected from these forces of nature, and they can stay put until outside conditions improve and the weather becomes more settled. In the things that matter most to them — temperature, moisture, and sunlight — they enjoy close to optimal conditions in a greenhouse. The one possible exception is sunlight, which will be in short supply on overcast days. (A greenhouse does afford the opportunity to use grow lights, a benefit under low-light conditions, but few organic growers resort to this measure.)

Of course, your seedlings can't stay in the greenhouse forever. In the same way that children grow up and venture out into the world, at a certain point, seedlings must take their chances in the field with Mother Nature. They need to be in good health when this occurs. Experienced growers will tell you that their own transplants are usually more robust, disease resistant, and dependable than those available through off-farm nurseries. But for this to be the case, you must be vigilant about maintaining a healthy and sanitary environment in your greenhouse. More on this later.

7. A Nice Place to Work

You're not likely to build a greenhouse for this reason alone, but once you have one, you'll appreciate the environment it provides. A greenhouse is a quiet place, untroubled by wind and other of nature's elements. When we begin seeding onions in mid-February, the outside temperature is frequently well below freezing. Inside the greenhouse on such days, as long as the sun is shining, it can be a balmy 65 or 70°F (18 or 21°C), without the heater running. Overcast days, of course, are another story. On such days an unheated greenhouse will be warmer than outside, but not a whole lot.

When heavy rain or thunderstorms threaten at any time during the growing season, we're likely to retreat to the greenhouse, where there is usually seeding to do or thinning to catch up on. During these times, the greenhouse enables us to stay productive.

I also enjoy our greenhouse because it offers a more microscopic view of the growing process. It's a lot easier to observe a seed germinating and passing through its cotyledon stage while it is living in a flat on a bench inside a greenhouse than it would be in the wide expanse of the field. And when your diminutive seedling is finally transformed into a mature vegetable ready for harvest, you have a sense of fulfillment from knowing you've been following its progress from the beginning.

What Type of Greenhouse?

Before the invention of plastics, greenhouses were made from panes of glass inserted into rectangular metal or wooden lattice frames. They were often called "glass houses" and were generally shaped like actual houses, with straight sides and pitched roofs. Many still exist and function quite well, but the panes of glass are easily broken and irksome to replace. These days, you can still find the old-style pitched-roof houses on the market, but most of them are designed for backyard gardens and are too small for commercial growers. Some of these small houses are still clad with glass, others with more durable polycarbonate panels, and some with polyethylene film.

Most modern commercial growers use Quonset- or Gothic-style hoop houses. For larger greenhouses, the Gothic style is preferred because it has a peaked central ridge that sheds snow and ice more effectively. The Quonset has a strictly semicircular shape. Both are relatively inexpensive and easy to build. Their primary elements are:

- Galvanized steel posts that are hammered into the ground

- Steel hoops that fit into the posts

- Ridgepoles, crossties, and braces to hold the hoops together

- Baseboards and endwalls (built with untreated lumber, if you are an organic grower) and a door for access

- A double layer of polyethylene film, inflated with a small squirrel-cage fan (the air between the layers of film acts as an insulator)

- Exhaust fan and louvers

- Heater (usually propane, oil, or wood) and chimney stack

- Thermostats for fan and heater

Building one of these is not rocket science. Any greenhouse supply company should be able to give you the specs, along with pros and cons, on the different houses available and everything you need to get started.

How Big?

Before investing in a greenhouse, you'll want to determine what size house will best fit the scale of your operation. It should be big enough to accommodate the maximum number of seedlings that you expect to be growing at any one time over the course of a season. The Weekly Planting Schedule in your farm plan (see chapter 3) will help you determine this number. Simply count how many plants you plan to start over your busiest 6- to 8-week period.

Each year on our farm, there are two periods when the greenhouse is operating at full capacity. The first of these is early April to mid-May. Through most of April, our fields are too cold and wet to plant much

The Expanding Greenhouse

> My Story

WE STARTED WITH A 17- by 48-foot hoop house with polyethylene covering, plywood endwalls, and a hinged door. We installed a propane heater and an exhaust fan with a thermostat and louvers, which served us well enough for about 10 years. Then, as our operation grew bigger, it became increasingly too small. Our solution was to remove one of the endwalls, extend the ridgepole, add seven more hoops, then reattach the endwall and staple on a larger sheet of plastic. (Note: A 4-foot spacing between hoops is standard, which means our new length needed to be divisible by four.)

This expansion brought us to 17 by 76 feet, or 1,292 square feet of covered space — definitely an improvement over the original house, though not quite as big as I would have liked. At that point we were constrained by the size of the existing heater and fan. If we had extended the length any farther, we would have needed to replace both heater and fan with bigger units. We opted instead to make do with a smaller house.

outside, other than peas, onions, and some cold-hardy greens. Even until mid-May, we are still worrying about late-spring frosts and are reluctant to move tender crops (such as tomatoes, peppers, basil, and squash) into the field. So through much of April and the first half of May, our greenhouse is pretty crowded.

Our second busy period is mid-June to mid-August, as we prepare for the fall planting of hardy greens (kales, collards, broccoli, chard, lettuces, and others). During these two periods, we need to be able to accommodate as many as 40,000 plants over a 6- to 8-week stretch to keep pace with our planting schedule.

Access space and work areas. Only a portion of a greenhouse will actually be used for plants. Our greenhouse is 1,292 square feet, but much of this space is needed for other uses. Aisles and crosswalks allow us to reach plants with a hose and watering wand, as well as perform general maintenance on the greenhouse. These walkways consume a lot of space. The propane heater takes up space, and so does the frost-free hydrant that brings water to the greenhouse. Some space is also needed for greenhouse activities such as seeding, thinning, and filling flats with potting mix. Together, these uses account for about 60 percent of our floor space. This leaves us with just 40 percent, or 516.8 square feet, of bench space for the actual plants.

We like to be able to move around our greenhouse easily. Many growers are probably able to get by with less access space, which would permit them to dedicate more space to plants, albeit in a more crowded environment.

Propagation system. The type and size of flats you use, along with the number of cells they contain, will affect how many plants will fit in your greenhouse.

There are numerous propagation trays, flats, or containers to choose from. Most are plastic; some are Styrofoam; others are made from coconut husks and other naturally occurring, biodegradable materials. There is also the option of using stand-alone soil blocks, which don't require any container, only a flat surface to sit on.

Most commercial growers use plastic flats because they are inexpensive, an efficient use of space, and, if handled carefully, reusable. Styrofoam flats are more expensive and require more space, but they do last longer than plastic flats (if you can keep them away from small rodents that like to build nests in them during winter). Biodegradable containers are bulky and a lot more expensive, but they are an obvious plus for the environment. Soil blocks also take up a fair amount of space and require a more exact potting mix (in terms of ingredients) that will hold together well when pressed into a block. Soil block enthusiasts claim that their method produces transplants with better roots that adapt more quickly when moved to the field, but in my experience, most growers who are farming more than an acre or two find them impractical.

We do most of our seeding in the industry-standard "1020" plastic flats and try to reuse them for as long as possible. These flats are approximately 10½ inches wide by 21 inches long, which means each has a footprint of 1.54 square feet. By dividing the amount of bench space available in our greenhouse by the size of a single flat, we can arrive at the number of flats we have room for:

516.8 sq ft ÷ 1.54 sq ft = 335.59 flats

Cells per flat. The next thing to consider is how many cells each flat contains. On our farm, we work mostly with flats containing 72, 98, 128, 144, or 162 cells. All these flats have the same 10½- by 21-inch footprint, which means that the 72s have the biggest cells and the 162s the smallest. Which cell size we choose depends on the crop being planted (more on this later). The 128-cell flat is the one we use most often and serves as a good average. For this reason, I'm going to use it in my example.

To determine how many flats we need for our maximum number of plants:

40,000 (plants needed) ÷ 128 (avg. # cells per flat) = 312.5 flats

Fortunately, we have space for 335.59 flats, so we can fit in another 23 flats if necessary.

If our average flat had 162 cells, instead of 128, we would be able to fit in more plants — a total of 50,625, to be exact — so the choice of flat (namely, its cell size and number of cells) does offer a grower some flexibility. But of course, flats with a smaller cell size would result in smaller plants with less substantial root systems. It's a trade-off.

Allow for expansion. Once you've determined the right size for your greenhouse, based on existing and projected crop plans, I suggest that you build it about one-third bigger, assuming you have the space and can afford the modest additional cost. I wish we had done this — it would have saved us labor and expenses down the road. Of course, a larger

The One-Third-Bigger Rule

These days, I apply the one-third-bigger rule to almost all construction projects on our farm. Over the years, we've put up several outbuildings — sheds, pole barns, worker housing, high tunnels, and a greenhouse. There's not a single structure that I wish we had made smaller, but there are quite a few that I wish were a little bigger. As your business grows, you will very likely acquire more tools and equipment, plant more crops, and need more help. You'll find that covered space of all sorts is an asset that will often pay for itself surprisingly quickly. This is especially true of greenhouses.

greenhouse means a larger area to heat during cold weather, so don't go overboard on size. We'll consider ways to keep down heating costs later in this chapter.

If you don't want to start out building a larger greenhouse, you could site the one you do build in a location where you could add to its length in the future. But be aware: When you expand the size of an existing house, you may need to replace your heating and ventilation units with larger ones — unless you had the foresight and capital to install somewhat larger units than were necessary in the beginning.

More than one greenhouse. Yet another approach is to start with a small house and build an additional house (or houses) as needed. Obviously, additional structures will cost more money, but they will also increase your management options. A second greenhouse will enable you to separate plants by type and adjust temperature (and possibly humidity) to meet different plants' needs.

For example, heat-loving plants such as tomatoes, peppers, and eggplants might be grouped together in one house where you maintain a high ambient temperature. Plants that can be quite happy in cooler growing conditions, such as brassicas, lettuces, and most other greens, could be grouped in a different house, where fans and heat sources are set to keep the ambient temperature lower. With a second greenhouse you may be able to save on heating costs by keeping one house at a consistently lower temperature than the other, or keeping one house shut down during periods when you have less need of space.

(137)

How to Size Your Greenhouse

1. Determine the maximum number of plants you will need in your greenhouse during your busiest 6- to 8-week period.
2. Decide on the type of flats you will be using, their dimensions, and the average number of cells per flat. (If you prefer to use soil blocks, calculate the amount of space they will require.)
3. Calculate how many flats or soil blocks will be needed to accommodate the anticipated maximum number of plants.
4. Calculate the amount of floor area needed for the necessary number of flats or soil blocks.
5. Draw up a floor plan and calculate how much space will be needed for access aisles, work areas, heater, hydrant, and anything else you would like to have inside your greenhouse.
6. Add the space needed for flats to the space needed for access aisles, work areas, and so on, and you will arrive at the approximate size for your greenhouse.

Other Factors to Consider

If you're still with me after this lengthy discussion of how big to make your greenhouse, there are a few other things to consider before you get to break ground. Choosing an appropriate site is an obvious one. Another is the greenhouse's orientation with respect to the sun's movement across the sky and the prevailing winds. It's also important to site your greenhouse in a location that has good access and will be easy to reach with water and electricity, and you'll need to think about the type of heating and ventilation systems you plan to use.

Choosing a Site

It's best to build a greenhouse on level ground, even if you have to move around some earth or bring in fill to achieve this. A slight slope along the length of the house is acceptable, but any slope from side to side should be avoided. Uneven ground makes it difficult to maintain level benches, and unlevel benches mean unlevel flats. Because water obeys the law of gravity, even on the mildest slope, unlevel flats will invariably result in uneven watering. Also, it is much harder to attach polyethylene film evenly and securely to a structure that is not on level ground.

Orientation. An east–west orientation is preferable (it will allow more light to reach your plants) but not crucial. Many greenhouses are oriented in other directions. Not all plants need full sun all day, so having some shady areas (for instance, near the endwalls) isn't necessarily a bad thing.

Shade. Stay well away from human-made structures (e.g., barns, silos, and other buildings) and natural features (e.g., hedgerows and large trees) that will block the sun. You don't want shade on your greenhouse that you can't control. This doesn't mean all shade is bad. In midsummer, we drape a large shade tarp over our greenhouse to reduce the inside temperature, with full knowledge that a little less light will reach the plants. But this is shade we have control over. In cool, overcast weather we have the option of removing the shade tarp.

Proximity to utilities and home. It's important to consider the greenhouse's proximity to water and electricity. You'll need to have both of these. Water comes into our greenhouse through a buried line and a frost-free hydrant (a hydrant that will not freeze in the winter). Our electric feed line is also buried.

All water lines need to be below the frost level, which might be 3 or 4 feet below ground, depending on where you live. Unless you're big on ditch digging, or live in a region with mild winters, you will almost certainly need to hire someone with a backhoe to install them. If you're starting from scratch, it may be easiest to bury your electric and water lines in the same trench. In most cases, the closer you locate your greenhouse to existing water and electric lines that you can easily access, the less cost and disruption you will incur.

Also consider the proximity to where you live. On a cold winter's day with snow or ice on the ground, you may not want to be trudging great distances to get to your greenhouse.

Ventilation

On a warm, sunny day, the temperature inside an unventilated greenhouse can get to be over 100°F (38°C) very quickly (uncomfortably hot for most plants). A large fan at one end of the house and louvers at the other end will keep things cooler. A thermostat will let you program the fan to come on when the ambient temperature reaches a certain point. Additional wiring will ensure that the louvers open whenever the fan starts up. We've put screening over our louvers so that unwanted insects are not easily drawn into the greenhouse. Also, during warmer weather we fit a screen into our greenhouse door for further ventilation.

Small greenhouses heat up faster than larger ones, so good ventilation is imperative. On the bright side, small greenhouses cost less money to heat. If you're on a very low budget, you might use roll-up sides and open doors for ventilation, but this approach will allow insects easy access and require that you make a lot more trips to your greenhouse to maintain an optimum temperature.

Front view of greenhouse · The louvers on either side of the doors open automatically when the thermostatically controlled fan turns on. Note the propane tank and the heater exhaust stack.

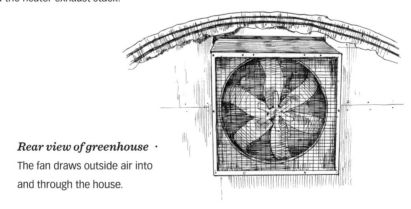

Rear view of greenhouse ·
The fan draws outside air into and through the house.

Heating and Cooling

After the labor and materials needed to build a greenhouse, the main expense is heat. For summer use only, a ventilated, unheated greenhouse will suffice. But a major reason to have a greenhouse is to start plants inside when it is cold outside. For this, you need artificial heat.

Greenhouse furnaces are most often fueled by propane, oil, or wood. Propane tends to be more expensive but is a cleaner fuel. Wood, especially from your own woodlot, is the least expensive, but you must feed it to the furnace on a daily basis.

Some growers outfit their greenhouses with temperature alarm systems that alert them if the heating system fails and the temperature inside the house drops to a dangerously low level.

As already mentioned, we drape a shade tarp over our greenhouse in the hot summer months of June, July, and August. This brings down the daytime temperature by at least 5°F (3°C), creating a better environment for the fall greens that are usually occupying the greenhouse at that time. Of course, a shade tarp also reduces the amount of light coming into the greenhouse. Choose a shade tarp with a light transmissivity rating of 80 percent or higher.

Inside the Greenhouse

You've sited and built your greenhouse. You've installed a fan and heater. You've brought in power and water. Or maybe you've been lucky enough to arrive at a farm with an existing greenhouse.

Now it's time to get to work and grow some plants. In this section I'll discuss what goes on inside a greenhouse. I'll cover greenhouse benches, potting soils, choosing the right flats, seeding methods, germination aids, and fertility — and a few other matters along the way.

Benches

It is possible to leave flats of germinating seeds and seedlings on the greenhouse floor — possible, but not advisable. If you do use the floor, be sure to put down greenhouse fabric first so the flats don't have direct contact with the soil. The fabric will reduce the chance of a soilborne disease making its way to your plants.

Plants will have better air circulation and better drainage if you set flats on benches. A couple of feet off the ground, they are less likely to contract diseases. Benches also make it easier to observe the progress and condition of plants and to move around the greenhouse with a hose and water wand. Most growers soon opt for a bench system.

There are several types of greenhouse benches on the market. Most are made from metal or plastic. Good-quality, commercial-grade benches can be costly, unless you pick them up secondhand. Many small growers choose to build their own.

Whichever type of bench you choose, the surface should be level and non-sagging; otherwise, water will pond in low spots. Also, because good drainage and air circulation are important, bench surfaces for seedlings should not be made from a solid material, such as plywood. Wire mesh, wooden lath, or a heavy-duty plastic grid will do the job and allow excess water to drain away. The one exception to this rule is when the seeds are first germinating and they need only a light misting of water. At this time, they may be set on a solid surface or on heat mats.

We constructed our own benches using old, salvaged 4×4s for the legs, 2×4s for the frames, and heavy-duty, galvanized wire mesh for the surfaces. They have served us well enough, though here and there we've found it necessary to put additional 2×4s under the wire mesh to prevent sagging.

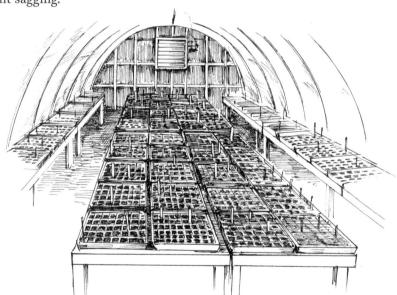

These homemade greenhouse benches are keeping seeded flats a couple of feet off the ground, allowing for better air circulation.

Greenhouse benches are usually at least 10 feet long. Those placed along the edges of the house should not be more than 3 to 4 feet wide, so you can comfortably reach over them. Benches situated in the middle of the house, with aisles on either side, can be double this width, but they will be difficult to move. A better and more versatile alternative is to set two 3- to 4-foot-wide benches side by side.

Potting Soil

There are many potting soils, or potting mixes, on the market, but most of them contain synthetic fertilizer and other substances that are not permitted in organic systems. Certified organic growers must use mixes that meet the standards of the NOP. These mixes usually include a fair amount of compost and are not sterile — meaning that they contain living organisms.

Some growers make their own potting soil using various combinations of compost, peat moss, sand, and naturally occurring forms of phosphorus, potassium, and calcium. In the early years on our farm, we made our own potting soil with mixed success. We had problems with consistency, and sometimes our compost contained viable weed seeds that germinated and competed with the vegetables we were sowing. Not a good thing. Now we find it easier and more cost-effective to let the professionals make the potting mix. We buy at least a ton of mix every year from a place called McEnroe Organic Farm in Millerton, New York, and start our seeds with confidence.

Choosing the Right Flats

I've already mentioned that all of our flats have the same dimensions but different numbers of cells. Our most commonly used flat has 128 cells that measure 1⅛ inches on each side and are 2 inches deep. These work quite well for lettuces and most of the other greens we grow. Swiss chard, broccoli, and cabbage seem to do better when they have a little more space, so we seed these crops in flats with 98 cells, each of which is a little bigger — 1¼ inches on each side.

Mesclun greens are seeded in 144- or 162-cell flats — we neither expect nor want them to grow into tall plants. We also use the 144s and 162s to start tomatoes and peppers, but once the seedlings have put on 2 or 3 inches of height and developed enough roots to fill their cells, we

Potting-on · A table knife is used to open a wedge-shaped hole for a tomato seedling.

Flat Sizes We Use for Different Crops

Don't feel compelled to copy the system below. Variations on it will work fine, and there are several other flat sizes to choose from. Just be aware that some crops need more growing space or more time to grow than others. Experiment with different flat sizes and see which work best for you. Remember, those plants that need to stay in the greenhouse for more than 6 weeks will usually benefit from the extra space and fresh growing medium that potting-on will give them.

A final note: When we run out of a certain flat size, we don't hesitate to go one step smaller or bigger and seldom notice much difference. On our farm, very few things are written in stone.

144s and 162s: Mesclun salad greens, scallions. For starting tomatoes, peppers, eggplant.

128s: Lettuces and most other greens, basil, parsley. Potted-on rosemary, sage, thyme.

98s: Swiss chard, broccoli, cabbages, celeriac, winter squash.

72s: Summer squash, zucchini, cucumbers. Potted-on tomatoes, peppers, eggplant. Also sometimes use for broccoli.

10s: Onions, shallots. Many herbs that will later be potted-on to 128-cell flats.

replant them into our larger 72-cell flats. This is called "potting-on" and always results in sturdier, healthier, and more vigorous plants.

We commonly use 10-cell flats for onions, shallots, and small seeded herbs. These flats have the same 1020 dimensions (10½" × 21") but only 10 large cells, each measuring 3½ by 4¾ by 2¼ inches deep. We seed onions and shallots in three shallow rows per cell, with an inch between rows. Each row receives approximately 15 seeds, for a total of about 45 seeds per cell. We've found these flats work quite well for our allium crops, even though some roots intermingle, making the seedlings a little more difficult to separate and transplant. Because we usually transplant our alliums in groups of two or three, this doesn't bother us too much.

The 10-cell flats are also used for certain herbs (sage, thyme, rosemary, marjoram, oregano). After germination we thin the tiny herb seedlings down to about 15 per cell and later pot them in 98- or 128-cell flats.

Seeding

If you own a greenhouse, you'll learn how to seed flats soon enough. Depending on the type of seed you're sowing, there are a few different approaches. Large seeds such as cucumbers, squash, and Swiss chard can be picked up easily between thumb and forefinger and pressed into the potting soil. Medium-size and small seeds are a bit trickier. Many small growers fold a sheet of stiff white paper in half to create a V-channel for their seeds. With the folded paper in one hand and a pointed object such as a pencil or a large nail in the other, they are able to coax the seeds — one or two at a time — into each cell of the flat they are working on.

Handheld seeders. There are a few simple devices on the market that claim they can help you with seeding. Most don't offer much of an improvement over the folded-paper-and-nail method, but we've found one to be quite useful. It's a vibrating device called the Vibro Hand Seeder from Gro-Mor. It runs off a 9-volt battery or a power converter cord. Seeds are placed in a shallow metal channel and dribbled into the cells of your flat, either slowly or quickly, depending on the vibration setting you choose. But a word of warning: These seeders work best with oval or elongated seeds and not so well with perfectly round seeds, which tend to roll out of the metal channel uncontrollably.

Vacuum seeders. If you want to go a step beyond a vibrating seeder, you might consider investing in a vacuum wand seeder. This will speed up the process considerably because it will enable you to drop an entire row

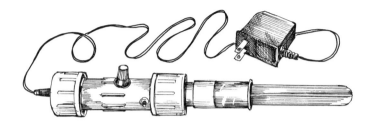

Vibrating hand seeder

of seeds into a flat at the same time; however, it is a lot more expensive (over $500 versus $60 to $70 for a vibrating seeder), and it is less versatile. If you're using a vacuum seeder, you'll want to standardize your flats, with regard to both size and number of cells. Large growers and commercial greenhouse operations use vacuum seeders that can seed an entire flat at the same time.

My advice would be to start out seeding by hand, or with a handheld vibrating seeder. If you find that you are specializing in just a few crops and seeding multiple flats (more than 10 or 20) of each every week, and you're willing to work with just one or two flat sizes, it's time to look into more advanced seeding equipment.

Watch Out for Older Seeds

In addition to having lower germination rates, older seeds often have less vigor. This means they may grow poorly and produce less vigorous plants, especially when they are subjected to the variable and more challenging conditions that generally prevail in the open field, as opposed to the controlled environment of the greenhouse. The result can be rows of disappointing plants that may not be able to properly complete their life cycle.

It makes sense to use new or relatively new seed whenever you can. If you do use seed that is more than a couple of years old, be aware that vigor usually declines before the germination rate does. There are ways to test older seed for vigor, but since I've never done this, I'll let the seed experts advise you.

How Many Seeds per Cell?

When you are deciding how many seeds to place in each cell, there are three factors to consider: Germination rate, cost, and the amount of thinning you're willing to do. The seed's germination rate should be given on the packet. The higher the germination rate — let's say, above 90 percent — the more likely it is that a single seed in its own cell will germinate for you.

The second consideration is cost. Most seeds are relatively inexpensive, but not all. For example, some fancy hybrid tomato seeds, with various disease resistances built into them, can cost as much as 30 cents apiece. Five hundred of those will cost you $150. That's expensive. A third factor to consider is how much time you're willing to spend thinning.

Taking these factors into account, we've come up with the following guidelines for all crops that will be transplanted into the field as single plants (singletons):

- If the seed is expensive or we don't have much left, just one per cell

- For most other seeds, two per cell

- When seed is older (1 to 2 years) or has a low germination rate (less than 70 percent) — three per cell

A final word on seeds per cell: When dealing with more expensive seeds, or seeds in short supply, we sometimes put two seeds in roughly 20 percent of the cells at one end of the flat, and one seed in the remaining 80 percent. If we don't get full germination in the 80 percent of cells that received just one seed, we can fill in the empties from the cells that received two seeds. We usually dig, separate, and replant the little seedlings a couple of weeks after germination. This requires gentle fingers, a small implement like a table knife, and a little water to set the roots.

Thinning

When you put more than one seed in a cell, you are normally assured of getting a flat in which 100 percent of the cells will give you seedlings, but you'll almost certainly need to do some thinning. In my mind, knowing that we will get close to or exactly the number of plants we want (and are planning for) justifies the time spent thinning. Also, by using flats at full capacity, we make more efficient use of our greenhouse space, which is in short supply during busy periods.

Crops like lettuces or cabbages, which are sold by the head, should always be transplanted as singletons, and therefore must be thinned. The same goes for most other greens that we grow, even though many of them are sold by the bunch. There are certain crops, though, especially Asian greens such as mizuna or mibuna, that can be transplanted singly or in groups of two or three. With these plants, you don't need to be so strict about thinning.

Germination

The three main variables influencing the germination of seeds are moisture, temperature, and light.

Moisture. This is pretty much a universal requirement for seed germination. At the most basic level, a watering can will do the job. Better than a watering can is a hose, with a watering wand, that comes to the greenhouse from an outside source. Still better is a frost-free hydrant installed in your greenhouse. A hose or some other watering system (overhead sprinklers and a timer, if you want to get really fancy) can be connected directly to the hydrant. For more information on watering, see page 153.

Temperature. Soil temperature has a big effect on how well seeds germinate. As long as they are kept moist, most vegetable seeds will germinate well when the soil or potting mix is between 65 and 85°F (18 and 29°C). Above and below this range, some seeds will fail to germinate, while others will take longer to germinate and their rate of germination will drop. Minor deviations from the ideal range may not be so consequential, but big deviations will have very negative results. Bear in mind that the longer it takes for a seed to germinate, the greater the chance that it will rot or meet with some other misfortune, such as getting eaten by a mouse.

Some varieties of lettuce prefer cooler soil — 65 to 75°F (18–24°C) — and germination of all lettuces drops off sharply above 85°F (29°C). This can pose a problem for growers trying to start lettuce in midsummer.

Light. When their moisture and temperature requirements are met, most vegetable seeds will germinate happily when covered with a thin layer (usually about twice the diameter of the seed being planted) of potting mix or soil. Others, especially very small seeds, some flowers, and certain varieties of lettuce, prefer being exposed to some light. These should be left uncovered after sowing or covered with an extremely thin layer of soil or a moist paper towel. A good seed catalog will tell you which seeds should be left uncovered.

Heat Mats

An excellent way to keep your potting medium warm, ensure rapid germination, and reduce the incidence of damping-off diseases, is to sit your flats on electric heat mats (also called plant propagation mats). The bottom heat provided by these thermostatically controlled mats will keep the potting soil at an optimum temperature.

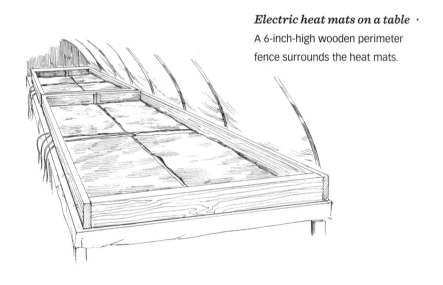

Electric heat mats on a table · A 6-inch-high wooden perimeter fence surrounds the heat mats.

Germinating flats on heat mats · Note the paper towels on the surface of some of the flats and the row cover on the right that will be draped over them at night.

For most seeds, we set our heat mats to turn on when the soil temperature drops below 75°F (24°C). On sunny days, even in winter, there is usually enough warmth in the greenhouse that heat mats are not needed. But on overcast days or at night, when temperatures drop, they turn on and do their important work.

In addition to hastening germination, we've found that heat mats can help us reduce fuel consumption in our greenhouse during a short but significant period of the year — when outside temperatures are low and only a small portion of the greenhouse is occupied.

Our method. We lay out the heat mats on a long table, the surface of which is two sheets of 4- by 8-foot plywood, end to end, covered with heavy-duty plastic (old greenhouse poly works just fine for this). On all four sides of the table, there is a 6-inch-high wooden fence or wall made from lengths of 2×6 lumber. See illustration on page 149. The seeded flats are set on the heat mats and watered or misted as needed during the day. At the end of the day, we drape a large sheet of heavy-gauge row cover over the tables and use spring clamps to secure it at each end.

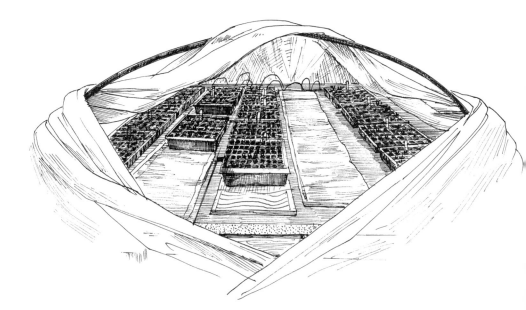

Hoops and row cover are used at night to retain the warmth provided by the heat mats (this is a variation on the wooden fence and upright post system described in the text).

The row cover is held up by the wooden fence or walls and an 18-inch-tall upright post, with a padded top, placed in the center of the two tables. In no place does the cover rest on the surface of the flats. The arrangement looks a bit like a low, elongated tent. The row cover acts like an insulating blanket and creates a microclimate for the seeded flats. It ensures that most of the heat generated by the mats stays with the potting soil and does not dissipate into the larger expanse of the greenhouse. This saves electricity and more to the point, it means we do not have to run our propane heater.

Hot-Water Germination Systems

Heat mats are not the only way to provide bottom heat to speed seed germination. Another approach is to install copper or PVC water pipes in shallow beds of sand or some other heat-conducting medium. Hot water, from a hot-water heater, is circulated through these pipes and keeps the sand at a steady warm temperature. A system of this sort is more elaborate and costly to put in place but should pay for itself over time. High-quality, durable heat mats and the thermostats to operate them are also quite expensive. But they, too, will pay for themselves within a few years.

If you want to reduce energy costs, you might consider installing solar panels to power your heat mats or hot-water germination system.

After the Heat Mats

Once seeds germinate, they are removed from the heat mats and placed on benches in another part of the greenhouse. Again, our goal is to avoid running the propane heater until absolutely necessary. To this end, we have constructed a simple rectangular box frame from 1.5-inch PVC pipe and corner fittings. Its dimensions are 10 feet long by 15 feet wide by 4 feet tall. It stands about 20 inches above the two large benches that it encompasses. On cold days and at night, when temperatures drop, we drape row cover over the PVC frame, creating another small, insulated environment. Spring clamps at both ends and on the sides keep the row cover in place and taut.

Underneath one of the benches, we place a small ceramic electric heater, which is set to go on when the temperature under the row cover drops below 60°F (16°C). On an average night in late winter or early spring, with temperatures well below freezing, the electric heater might use a few dollars' worth of electricity. The propane heater, on the other hand, if it were turned on and asked to maintain a temperature of 55°F (13°C) throughout the entire greenhouse, would gobble up several times as much fuel every night. Not good for the wallet or the environment.

By creating these small, insulated zones within the larger greenhouse and applying heat locally, we are able to speed germination and maintain excellent growing conditions for at least a few weeks without having to heat our entire greenhouse, which would certainly require burning more fuel.

Our insulated and individually heated zones can accommodate about 75 flats. Once we exceed this number, we begin using the propane heater to maintain the entire greenhouse at or above 55°F (13°C), though, for a few weeks, we may continue to use the small ceramic heater under the benches blanketed with row cover for heat-loving plants. We also continue to use the heat mats for germination, keeping them set to go on at about 75°F (24°C). When no longer needed, the PVC frame is easily dismantled and stored.

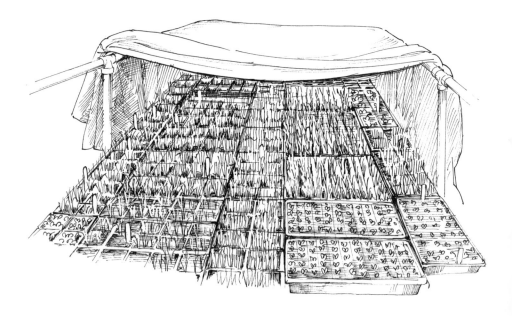

PVC frame and row cover sit over flats on greenhouse benches. You might also place a small ceramic heater underneath the flats.

Fertility

You don't need to provide any supplemental feed to seeds while they are germinating. Also, in the first couple of weeks after germination, there should be adequate nutrition in a compost-based potting mix.

But seedlings that are rooted in 1 cubic inch or less of potting soil soon use up the nutrients available to them. Furthermore, frequent watering of such a small volume of soil can cause nutrients to leach out through the drain hole in the bottom of the flat. Your plants are not likely to die from malnutrition during their relatively short stay in the greenhouse, but if you want to keep them in peak health, it's best to give them some supplemental feed in the few weeks before transplanting. You can do this by spraying a fish emulsion and kelp mix every couple of weeks, a practice known as "foliar feeding" (the emulsified fish parts provide nitrogen and phosphorus, and the kelp furnishes micronutrients). Alternately, you can water them periodically with a diluted fish and kelp mix.

pH. Water used in a greenhouse should have a pH between 5.2 and 6.8. Water with a pH outside this range, especially above 6.8, will severely limit the availability of certain nutrients (most notably, manganese, iron, copper, zinc, and boron), resulting in poor plant growth. This can be a big problem when seedlings are growing in very small cells. If you suspect your water is highly alkaline, you should have it tested. The problem can be corrected by adding citric or sulfuric acid to the irrigation water, but seek professional advice before doing this.

Watering

As I already mentioned, water comes to our greenhouse through a frost-free hydrant fed by an underground line. A hose long enough to reach all corners of the greenhouse is connected to the hydrant; then a 3-foot water wand, with a shutoff, is attached to the hose. At the end of the water wand, we attach a Y divider that allows us to use two different sprinkler heads on the same wand. One has very fine nozzles for misting flats of ungerminated or just germinating seeds. The other sprinkler emits a heavier flow, like falling rain, and is used to more thoroughly penetrate the potting soil as plants grow and develop root systems. Each sprinkler head has its own shutoff, allowing us to use just one at a time.

Watering a greenhouse is no simple matter, especially when you are dealing with a variety of crops at different stages of development in flats with cells of different sizes and under changing weather conditions. The fact that these variables often influence and interact with each other only adds to the challenge.

I view greenhouse watering as more of an art than a science, but an art that is grounded in common sense and easily understood principles. A good greenhouse waterer/manager needs to develop an observant eye and always be on the lookout for telltale signs that the plants under his or her care are not entirely happy.

To start out, you need to get a handle on the following five basic variables:

- **Temperature.** On hot, sunny days, plants need more water.

- **Cloud cover.** On overcast days, plants need less water.

- **Humidity.** On humid and rainy days, plants need less water.

- **Size of plant.** Big plants need more water than small ones.

- **Type of plant.** Plants that are adapted to dry climates — such as the Mediterranean herbs, sage, thyme, and rosemary — need less water. In fact, too much will often kill them.

The above five are the basics to keep in mind, but there's more to it than that. Here are some other important guidelines to observe when watering your greenhouse:

Water according to stage of growth. Flats with germinating seeds should be kept moist at all times but not saturated. Try to keep the top 0.5 inch of potting soil damp, especially during the day.

As seedlings grow and develop roots, there should be water throughout the entire depth of the potting soil, especially in the lower portion, where the roots are. If only the top portion of each cell is moist, the roots will stay there. You want them to branch out in search of water and ultimately fill the entire cell.

Allow flats to dry down. After seeds germinate, you should let the flats get moderately dry at least once a day. It is better to water thoroughly and less frequently. By allowing the soil to dry before rewatering, you encourage root growth. If there is a constant supply of water, plant roots have no incentive to go in search of moisture. Moreover, constantly wet plants are more vulnerable to disease.

Pay attention to time of day. Adequate moisture is important during the warmest and sunniest part of day; that is, midmorning to late afternoon. Toward the end of the day, flats should be allowed to dry off somewhat. Avoid watering plants late in the day, unless they are drooping and clearly in stress. Plants need minimal water at night, and wet leaves at night will make a plant more vulnerable to foliar diseases.

Establish a regular schedule. Develop a regular watering schedule, or at least a schedule to determine whether watering is needed. The first watering of the day is usually best done 2 or 3 hours after the sun rises. But the condition of the plants and the amount of moisture in the flats are the best indicators of when to give water.

Check weights. Periodically check the weights of a few sample flats to determine moisture content. Flats should be lightest in the morning before the first watering. Be aware that flats with different cell sizes can have very different weights. It helps to become familiar with the average weight of each type of flat — when it is saturated and when it is fairly dry.

Organize flats. Flats with small or shallow cells dry out faster than flats with big cells, and therefore require more frequent watering. Separate and organize flats in the greenhouse according to cell size. This makes watering easier.

Water evenly. Keep flats level and water them evenly. There is a strong tendency to apply more water to the center of a flat than to the edges. Always take this into account.

Remain flexible. Remember, every day is a new day and will call for its own approach to watering. Don't mindlessly repeat what you did the day before, unless everything — temperature, sunlight, humidity, condition of plants — is just the same.

Use one waterer. Avoid having several different people water your greenhouse. If at all possible, have the same person do it every day, and that person should have an eye for detail.

Bottom Watering

Bottom watering involves setting a flat of seedlings in a larger tray containing an inch or two of water and allowing the water to enter the flat from below. This method is time consuming but sometimes necessary.

Excessive top watering and prolonged surface wetness can cause a crusty, green mold to form on the surface of your flats. This condition is unhealthy for plants. It severely reduces soil aeration and makes it harder for water to penetrate down to the lower plant roots, where you need it. Allowing water to rise up, by capillary action, through the bottom half to three-quarters of the flat can mitigate this problem. The process shouldn't take more than 2 or 3 minutes. Don't leave the flats immersed so long that the water rises to the soil surface.

Bottom watering is more likely to be necessary on slow-growing plants that spend a couple of months or more in the greenhouse — onions and celeriac, for example, or Mediterranean herbs such as sage, thyme, and rosemary. If you do have to resort to this method, consider adding some fish emulsion to the tray of water, so that your plants get a nutritional boost as they receive their water.

Greenhouse Hygiene

The protected and controlled environment of a greenhouse is ideal for starting and growing tender young plants. But it can also provide a haven for certain pests and diseases, which, once established, can spread rapidly. Good hygiene and a vigilant eye are your best defenses. Healthy plants that are getting the attention, water, air movement, and nutrients they need are much less likely to succumb to pests and diseases. Here are some simple steps you can take to minimize problems.

- **Clean and sweep out your greenhouse periodically.** Debris of any type provides hiding places and cover for insect pests and bacterial and fungal organisms.

- **Cover the greenhouse floor with a porous weed barrier** to make sweeping easier and to block weeds. Try to avoid tracking in dirt from the outside.

- **Remove any weeds or grasses that creep in from the sides.** They can harbor pests and diseases.

- **Remove any plants that show signs of disease or insect infestation.** Never bring diseased or insect-infested plants into your greenhouse.

- **Always use clean flats,** and sweep table surfaces after seeding, thinning, or potting-on.

- **Avoid heavily watering plants late in the day.**

- **Use screens on the greenhouse door and louvers to keep insects out.**

- **Don't allow smokers to handle plants in the nightshade family,** such as tomatoes, peppers, and eggplant. If this is unavoidable, make sure the smoker has thoroughly washed his or her hands and wears gloves. Even tiny amounts of tobacco can transmit a disease called "tobacco mosaic virus."

- **At the end of each season, remove all flats, trays, and containers.** Pressure wash or hose down all benches and work surfaces.

- **Clean flats and trays outside of the greenhouse.** Be sure to thoroughly wash off all potting soil and other organic material outside of the greenhouse; then use a disinfectant approved for organic use. Many growers soak their flats in a 10 percent bleach (such as Clorox) solution (1 part bleach to 9 parts water) for 30 minutes and follow this with a water rinse. The active ingredient in bleach is sodium hypochlorite.

Two of the more common problems you will likely deal with are aphids and damping-off.

Aphids are most likely to show up in midsummer, when the greenhouse is crowded and receiving frequent waterings. Aphid populations can surge quickly and do significant damage. They can also spread diseases. If you catch them early, they are relatively easy to control by spraying insecticidal soap. Another approach is to release parasitic wasps or lady bugs into your greenhouse. But again, this should be done when the infestation is still light.

Damping-off is a complex of fungal diseases — most commonly, *Rhizoctonia* and *Pythium* — that can be deadly to greenhouse seedlings. *Rhizoctonia* is active at the soil surface. It creates a girdling, hourglass impression near the base of a plant's stem, soon causing it to fall over and die. *Pythium* operates farther down in the soil. It attacks and kills a seedling's roots.

Damping-off is most likely to occur under cold, wet, crowded

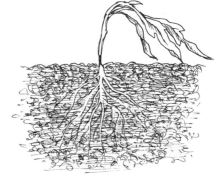

Seedling damping-off · Note the hourglass-shape thinning of the seedling's stem, just above the soil line.

conditions and will often take hold at night. Using clean flats, keeping plants well thinned, and maintaining good air circulation are your best defenses. If you detect damping-off, rogue out affected plants immediately.

Hot and Cold Treatment

Allowing a greenhouse to experience extremes of heat and cold can be a good way to eliminate certain pests and diseases. An unheated greenhouse in a northern winter will get cold enough to knock out some of the "bad guys." Similarly, in the full heat of summer, closing down a greenhouse and shutting off ventilation for a couple of weeks can put the kibosh on some troublemakers. Replacing the weed barrier on the floor with clear plastic will cause the soil underneath to get very hot. This process is known as "solarization."

There are numerous other pests and diseases that can invade your greenhouse. You'd be wise to acquire a book that tells you more about them and provides information on prevention and control. But don't get too anxious. By taking the simple precautions outlined above, you should encounter very few problems. At least, that's been our experience.

Hardening Off

Before seedlings are transplanted to the field, they should be hardened off. This is done by moving them from the controlled and somewhat pampered environment of the greenhouse to the outside world and placing them on something we call "hardening-off structures." The plants remain in their flats but are more directly exposed to nature's elements, in the form of sun, wind, rain, and lower temperatures. On the hardening -off structures they get a taste of what lies ahead and some time to prepare themselves.

Our hardening-off structures resemble greenhouse benches, in that they have legs that hold the flats a couple of feet off the ground and a surface of galvanized wire mesh to ensure good drainage. But they differ in that they are built with a sloping, partial canopy a few feet above the plants that provides some shade and a little protection from heavy rain.

The hardening-off period can last anywhere from a few days to a week or more. During this time, plants continue to receive water but somewhat less than they are used to. The idea is to toughen them up a little but not expose them to severe weather changes that might be harmful or fatal. Sometimes, if a cold spell is headed your way, it might be necessary

The hardening-off structure at left has flats of tomato seedlings; the one below holds onion seedlings.

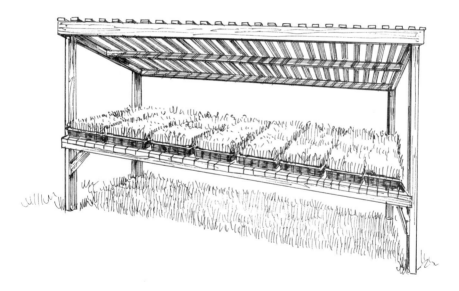

to interrupt the hardening-off process and return tender plants such as basil or tomatoes to the greenhouse for a night or two.

Hardening off is most important early in the season or at any time when plants are going from a heated greenhouse to outside temperatures (especially nighttime temperatures) that might be a lot lower than they are accustomed to. In midsummer, when outdoor temperatures remain well above freezing both day and night, hardening off is less critical, though still advisable.

Greenhouse Dos and Don'ts

As with managing your entire farm, managing a greenhouse requires some organization. With a little experience under your belt, you'll come up with an arrangement that works well for you. In the meantime, here are a few suggestions to get you started:

▶ **Set flats on benches,** rather than the greenhouse floor.

▶ **If you use heat mats, don't leave flats on them after seeds have germinated.**

▶ **Group similar plants at similar stages of development.** For example, if you have several flats of lettuces that were started 3 weeks earlier, keep them together. Don't mix them with lettuces that were seeded just a week before. The latter will probably need less water.

▶ **Don't mix widely different types of plants,** even if they were seeded at approximately the same time, unless their water requirements are similar. Flats of thyme and sage, for example, need to be watered less often than flats of broccoli or summer squash.

▶ **Identify flats that need thinning and possibly group them together.** Make sure that thinning is not overlooked and occurs in a timely manner. Plants should be thinned soon after germinating and certainly well before transplanting.

▶ **Move flats of seedlings that are nearly ready to go to the field closer to the greenhouse door.** We've found that this conveyor belt approach helps us see clearly which plants are coming up next for transplanting.

▶ **Follow the guidelines for greenhouse hygiene** and pest and disease control outlined in this chapter.

▶ **Harden off plants before transplanting.**

Chapter 7

···

Growing Under Cover

L ET ME START BY SAYING what a high tunnel (or low tunnel, for that matter) is not. A high tunnel is not a greenhouse, though from a distance it may look like one and, in fact, may be made from the same materials. A greenhouse, at least as we have defined it in chapter 6, is a place to start plants from seed — usually in flats or pots that rest on benches. A standard greenhouse will have a heater for when the temperature inside drops too low and an exhaust fan for when it gets too high.

Your typical tunnel, however, whether it be high or low, has neither a heater nor an exhaust fan. It is a very low-tech, eco-friendly structure because it relies only on the sun for its warmth and on roll-up sides and open doors to keep it from getting too hot. Plants in a tunnel are usually grown in the ground, just as if they were in the field. In fact, you could say they are grown in the field — it's just that they have a tunnel over the top of them.

We have two high tunnels on our farm. The smaller one is 21 by 96 feet and 11 feet high. The larger is 30 by 96 feet and is 14 feet high. I see more high tunnels in our future.

High tunnels are passive solar structures. Plants are grown directly in the ground without artificial heat. Note the end doors and roll-up sides, which are used for ventilation.

Nine Advantages of High Tunnels

High tunnels have become increasingly popular with organic vegetable growers for several very good reasons. The one most often cited is season extension in both spring and fall and, in some cases, even into the winter. Other noteworthy benefits include: Protection from the elements and most herbivores; better-looking, higher-yielding, and better-storing crops; and a good place to work when it's cold or wet outside. Here, we'll consider these and a few other good reasons to invest in one or more high tunnels. For many growers, it's an investment that will soon pay off.

1. Season Extension

Most farmers in North America are constrained by the length of the growing season, which is usually defined as the number of frost-free days. Tender field-grown crops, such as tomatoes, cucumbers, and basil, cannot be planted until after the last frost of spring and end their lives when the first frost arrives in fall. Hardier plants such as lettuces, kale, and spinach can survive and even prosper in the field much longer at both ends of the season, but eventually, their days of reckoning come.

A high tunnel can extend the season for any of these crops by a few weeks or even a month or more in both spring and fall. You can provide additional protection by creating a low tunnel within a high tunnel — hoops and a blanket or two of row cover will keep many cold-hardy plants alive through most or all of winter. It should be noted, however, that during the coldest and darkest months, plant growth will slow almost to a halt. Whether you opt for the protection offered by a high tunnel alone, or use the "tunnel-within-a-tunnel" approach, you will extend your harvest season and increase yield.

2. Wet Conditions Can Delay Field Planting

Wet outdoor conditions often delay spring planting, even though temperatures may be high enough. You can't plow or rototill a field while it is still saturated with water. But even in the wettest spring, the soil inside a high tunnel will be relatively dry, except perhaps for some seepage along the edges. This means you can plant as soon as you feel the days and nights are warm enough for your crop. Regardless of the time of year, you seldom have to worry about the ground being too wet under a high tunnel.

3. Protection from the Elements

As is well known, and often stated in this book, many of the problems that beset field-grown plants are caused by the weather. Excessive wind, hail, and rain can damage and stress plants. Wind desiccates many vegetables and slows their growth; hail can destroy plants, or at least cause injuries that make them more vulnerable to invading insects and diseases.

And rain, because it is so common, is the biggest worry of all — wet foliage makes it easier for diseases that might be lurking in the neighborhood to gain a foothold and spread. Foliar diseases, in particular, are a real problem for organic growers who don't have chemical fungicides in their tool chests.

Inside a high tunnel, plants are largely protected from these three forces of nature.

4. Better Crops

Crops grown under high tunnels are often better looking, better tasting, better storing, and higher yielding. Field crops that bear fruit, such as tomatoes and peppers, do fine when the weather cooperates. But if they receive prolonged, heavy rain or an unexpected spell of cold weather, their growth slows down, and their quality and appearance can take a quick turn for the worse. With tomatoes, especially, splits and cracks are common when the plants absorb too much water within a short period of time. A promising crop can soon turn into a harvest of "seconds" with a short shelf life and disappointing flavor.

Heavy rain causes soil to splash up on lettuces, escarole, and many bunching greens, making them less attractive, unless you go through the laborious process of washing individual heads and bunches. Saturated greens, as just about every other vegetable, don't taste as good or last as long. This is not to suggest that cool weather greens are a good choice for a high tunnel during summer months. They are not. But in early spring and late fall, they can do very nicely indeed.

An earlier planting date, more warmth when needed, and protection from the sometimes destructive forces of nature will combine to produce significantly higher yields in high tunnels.

5. Protection from Herbivores

A couple of years ago, we found a nest of baby rabbits in one of our tunnels. We evicted them as soon as they were big enough to hop off on their own. That's about the only time I've seen four-legged herbivores inside a tunnel. To date, no deer or woodchucks have ventured in, so we haven't had to worry about fences or other deterrents, which is certainly not the case with our field-grown crops. Sometimes, birds fly into the tunnels and dine on any insects that have taken up residence, but this we don't mind.

Late Blight

> My Story

IN JULY 2009, after a long spell of cool, wet weather, we got late blight in our field tomatoes and lost the entire crop (more than 2,300 plants) in a couple of weeks. Of course, we didn't just lose the plants. We also lost a considerable amount of labor and resources. Beds in three different fields had been carefully prepared, black plastic and drip tape laid, each transplant set in the ground by hand, and each given spoonfuls of rock phosphate and organic fertilizer. Some eight hundred wooden stakes and a few days of vigorous work had gone into trellising the young plants using the Florida weave (basket weave) system. All of this time and resources amounted to nothing.

That year, we had also put some 50 tomato plants in our one high tunnel, figuring they would give us a modest amount of ripe fruit before the field tomatoes came in. This they did, but more importantly, they continued to grow and produce fruit for a couple of months after the scourge of late blight hit us.

Our tunnel tomatoes did eventually contract late blight, but because their foliage stayed dry, the disease was not able to spread as rapidly as it did in the field. We harvested at least 500 pounds of fruit before the plants finally succumbed. Not a bountiful haul, I'll admit, but enough to bring in a couple of thousand dollars and convince me to invest in a second tunnel. This second, larger tunnel now houses about three hundred tomato plants, along with lesser numbers of peppers and herbs. It has nearly paid for itself in 2 years.

6. Moderate Cost

Tunnels cost a lot less to build and maintain than greenhouses, largely because they require only one sheet of poly and don't need a heater, fuel, fan, louvers, or the wiring and electricity that greenhouses call for. Most growers, with a good set of instructions and a couple of willing helpers, should be able to construct a high tunnel without professional assistance. The cost and the amount of setup time will vary depending on the size of the tunnel.

Cost estimates range from 75 cents to $1.50 per square foot for the basic hardware and plastic needed; therefore, a 21- by 96-foot tunnel would run you somewhere between $1,500 and $3,000. The heavier the gauge of steel, the higher the price. Then, you will need lumber for baseboards and to frame out endwalls and possibly doors. It may also be necessary to do some site preparation, such as leveling with a bulldozer.

There is also the matter of bringing water to the tunnel. If you're not planning to grow plants in subzero temperatures, garden hoses and shutoff valves might be adequate. Most growers, though, will want the option of irrigating their tunnel during the winter months. A buried water line and frost-free hydrant will make cold-weather irrigation a lot easier but will add to the price tag. When all is said and done, your final cost could go up by 20 to 40 percent; therefore, you might end up spending $4,000 or more for a 21- by 96-foot tunnel. Not exactly small change, but still a good investment. Many farmers find that the larger and more reliable yields they gain from a high tunnel cover the cost of their investment in 3 or 4 years.

We bought our two tunnels from Ledgewood Farm in New Hampshire. Ed Person, the proprietor of Ledgewood, was very patient with us and offered plenty of guidance and advice. The tunnels are sturdy and can handle severe winter conditions. Other high-tunnel merchants include FarmTek (Iowa), Agra Tech (California), A.M. Leonard (California), Poly-Tex (Minnesota), Zimmerman (Missouri), and Harnois (Quebec). The English company Haygrove offers a wide selection of single-bay and multibay tunnels at very competitive prices.

If you're on a very low budget, you might look into building your own smaller and less sturdy tunnel with rebar ground posts and 1.5- to 2-inch schedule 40 PVC.

7. A Pleasant Place to Work

Like greenhouses, high tunnels provide a nice change of work venue — except in the middle of a very hot day. The environment inside a tunnel, even when the sides are rolled up, is quieter, less windy, and often more soothing and contemplative than the rough-and-tumble of the open field. On cold days, when temperatures are below freezing, it can be quite pleasant inside a tunnel as long as there is some sun in the sky. Even in overcast weather, it will be noticeably warmer inside a high tunnel.

Before we had high tunnels, it was always a challenge to find productive work for the crew and me on seriously rainy days. Now when I make up a work plan for the week ahead, I schedule tunnel chores — such as pruning, trellising, harvesting, or weeding — for any wet days. This way, all of us stay dry, and the fields are spared the compacting effects of human traffic on wet soil. On cold days, too, the tunnels are always the preferred spots to be.

The tunnels are especially important on rainy harvest days. Really bad weather can seriously interrupt picking in the field. This is never a problem in a tunnel.

8. Season Extension for Your Workers

Many farmers have found that high tunnels not only extend the season for certain crops and therefore expand their income-producing window, but they also make it possible to retain employees who otherwise would have to be let go when the season winds down. When you can offer only 6 or 8 months of employment, there's a good chance you won't see that season's workers ever again. Holding onto productive employees for an extra couple of months and possibly the entire year is very appealing to many farmers.

9. Vertical Space

Because tunnels require a significant investment of capital and labor, most growers want to get as much production out of them as possible. One way they do this is by exploiting the vertical dimension. With tomatoes, for example, farmers often use a trellising system that allows the plants to grow to a height of 8 feet or more on a single stem, producing fruit all the way to the top. The fruit stays clean and is easy to pick. This use of vertical space would be very difficult to achieve in the open field. For more details on growing and trellising tomatoes in a high tunnel, see chapter 13.

Some growers exploit the vertical space in their tunnels by suspending hanging plant baskets above low-growing crops such as peppers or cucumbers.

The trellised tomatoes growing in this high tunnel are pruned to a single stem.

Some Vegetables That Do Well in High Tunnels

IN SPRING, SUMMER, AND EARLY FALL

▸ Tomatoes
▸ Peppers
▸ Eggplant
▸ Cucumbers
▸ Basil

IN FALL AND WINTER

▸ Lettuces
▸ Kale
▸ Collards
▸ Asian greens

YEAR ROUND

▸ Perennial herbs — rosemary, thyme, lavender

Special Considerations for High Tunnels

Before you set out to construct your first high tunnel, you need to choose a good site for it. Orientation with respect to the sun and ventilation through prevailing winds are important factors to consider. Other variables that will come into play, but are not necessarily site specific, are provisions for pollination of fruit-bearing crops and the methods you will use to maintain fertility. You also want to consider your long-range plan for crop rotation.

The site. Like greenhouses, tunnels are best located on level ground away from any objects that might throw shade. A slight slope is okay if it is along the length of the tunnel. Avoid any slope from one side to the other.

Orientation. Because air movement is so important, many growers choose to orient their tunnels perpendicular to the prevailing winds so that during the hot months, they can capture as much free air as possible through the roll-up sides. If, on the other hand, you're trying to maximize the amount of sunlight reaching your tunnel during the winter months, you might choose a different orientation.

For maximum sunlight, the rule of thumb is this: In the Northern Hemisphere, above 40° latitude, the best orientation is east–west (it captures more low-angle sun). Below 40° latitude, a north-to-south orientation is better (the sun is much higher in the sky at these latitudes). For a more detailed discussion of orientation, see the website Hightunnels.org.

Ventilation. Tunnels with widths of 21 feet are usually ventilated well enough by natural airflow. Wider tunnels (30 feet is generally the widest for single-bay tunnels) can have ventilation problems, especially in their centers. Some growers install overhead fans in wide tunnels to keep plants in the middle rows drier. See page 333 for more on ventilating high-tunnel tomatoes.

Pollination. In the absence of insects, many crops, such as tomatoes, rely on the wind for pollination. So if you want plenty of fruits to develop, and are short on insects, you need to have a good flow of air through your tunnels after flowers have set. You might also try physically shaking the plants to help them self-pollinate.

Fertility. Because you want your high-tunnel plants to produce as much as possible, you need to maintain a high level of fertility in the soil in which they are growing. This usually means adding ample amounts of compost and possibly other amendments after each crop. Fallow periods and green manures, especially leguminous ones, are good for building soil health and fertility, but they temporarily remove valuable real estate from production. For this reason many growers don't use these soil-building measures as often in their tunnels as they do in the open field.

Crop rotation. Rotating crops in high tunnels can be a major challenge because of the limited number of crops. In the absence of a good rotation plan, it is helpful to maintain high fertility and use disease-resistant varieties. Tomato growers may use grafted plants to reduce the chance of disease.

Of course, the more tunnels you have, the more you will be able to rotate different crops through them. You could also use smaller tunnels that can be easily dismantled and periodically moved to different locations. Some tunnels are actually designed to be moved on sleds, with the aid of a couple of tractors.

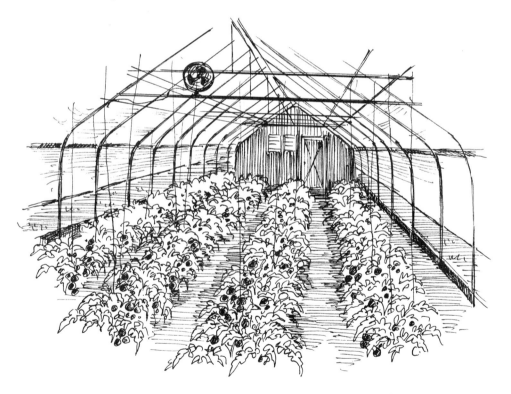

In addition to its roll-up sides, this high tunnel has an overhead fan to distribute air around the plants in the center.

Low Tunnels and Caterpillars

Low tunnels and caterpillars offer many of the same advantages that high tunnels provide, but not all. They are generally temporary structures used for season extension and protection from pests, diseases, and the elements.

Low Tunnels

The big difference between a high and a low tunnel is that you can stand up in one but not in the other. Another difference is cost — low tunnels cost a lot less. Once you have the hoops and the covering, you can put up low tunnels very easily. They are usually 2 to 3 feet high and 4 to 5 feet wide; length varies depending on need. Simple low tunnels are often used to keep insects away from vulnerable crops in summer. In the colder months, they are used for season extension and sometimes even overwintering.

Low-tunnel hoops can be made from 10-foot lengths of galvanized electrical conduit that is ½-inch-diameter, 1-inch PVC pipe and rebar posts, or 10-gauge galvanized wire. The conduit-style hoops are the strongest and most able to withstand a substantial snow load in winter. An inexpensive hoop bender will let you make your own uniform hoops. Wire and PVC hoops are less sturdy but will do a good job of protecting crops in spring, summer, and fall.

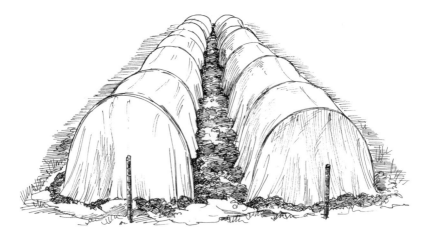

The row cover on these low tunnels is draped over hoops made from ½-inch galvanized electrical conduit.

Regardless of the material used, hoops are placed 4 to 5 feet apart and set well into the ground on either side of the bed you are planning to protect.

To protect crops from insects, drape lightweight row cover or gauze over the hoops and weigh it down with sandbags, bags of gravel, or any other heavy objects that you have on hand. Shoveled soil can also be used to hold down row cover but is impractical if the row cover is going to be repeatedly removed and replaced, as is often the case when harvesting. Row cover buried under soil also degrades more rapidly. For season extension, use heavier weights of row cover with greater insulating value.

Growers wishing to protect crops through winter often drape a layer of greenhouse poly over the row cover. This can work quite well but only when the hoop structure is sturdy and sufficiently braced to withstand the vicissitudes of winter. Use cross ropes to hold the plastic in place, along with sandbags or other heavy weights. Stakes at either end of the tunnel will keep both plastic and row cover taut and prevent them from caving in under a heavy snow load.

Different designs for low tunnels can be readily found online. Check out Eliot Coleman's "double-covered low tunnels." You can read about them in an online article by Jean English titled, "Extending the Growing Season with Coleman's Double-Covered Low Tunnels." Or get a copy of Coleman's *Winter Harvest Handbook* or his *Four-Season Harvest*. Eliot is a pioneer of season extension in the United States.

Tunnel Resources

There's much more to learn about high tunnels than what I've covered in this chapter, and a great place to start learning is the website Hightunnels.org. This site is maintained by an assortment of Midwestern Extension specialists, college professors, researchers, growers, and students, all with a strong interest in sharing their knowledge and experience of high tunnels.

The *Hoophouse Handbook* (see Resources) is an excellent collection of essays covering high-tunnel design, construction, and uses. The essays are written by growers with firsthand experience.

Caterpillar Tunnels

A caterpillar tunnel is a low tunnel on the way to becoming a high tunnel. The typical caterpillar is 5½ to 6 feet high, about 10 feet wide, and any length up to 300 feet. A person of average height should be able to walk down the center, as long as he or she avoids a bouncing gait. Inside a tunnel of this size, there should be enough room for two 4-foot-wide beds.

Caterpillars are easy-to-erect temporary structures that don't require perfectly level ground. They can be built with some of the same materials one might use to build a sturdy low tunnel; namely, rebar posts driven into the ground and 1-inch-diameter PVC hoops fitted over the top of them. Or, you might purchase 10-inch-wide hoops from a greenhouse vendor. As with the Eliot Coleman–style "double-covered low tunnel" mentioned above, a rope is often wrapped around the top of each hoop along the length of the tunnel, then tied to stakes firmly set into the ground several feet beyond either end of the structure. This added reinforcement gives better protection against wind and snow.

Caterpillar tunnels may be covered with row cover or greenhouse poly, depending on the time of year and the inside temperature desired. Cross ropes, in addition to sand bags or other weights, should be used to hold the row cover or greenhouse poly in place and prevent billowing.

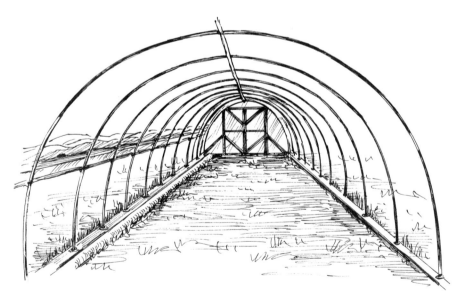

This caterpillar tunnel frame is made with 10-inch-wide hoops.

Caterpillars may be used to extend the growing season well into winter, or to provide heat-loving plants with a warmer environment during spring and early summer. When temperatures rise, however, caterpillars that are covered with greenhouse poly will need to be ventilated by rolling back the poly or opening the ends, or both.

Nine Points in Favor of High Tunnels

1. They extend the growing season.
2. They keep the ground dry, so you can plant when the weather is warm enough for your crop.
3. They protect plants from the elements.
4. Plants grown in tunnels often look, taste, yield, and store better.
5. Herbivores are less likely to eat your plants.
6. Tunnels cost less than a greenhouse.
7. They provide a pleasant place to work, especially on cold and rainy days.
8. They can extend the season for employees as well as crops.
9. They enable you to exploit vertical space, increasing your growing capacity on a small piece of land.

Snug Haven Farm

Owners: Judy Hageman and Bill Warner

Location: Belleville, Wisconsin (southwestern Dane County)

Year Started: 1989

Farm Size: 50 acres, with 1 acre of high tunnels and less than ½ acre of field crops in production

Crops: Spinach, arugula, tomatoes, cut flowers, small amounts of herbs and potatoes

WINTERS IN DANE COUNTY, WISCONSIN, ARE SEVERE. Nighttime temperatures of −15°F (−26°C) are not uncommon, and they can drop even lower. Well before it gets this cold, most growers take a break and put in some time with the seed catalogs. But not Bill Warner and Judy Hageman of Snug Haven Farm. They see the cold weather as a marketing opportunity. Bill and Judy's specialty is spinach. They say it tastes sweeter and crunchier when it has gone through a few good freezes.

Apparently, their customers agree: Bill and Judy sell a lot of spinach. Their outlets include the Dane County Farmers' Market in Madison (the largest producer-only farmers' market in North America); their own winter share program, which is like a mini-CSA (shareholders receive a 1-, 2-, or 3-pound bag of spinach every other week, for 11 weeks); and some of the finest restaurants in Madison and Chicago. Winter spinach grows more slowly than spinach planted for spring or fall harvest, but it is thicker and hardier, and by all accounts, it tastes better.

The bulk of the growing at Snug Haven Farm is done in high tunnels, or hoop houses, as Bill calls them. The farm has a total of 13 of these season-extension structures. They cover about 1 acre of land and range in size from 30 by 96 feet to 32 by 145 feet. All the spinach is planted in September. Harvest takes place weekly, from late October or early November until late April, but never before the plants have experienced freezing temperatures.

When it drops well below freezing outside, the spinach gets an extra blanket, in the form of a second layer of greenhouse polyethylene, to keep it warm at night and on seriously cold, overcast days. This layer is draped over what Bill Warner calls a substructure made from 0.5- and

¾-inch electrical conduit. The two sizes of conduit are cut and fit together to form 4-foot-high by 30-foot-wide support structures. The assembled pieces are spaced 6 to 8 feet apart. They reach across the entire width of a hoop house and are flat across the top, rather than bowed.

On warmer days, the sides and ends of the tunnels are opened to prevent the spinach from getting too hot and to allow excessive condensation to evaporate. Sometimes, Bill uses large sheets of row cover instead of polyethylene, though he prefers the poly for a few reasons: It seldom tears, it warms the spinach a little faster when the sun comes up, and it enables him to get a second use out of material that has already had a first life covering his hoop houses. But because the poly doesn't breathe as well as row cover, it requires more careful management. Even under its double covering, the spinach freezes repeatedly, but the soil inside the hoop houses rarely freezes. When it does freeze, it freezes no more than ¼ inch deep for a few hours in the morning.

Snug Haven Farm's hoop houses are not heated in the conventional sense, but on cold days, they may be warmed up for an hour or two with propane furnaces, to create a more comfortable harvesting environment. It's tough to cut spinach for any length of time when the temperature is much below 45°F (7°C). When necessary, the heaters are also used, for short periods, to thaw the spinach before harvesting.

Snug Haven Farm follows organic methods, but it is not certified. To Bill, the federal government's involvement and all the paperwork are big disincentives. Still, he hasn't altogether ruled out organic certification. The farm employs three or four part-time workers to help harvest the spinach, pick tomatoes in spring and summer, and do other chores. Each year, one of the hoop houses is converted into a greenhouse for starting tomatoes and flowers.

By early April, it's getting a bit warm to grow the kind of spinach that Snug Haven Farm likes to grow. This is when Bill and Judy switch to hoop-house arugula, tomatoes, flowers, and, from time to time, a few other assorted vegetables and herbs. Their goal is to have their wares at the market before the flood of fresh field-grown produce arrives. It's sometimes chilly work, but it pays off.

Chapter 8

··

Irrigation

A S A RULE OF THUMB, most crops will prosper when they receive the equivalent of an inch of rain every week. But this is just a general rule and should not be blindly followed. Here are 12 factors to take into account when assessing irrigation needs:

1. Plants will benefit from more water during the hot (and often dry) summer months, which in most parts of the Northern Hemisphere are June, July, and August. During these times, the equivalent of 1¼ to 1½ inches of rain each week, especially during periods of fruit set and development, would be fine for most vegetables.

2. Conversely, in cooler and sometimes more humid spring and fall weather, plants generally need less water.

3. Flowering and fruiting vegetables will be more productive if they receive additional water (about the equivalent of 1½ inches of rainfall every week) just before and during their flowering and fruiting stages. Examples include tomatoes, peppers, eggplants, and squash.

4. Potatoes need more water (again, 1½ inches per week) when they are flowering and setting tubers — that is, if you want good-size tubers.

5. Larger plants with lots of stem and leaf tissue to maintain, such as zucchinis and summer squash, usually need more water than small ones like lettuce, for example.

6. Plants need less water and also less deep watering in their early growth stages, when their roots and tops are still small. As they put on size and their roots spread out, they need more water.

7. Soft-tissued, leafy greens need more water than plants with dry stems and leaves, such as the Mediterranean herbs, which can be seriously harmed if given too much water.

(176)

8. Plants growing in well-drained, sandy soils need more water than plants growing in heavier clay soils, which don't drain so easily and therefore hold their moisture for a longer period of time.

9. Organic matter in a soil holds moisture, so soils with high organic matter content (4 percent or above) will need less irrigation during droughty periods than soils that are low in organic matter (3 percent or below).

10. Plants growing in a mulch, especially an organic mulch like straw, will require less water than plants growing in bare soil because the mulched soil will lose less water to evaporation.

11. Topography plays a role in plants' water needs. Sloping ground, and especially steep slopes, will lose more heavy rain water to runoff than level or gently sloping fields. Also, south-facing slopes, which receive more direct sunlight, will dry out faster and therefore need more water than north-facing slopes.

12. Plants may require more water in the week or so after mechanical cultivation because cultivation can prune the shallow roots of your vegetables, making it more difficult for them to obtain water until the damage is repaired.

Determining Moisture Level

You need to know how much water is available in the soil at any given time to make the best irrigation decisions. There are a few ways to assess soil moisture content. One method is to simply dig down several inches and observe the color of the soil. Darker soil below the surface generally means that some moisture is present.

Ball test. You can also take a handful of soil, from several inches below the surface, and try pressing it into a ball or clump to gauge moisture content. If your soil is a loam (roughly equal parts sand, silt, and clay) and you are able to form a ball that maintains its shape when you open your hands, there is plenty of moisture present. If the ball of soil falls apart in your hands, its water content is probably low. This method doesn't work quite so well with soils that have a high sand or high clay content — sandy soils don't clump easily, but clay soils will clump readily.

Don't be deceived by a wet soil surface, such as may exist after a light rain. Water in the top few inches may be good for seedlings or germinating seeds, but it won't do much for existing plants that already have roots that have reached down a foot or more. What you really want is moisture at these deeper levels.

Soil moisture meters. Soil moisture meters usually measure percent relative saturation. Relative saturation, or field capacity, will vary depending on the type of soil being tested. After complete saturation and drainage, a sandy soil will have a field capacity of about 30 percent, a loam about 45 percent, and a clay soil about 65 percent. In other words, these are the amounts of moisture that the three different soils are capable of holding in their open pore spaces.

At the above levels of saturation, a soil moisture meter would read 100 percent, indicating that the soil has reached its field capacity, or maximum ability to hold water. Additional water will cause runoff, ponding, or flooding. Generally speaking, a soil moisture meter reading of 50 percent or above indicates that the soil has sufficient water to meet plant needs. As levels drop below 50 percent, irrigation may be called for.

Field Capacity of Soil Types

Sandy soil – 30%
Loam – 45%
Clay – 65%

High-quality soil moisture meters are expensive and will give reliable readings only when they are carefully cleaned and maintained. I would not urge most growers to rush out and buy one. Using a soil probe, or digging down several inches with a shovel or trowel and inspecting the soil, both visually and by touch, should tell you approximately what you need to know. The more often you perform this assessment, the better you will understand your soil and its moisture-holding capacity. This knowledge will enable you to make more informed irrigation decisions.

Moisture loss through evapotranspiration. Soil loses its water in two ways: direct evaporation from the soil surface and transpiration (giving off water vapor through the stomata) from plant leaves. The hotter and drier it gets, the more water is lost to these two processes, which are collectively referred to as "evapotranspiration." For example, in humid and cool (below 70°F [21°C]) weather, the equivalent of about ⅒ inch of rainwater is lost each day. Under dry conditions, with temperatures above

80°F (27°C), approximately ⅓ inch of water is lost daily. These differing rates of water loss need to be taken into account when deciding how much to irrigate.

Wilting leaves. Wilting leaves can be an indication that plants are in need of water, but they can also indicate stress caused by disease or an insect pest such as vine borers. On very hot days, some plants slow down photosynthesis and allow their leaves to wilt even when there is adequate moisture in the soil. So don't make irrigation decisions based on wilting leaves alone.

Irrigation Options

Crops can be supplied with water in different ways. I'll focus on two methods: Drip irrigation and sprinkler irrigation. I'll also consider the use of a water-wheel transplanter to get young plants started, as well as the more random approach of relying on rainfall.

Drip Irrigation

Though it requires somewhat expensive materials, drip irrigation is by far the most efficient way to deliver water to plants. It does not require high water pressure or a high volume of water. It is unlikely to cause puddling or runoff. With drip irrigation, plants receive water exactly where they need it — in their root zones — rather than on their leaves. Areas between widely spaced rows of vegetables, where weeds are likely to grow, receive little or no water, so you don't help the opposition.

Drip irrigation also makes it possible to fertilize and irrigate plants at the same time, a procedure called "fertigation". Fertigation involves running irrigation water through a soluble organic fertilizer, such as a fish or kelp mix, before it reaches your crops. To do this you'll need a dispensing unit or mix tank and some type of injection system.

Drip lines and plastic mulch. We use T-tape drip lines whenever we plant into plastic mulch. Our mulch-laying implement installs the drip tape under the plastic, in line with where the vegetables will be planted. In our loamy soil, the water spreads out about 12 inches on either side of the drip line. This means we need only one line for each 4-foot strip of plastic, even when we are planting two rows of a crop such as peppers, as long as they are not more than 18 to 20 inches apart.

Plastic mulch and drip lines are a good choice for heat-loving crops that will be in the ground for at least 3 months. On our farm, these include tomatoes, peppers, eggplant, cucumbers, zucchini, summer squash, and basil. Parsley is another crop we like to plant into plastic and irrigate with drip lines. It's not a heat lover, but it does stay productive for at least 5 months, and the plastic mulch blocks weeds the entire time. Some growers report good success planting onions into plastic mulch, primarily for weed control.

Drip lines stay in place throughout the life of the plants they are irrigating. Unfortunately, though, both drip lines and plastic mulch are usually good for just one season. Soaker hoses can last longer, but these are not practical in most commercial vegetable applications.

Drip lines and bare soil. Drip lines may also be used to water crops that are planted into bare soil. In these instances the drip lines are simply laid on the soil surface, as close to the crop as possible. When it is time to cultivate, they must be moved aside.

As already noted, we often use drip lines to moisten the soil over the top of direct-seeded crops such as carrots, turnips, radishes, or beans to facilitate germination. These lines are usually moved after they have served their immediate purpose and reused in a different location.

These peppers are planted in 4-foot-wide black plastic, with a single drip irrigation line for each bed.

Clean water. Drip irrigation requires clean water with no solid particles in it. Solid particles, however small, can easily clog the microscopic emitters that are evenly spaced along the length of drip lines. This will result in uneven watering. If your water source is a pond or stream, you need to have some kind of filtration system to remove solids. If you're using a well that has clean water with no sand or sediment in it, you may not need a filter. We've used the unfiltered water from the well on our property for drip irrigation for many years.

Rodents. In our area, small rodents, probably meadow voles or field mice, sometimes bite into drip lines, either for their amusement or to take a drink. Either way, it creates a problem, causing the somewhat pressurized water to gush out at a rate far beyond what is desirable. A leak of this sort can be repaired using a simple and inexpensive union — something called a "tape loc to tape loc coupling."

A greater challenge is detecting the location of such leaks, especially when the drip lines are hidden from view under plastic mulch. The best way to do this is to monitor the lines regularly. After water has been running through them for several minutes, they should be fully engorged and firm to the touch throughout their entire length. If they are not, you need to walk up and down the rows to find out where the leak is. Usually, you can hear the water gushing out underneath the plastic.

Pressure regulator. Most drip tape products cannot handle a water pressure higher than 10 pounds per square inch (psi). If the pressure where the water enters the drip lines is greater than this, you could pop the lines out of their fittings. To prevent this from happening, you need to install a pressure regulator. A simple pressure gauge will tell you the water pressure.

Ask the experts. If you think drip irrigation is right for you, consult with two or three of the companies (Trickl-Eez, DripWorks, Belle Terre Irrigation, and others) that carry the products you are interested in. Tell them you want expert advice and help in designing a system for agricultural use.

Keep Records

Whether you are irrigating with sprinklers or drip lines, keeping good records and carefully monitoring your plants' progress will point you in the right direction. As always, work with the resources you have, be flexible, and stay light on your feet, even when they get a bit wet.

True, they're in business to sell their products, but they also want to establish good, long-term relations with farmers who may end up buying from them year after year. They can do this by providing solid advice.

The more information you give them, the better they will be able to help you design a system that's right for you. Here's a list of information and data you should have available:

- **Dimensions of the growing area,** or areas, you wish to drip-irrigate
- **Principal crops** you will be growing and whether you plan to drip-irrigate all of them
- **Type of soil** — loam, sandy loam, clay loam
- **Water source** — farm well, pond, river, municipal water — and how close it is to your fields
- **Amount of water flow** (in gallons per minute) you can count on over an extended period of time
- **Water pressure** in psi
- **Length of your rows** and the spacing between them
- **Topography of your terrain** — whether it is level or sloping

Sprinkler Irrigation

Unlike drip irrigation, a sprinkler system comes close to simulating rainfall. It furnishes water that you can see, so it's easy to tell when something is amiss, such as a blocked nozzle, a sprinkler head that has come loose, or a leak in the feed line. If you have an adequate supply of clean water, sprinkler irrigation can cover a large area at one time and put down water more rapidly than drip irrigation. But for big areas (say a third to a half acre or more), you need a lot of water and high pressure — much more of both than you would need to drip-irrigate an area of similar size.

Sprinklers irrigate everything, including empty spaces between rows where weeds can set up shop.

Water quality. For many years, farmers pumped water from ponds, streams, and rivers and used it, in an untreated form, to irrigate their crops. This is no longer recommended. It is important that you use clean water for sprinkler irrigation. Potable well water is fine, but untreated surface water can cause problems. In fact, depending on where you live and how you market your produce, it may be outright prohibited. This is

Plant Diseases from Untreated Water

The same pathogens that can cause a food-borne disease can also, under certain conditions, transmit bacterial and fungal diseases to plants. In the early years of our farm, we believe our onions contracted a bad case of bacterial soft rot after receiving a dose of sprinkler irrigation on a hot, muggy day. That was the end of those onions. Yet another reason not to sprinkler irrigate with untreated surface water.

because surface water can contain pathogens, including a variety of bacteria and even protozoa, and these organisms have been known to find their way onto vegetable crops, causing people to get sick and, in rare cases, die. The problem can be overcome by treating surface water with a sanitizing solution before it makes contact with growing plants.

You may safely apply untreated surface water, through drip irrigation, to the root zones of plants whose edible parts are above ground only. For example, it would be okay with staked tomatoes or peppers, but not carrots or onions or turnips, which are in full contact with the soil. But in all probability you would need to filter the water to remove solids that would clog the microscopic drip emitters. You may also sprinkle untreated surface water onto a field before or just after seeding a crop. It's just the edible portion of growing plants that you should keep it away from.

Layflat hose. The sprinklers on our farm are fed by a pump that draws water from a half-acre pond. The water is transported to the sprinkler heads through 1½-inch VinylFlow (Layflat) discharge hose. This is a nonrigid hose that lies on the soil surface. When drained of water, it lies flat and can be easily moved. It is durable enough that it can occasionally be driven over with a tractor or truck.

Ask the experts. As is the case when designing a drip system, it would be wise to consult with the experts if you're planning to install sprinkler irrigation on your farm. You'll need to provide a lot of the same information as you would for drip (usually, the same companies carry both). There are a number of options. The experts should help you choose products that are scaled to the amount of water you have and the size of your operation.

Rainfall

Depending on the region in which you farm and its annual weather patterns, rainfall can make up anywhere from a small to a large percentage of your crops' water needs; however, an ample rainfall can be a mixed blessing. True, it will cut down on your need to irrigate, and this can save time and money. But unfortunately, you don't get to decide when the rain is going to fall or how much you're going to get. Sometimes, the rain comes when you don't need it and positively do not want it.

Too Much Rain

Too much rain can drown beneficial soil organisms and leach nutrients from the soil, especially nitrogen and potassium. And excessive rain can make it difficult to conduct fieldwork such as planting, cultivating, and harvesting. Plant diseases are also more likely to gain a foothold and spread under wet conditions. This can make life especially difficult for organic growers, who cannot use chemicals to control diseases.

Like it or not, the rain will fall when it will fall. Sometimes it will make you happy, and sometimes it will make you sad. The best you can do is keep an eye on weather forecasts and learn to work around rain events. If a forecast tells us that a significant rain event is on the way, we try to do as much planting and cultivation as possible before the rain arrives. After a heavy rain, we stay out of our fields and cut back on irrigation.

A Rain Gauge

Use a rain gauge to measure rainfall amounts, and record these in a rain log. Note the date and amount of rain, its intensity, its approximate duration, and whether it came during the day or night. A slow, steady rain falling at night will soak deeply into your soil, and very little moisture will be lost to evaporation. A violent thunderstorm on a hot summer day can drop an inch or two of rain in an hour, but a lot of it may be lost to evaporation or runoff.

By knowing exactly how much rain has fallen over the previous week or two, and the conditions under which it fell, you can make informed decisions about when to irrigate. If your rain log tells you that you've received

Rain gauge

2 inches of precipitation in the previous 5 days, you probably should not run your irrigation system for a week or more.

Be sure to place your rain gauge in an open area, away from trees or buildings that might affect the reading, and bring it inside when freezing weather threatens. Over the years we've lost a couple of gauges that we had forgotten to empty before the first hard frost. If you're looking for more precise readings, choose a large rain gauge over a small one.

Rain on Plastic Mulch

Most of the rain falling on plastic mulch (with drip lines underneath) runs off to the sides and doesn't find its way to the root zone of your plants. This is particularly true for small amounts of rain — say a half inch or less. If we get an inch of rain, we usually cut our drip irrigation time by about 25 percent. If we get 2 inches or more, we might cut it in half, or even eliminate a session altogether. But before making such a decision, we take a soil probe or trowel and go down a foot or so under the plastic to see how much moisture is present.

Water-Wheel Transplanter

Most people probably do not consider water-wheel transplanters to be irrigation devices, but they do provide plants with water at the moment they are set in the field — a very crucial time. This water helps to set the plants' roots in the soil and provides start-up moisture for a week or so thereafter. You can easily adjust the amount of water you give to plants. If the soil is already moist, you might provide less water. If the soil is dry, and especially if no rain is expected, it makes sense to give your transplants a generous amount of start-up water. Fish emulsion, kelp, or other organically approved and water-soluble fertilizers can be added to the tanks of your water-wheel transplanter so that plants receive both irrigation and a nutritional boost at the same time.

Watering Specifics

Depending on whether you are using sprinklers or drip lines, it is better to irrigate during certain times of the day than others. And there are times when sprinkler irrigation is a plain bad idea. You also need to decide how much water to provide and how deeply you want that water to penetrate into the soil. The development stage of your crop and the depth of its root systems will greatly influence these decisions.

When to Water

Under stable weather conditions, sprinkler irrigation should be carried out early in the morning — soon after sunrise. As the day heats up, the evaporation rate increases dramatically. Up to 50 percent of the water applied by sprinklers can be lost in the middle of the day, especially if a dry wind is blowing. Less water is lost through evaporation on windless, cloudy, and overcast days.

Avoid irrigating with sprinklers at the end of the day or at night. The water will soak into the soil nicely, but the plant leaves will very likely remain wet throughout the night, which makes them more susceptible to fungal and bacterial diseases.

If you discover that any of your plants are severely water-stressed, don't wait until the following morning to irrigate — give them water as soon as possible, regardless of the time of day. But try to avoid finding yourself in this situation. Severe water stress can have bad consequences. It retards growth, causes flowers to abort, and leads to premature flowering and fruit set. It can also kill plants outright.

You may drip-irrigate under plastic mulch any time of day or night, because there will be minimal evaporation under the mulch and there is no risk of the plants' leaves getting wet. If you drip-irrigate on the soil surface (without plastic mulch) during the heat of the day, some water will be lost to evaporation but nowhere near as much as if sprinklers were used.

How Deeply to Water

The top 18 to 24 inches of most soils is where the major action takes place. It's where most of the soil flora and fauna hang out, and where most organic matter decomposes. It's also where plant roots get the bulk of their water and nutrition. This is largely because the upper levels of the soil have more pore space, which means more room for air and water. Under these aerobic conditions, microbial life flourishes.

At deeper levels — below 2 feet in most soils — the environment is more anaerobic (contains less oxygen) and less hospitable to the majority of soil organisms. This is not to say that there is no water or nutrients at these deeper levels to nourish plants. There certainly can be, but the major feeding grounds are higher up.

Don't Water Too Much

You've probably heard of soils being described as waterlogged after periods of extended rain. This is not a good thing. Too much water can be more harmful than not enough. The more water there is, the less room there is for air. Both plant roots and aerobic soil organisms need air just as they need water, and the lack of air can be stressful or fatal. At the very least, prolonged wetness will impede a plant's ability to absorb nutrients. Excessive rainfall or irrigation also often results in the loss of plant nutrients through leaching.

For young seedlings in the early stages of root development, it's okay to water to a depth of just 6 to 9 inches, but this is not sufficient for most mature plants. Shallow watering creates no incentive for mature plants to send their roots well below the surface, where they could tap into deeper reserves of moisture when a dry period comes along. It's important, therefore, to encourage plant roots to go down at least 18 to 24 inches, if your soil is deep enough to permit this. So occasional watering to greater depths should be part of the equation. In our region, periodic heavy and prolonged rainfall is usually enough to meet the occasional deep-water needs, so we focus our attention on maintaining adequate moisture in the top 18 inches.

How Long to Water

We know that the equivalent of an inch of rain each week will meet the water needs of most plants, most of the time. Incidentally, 1 inch of water across 1 acre of land equals approximately 27,000 gallons — a fair bit of water, especially if you're drawing it from a well. Putting aside other variables, the question then arises: How much irrigation water should we apply to meet the 1-inch rule?

When Using Sprinklers

If you're using sprinklers, it's easy to arrive at an approximate answer. Simply stand up several empty coffee or soup cans in the area that is covered by the sprinklers. The cans should be the same size and should be standing straight, not at an angle. Place them in spots that you think are receiving an average amount of water.

Note how long it takes for an inch of water to accumulate in each can. Some may fill up a little faster than others, depending on how evenly your sprinklers are spreading the water. If this is the case, take an average. Let's say it takes 2 hours. This is how long you should run the sprinklers to get the equivalent of an inch of rainfall. If the spray pattern is uneven, you'll need to compensate for this by moving the sprinklers more frequently and allowing for some overlap. In general, more overlap will result in more even watering and enable you to get the job done faster.

When Using Drip Irrigation

It's a little harder to determine how long you should run a drip irrigation system to receive the equivalent of an inch of rainfall. You can start with a formula, which I've borrowed from the catalog of an irrigation company called Trickl-Eez, but you'll also need to take into account your particular crop and growing conditions.

To apply the formula you will need to know:

- Drip tape flow rate

- Distance between drip tape emitters

- Row spacing or distance between drip lines

Drip tape flow rate is measured in gallons per minute. Low-flow emitters (0.22 gallons per minute [gpm]/100 ft) are recommended for heavy clay soils; medium-flow emitters (0.34 gpm/100 ft) are good for medium-textured soils; and high-flow emitters (0.45 gpm/100 ft) are suitable for light-textured, well-drained soils. If you use high-flow emitters in a clay soil, you might put out more water than the soil can absorb, resulting in runoff.

Emitters can be spaced 8, 12, or 16 inches apart. Choose drip tape with emitters that are closer together for light, well-drained soils and farther apart for heavy clay soils.

Now here's the formula:

$$(0.052 \times \text{feet between rows} \times \text{emitter spacing in inches}) \div$$
$$\text{flow rate of each emitter in gallons per hour} = \text{Number of hours}$$
$$\text{to irrigate per week}$$

Applying the formula. Let's say we have five rows of tomatoes in black plastic, each 200 feet long and each with its own drip line. There are 6 feet between rows and 12 inches between the drip line emitters. (Let's assume we have enough water to maintain good pressure across 1,000 feet of drip line.)

The drip lines we are using have a flow rate of 0.45 gpm, per 100 feet. This amounts to 27 gph (0.45 × 60), per 100 feet, or 0.27 gph (27/100) per foot. The arithmetic, then, would look like this:

$$(0.052 \times 6 \times 12) \div 0.27 = 13.87 \text{ hours per week (round off to 14)}$$

According to this formula, we need to drip-irrigate our tomatoes for 14 hours each week. If we are okay with this, we would then break the 14 hours into three sessions of 4 hours and 40 minutes each. The first we'd apply on Monday, the second on Wednesday, and the third on Saturday.

Adjusting the numbers. In my mind, however, 14 hours of drip irrigation per week is a bit high. Taking into account the nature of our soils, the large amount of organic matter they contain, and the fact that our water supply is limited, we usually irrigate our field tomatoes for 9 hours a week (divided into three 3-hour sessions). But even these three 3-hour sessions may be adjusted — either raised or lowered — depending on the following four variables: (1) the size of the plants, (2) their stage of growth, (3) how hot it is, and (4) recent rainfall amounts.

The point I'm trying to make here is that the above formula should be viewed as a starting point and need not be slavishly adhered to. Conditions will vary from region to region and even from farm to farm. Sandy, well-drained soils are going to need more water than heavier soils with high clay content. The goal is to provide just enough water to grow a healthy and productive crop. Over time, and with a little experimentation, you'll discover what watering regimen works best for you. A cautionary note: One thing we've learned the hard way is that it's unwise to stint on water during the late stages of flower set of tomatoes and peppers. Doing so can cause a bad case of the physiological condition known as blossom end rot, the symptoms of which are dark leathery patches near the blossom end of the fruits. On the other hand, tomatoes that absorb more water than they need usually don't taste that good. Ditto for strawberries.

Drip Irrigation Schedule

If you are drip irrigating more than a handful of crops (especially those planted into plastic mulch) you'd be wise to draw up a Weekly Drip Irrigation Schedule, rather than rely on instinct or memory to determine when your plants may need water. This schedule should list the days of the week, the crops to be irrigated on each day, and the length of time the drip irrigation will run. A Drip Irrigation Schedule need be no more than a sheet of 8½- by 11-inch cardstock with a column and entries for each day of the week.

If you have a crew and want to divide up drip irrigation duties, display the schedule in a prominent place. You can list the person responsible on any given day and ask him or her to confirm that the job is done. For example, Worker A might be assigned to drip-irrigate the tomatoes in Field 05 on Monday morning for 3 hours, and then the summer squash and basil in Field 06 for 3 hours on Monday afternoon. Worker B might do the peppers and eggplants in Field 07 for 3 hours on Tuesday morning and then the parsley in Field 01 for 2 hours in the afternoon, and so on. Or you might prefer to have just one person in charge of drip irrigation.

Depending on the size and water needs of the plants you are irrigating, you will need different schedules at different points in the season. For example, you might create a drip schedule for late spring and early summer, another for midsummer, and a third for late-summer and early fall. In the event of excessive rainfall, you can adjust down irrigation times. In periods of extreme heat and drought, you might increase them.

You might also draw up a schedule for sprinkler irrigation, but this will be affected more directly by rainfall.

CHAPTER 8 RECAP

Irrigation Methods and Tips

▶ The two primary methods of watering crops are drip irrigation and sprinkler irrigation. Though more expensive, drip irrigation is the most efficient watering method. Sprinklers can cover a large area in a shorter space of time but require plenty of water under high pressure.

▶ When the weather is stable, irrigate in the morning. If you are drip irrigating under plastic mulch, you can water at any time of day or night.

▶ Water to a depth of 6 to 9 inches for young seedlings, and 18 inches for established plants.

▶ For mature, deep-rooted plants in areas of low rainfall, occasionally water to depths of 18 to 24 inches.

▶ Use coffee cans to estimate the amount of water your sprinklers are providing and run the sprinklers until they put out the equivalent of an inch of rainfall.

▶ Use a formula to estimate how long to water with drip lines.

▶ Keep good records and be flexible.

Tamarack Hollow Farm

Proprietors: Amanda Andrews and Mike Betit

Location: Burlington, Vermont

Year Started: 2009

Acreage: 88 acres, with 6 acres in production and the balance in pasture or woods

Crops: Mixed vegetables; livestock

AFTER GRADUATING FROM RUTGERS UNIVERSITY with a degree in English literature, Amanda Andrews put in a year with a nonprofit working on justice issues and wrestled with what to do next. Her family and teachers encouraged her to apply to a doctoral program in literature (she had excelled in the field), and she almost did. But there was something about the prospect of many more years in academia that didn't feel quite right. Amanda wanted to try a different path; she wanted to be outside, she wanted to experience physical work, and she had environmental leanings. In 2007, almost on a whim and with very little knowledge of farming, she came to work at our place for a season.

Amanda is slight of build. Within the first couple of weeks in the field, she pulled her shoulder and needed physical therapy. I was doubtful that she would last. But in time, the shoulder healed, Amanda got stronger and more limber, and she soaked up every bit of information about farming that she could. She worked alongside everyone else and did a good job of looking after our greenhouse. She even decided to stay on for a second season, during which, among other things, she drove our big truck to New York City and managed our Wednesday market at Union Square.

After our place, Amanda went on to explore other types of organic and sustainable agriculture. She worked for a dairy farm and cheese maker. She spent time at a small sheep farm, then a 350-member vegetable and egg CSA. She also worked at a farm that used horse traction. During these experiences, Amanda discovered that she enjoyed working with animals as much as she enjoyed growing vegetables and that the two nicely complemented each other.

In 2010, Amanda teamed up with Mike Betit of Tamarack Hollow Farm in Vermont and planted 1½ acres of vegetables on his 88-acre, leased livestock spread. She and Mike sold the vegetables and the meat he was producing at the Union Square Greenmarket in New York City (they had met there when she was working at our place). The next year, the two of them decided to expand the vegetables. They planted 6 acres of crops, including tomatoes, peppers, eggplant, potatoes, squash, onions, garlic, and assorted leafy greens. They did all of this in addition to looking after a hundred hogs, six hundred meat chickens, two hundred laying hens, and five Scottish Highland cattle.

Amanda and Mike continue to sell meat, eggs, and produce at the New York City Greenmarket. They also operate a 35-member CSA in the Burlington area. They use an ox (a castrated Holstein bull called Lucky) and animal traction equipment to hill potatoes and cultivate any row crops that are in the ground for more than 8 weeks. They also have a John Deere 5520 90-horsepower 4WD tractor with a bucket. The tractor, along with a five-bottom plow and a 10-foot harrow, are used for field preparation — that is, after the hogs have had some time to root around and loosen and manure the soil.

The year 2011 was tough for Tamarack Hollow and many other farms in Vermont. A record winter snowfall and spring rains caused flooding, which delayed planting of most crops until June. Then, at the end of August, just when prospects for a bountiful harvest were looking good, Hurricane Irene arrived in Vermont and was followed soon after by Tropical Storm Lee. Together, they brought more flooding and the loss of most of the farm's late-planted crops.

Amanda is enthusiastic about Tamarack Hollow's good alluvial soil but worries about its location on an oxbow of the meandering Winooski River. She notes that the extreme and erratic weather patterns that seem to be the new norm make farming "bottomland" a dicey proposition these days. Amanda and Mike have a 5-year lease on the land they're working. They are not sure they'll be renewing it. They've got an eye out for higher ground.

PART

III

LOOKING AFTER THE LAND

Chapter 9

....

Managing Your Soil

EVERY FARMER KNOWS, OR SHOULD KNOW, that soil is the cradle of terrestrial life and therefore the most vital of planetary resources. Without it, the earth would look very different. There would be no plants and no animals. Rocks and sandy deserts would prevail. Some scientists now believe it is likely that the earliest life on Earth originated in the changing forms of rocks and minerals that are the precursors to soil, rather than in some primordial soup of the sea.

One thing we do know is that healthy soils grow healthy plants, which in turn, nourish animals, humans, and many other forms of life. It doesn't get much more basic than that. Nations have risen and fallen by how they have used their soils. Wars have been waged over productive land. Great migrations have been undertaken in quest of fertile ground. Soil is, quite literally, the foundation on which earthly life exists. As a society, we abuse, ignore, and neglect — at our very great peril — the soils we have inherited.

The conventional approach. Millions of tons of synthetic chemicals have been applied annually to our agricultural land for more than half a century. Some conventional farmers have reached the point where they view the soil as little more than a convenient medium for plants to stand up in, to which they apply prescribed fertilizers and pesticides.

The relentless use of chemicals and heavy equipment and the continuous monocropping of corn, soybeans, rice, and wheat have taken a heavy toll on much of North America's once highly fertile farmland. These practices have damaged the soil's structure, depleted its organic matter, and reduced its ability to hold and retain moisture — all of which leave a soil more vulnerable to erosion by wind and water and are destructive to the soil's vital biotic community.

The organic approach. Organic farmers approach things differently. You'll often hear them say, "Feed the soil, not the plant," which is pretty much the opposite of what conventional farmers do. Organic farmers focus on keeping the soil healthy, fertile, and alive. They do this primarily by using composts, cover crops, animal manures, and mineral amendments and by rotating crops and allowing land to lie fallow for periods of time.

There is a lot to learn about the soil. Scientists spend their professional lives trying to unravel its secrets and at the end of the day are humbled by how little they know. Most organic farmers are not soil scientists, and we don't need to be. What's important is that we view the soil as a dynamic living resource that requires careful and nourishing management if it is to remain productive in a system of continuous, or near continuous, cropping. We must always resist any temptation to view the soil as a resource to be mined for short-term profit.

It takes years of farming the same piece of land to get a good feel for your soil — its characteristics, qualities, limitations, and variability. The last of these is especially true if you are working with more than one soil type, or on sloping and uneven ground with different drainage characteristics. The goal is to learn how to work with your soil in such a way that it will remain healthy and productive during your tenure on the land and be in as good or better condition for the growers who come after you. Learning as much as possible about your soil is one of the great joys of being a farmer.

There are several excellent books on soil management, written for organic and sustainable growers (see Resources). In this chapter, I provide basic information about soils that will help a farmer get started.

Soil Basics

Soil is the part of the earth's surface that will support plant and animal life. It is composed of layers or horizons that start above bedrock and progress upward through subsoil to topsoil. Farmers are mostly concerned with the topsoil. A healthy topsoil provides an environment that is rich in nutrients, minerals, and organic matter and is hospitable to plants and their roots. It will have good structure or tilth and plenty of pore spaces to hold air and water. The topsoil is where most of the soil's biological activity takes place, but subsoil and even the bedrock from which it derives are not irrelevant. Their depth and composition will in large part determine the type of topsoil that lies above.

Composition

Soils differ widely across the planet. Factors that account for the various soil types include the underlying parent material from which the soil is derived; the climate, rainfall, and weather patterns the soil has been exposed to over long periods of time; and natural phenomena such as flooding, volcanic eruptions, and glaciation. More recently, human activities like deforestation, agriculture, and mining have altered the nature of many of the earth's soils, usually for the worse.

Despite their differences, all soils are made up of the same four components: solid materials, water, air, and organic matter. The respective amounts of these four constituents may surprise you. We tend to think of soil as a primarily solid medium, but 50 percent of most agricultural soils are air and water. The composition of a typical, healthy farmland soil, by volume, looks like this:

Soil Types

The USDA's National Cooperative Soil Survey has identified and mapped more than 20,000 soils (also referred to as "series") in the United States. Most of them have been given names that usually derive from the area in which the soil was first mapped. The Mardin series, which appears on my farm, is one of them.

- **Solids (small pieces of rocks and minerals):** 45%

- **Water:** 25%

- **Air:** 25%

- **Organic matter (living and dead organisms and plants):** 5%

(Note: These percentages are for typical mineral soils in temperate regions. Not all soils are so blessed. In poor, arid-land soils, organic matter content can be as low as 0.5 percent. On the other hand, certain muck soils can have very high levels of organic matter — as much as 30 or 40 percent.)

All of the above components (solids, water, air, and organic matter) are vital for plant growth. Soil solids provide essential minerals and nutrients. They derive from the breakdown and weathering of the parent material and bedrock below. Air and water keep a soil alive. They occupy the very important pore spaces that enable plant roots to spread and soil organisms to move around and conduct their affairs, often in symbiotic relationships with plants. Soil air furnishes plant roots with oxygen and helps get rid of surplus carbon dioxide. Most of the nutrients that plants need are obtained from soluble compounds in the soil water.

Organic matter, though it makes up only a small percentage of a typical soil's volume, is hugely important—well beyond its proportional representation — but more on that later.

Texture

The solid, or mineral, component of soil is made up mostly of silicon, potassium, magnesium, calcium, and aluminum. Depending on particle size, these minerals are classified as sand, silt, or clay — sand being the largest, clay being the smallest, and silt lying in between. A soil's texture is determined by its relative proportions of these three particle sizes. If you farm in the northeastern United States, as we do, or other parts of North America that experienced the movement of glaciers, you're likely to have a fair number of rocks and stones as well.

Texture is important because it determines to what extent a soil can hold nutrients, water, and air — all of which are critical for healthy plant growth. For example, soils with a high proportion of sand (larger-size particles) will drain more quickly than soils with a high proportion of clay (smaller-size particles); however, soils with large amounts of clay tend to have a greater capacity to retain essential nutrients.

Soil experts have gone to the trouble to measure particle sizes and have come up with precise definitions for sand, silt, and clay:

Sand: Particles of 0.05 to 2 millimeters in diameter
Silt: Particles of 0.002 to 0.05 millimeters in diameter
Clay: Particles of less than 0.002 millimeters in diameter

The experts have also created something called a soil texture triangle (see illustration on next page). This is a simple, graphic way of classifying a soil's texture, based on its relative proportions of the three particle sizes. Clay is represented at the apex of the triangle and sand and silt at the left and right corners, respectively. The percentage of each particle size is represented along the sides of the triangle.

Soils that have all three particle sizes — clay, silt, and sand —are called loams. Using the triangle, a soil with 50 percent sand, 30 percent clay, and 20 percent silt is called a sandy clay loam. The USDA defines a total of 12 textural classes:

- Clay
- Silt
- Sand
- Loam
- Silty clay
- Sandy clay
- Clay loam
- Silt loam
- Sandy loam
- Loamy sand
- Silty clay loam
- Sandy clay loam

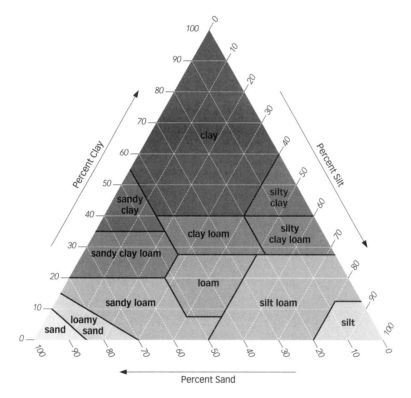

Soil texture triangle · The percentage of clay is read from the bottom left corner to the apex of the triangle; the percentage of silt from the apex to the bottom right corner; and the percentage of sand from the bottom right to the bottom left corner. The intersection of the three particle sizes within the triangle defines the textural class.

Balance is key. When it comes to soil texture, as with most things in life, balance is desirable and extremes can be problematic. Soils that are heavily weighted with clay will hold water and nutrients well but drain poorly, which can be a big problem during spring thaw and periods of heavy rain. They can also be difficult to work with tillage equipment such as plows, rototillers, and spaders. If tilled when too wet, soils with high clay content can bake in the sun and form hard, concretelike clumps that impede the penetration of water and air and make planting very difficult.

Soils with a high proportion of sand are easier to work but do not retain moisture or nutrients well. Usually, the most desirable soils are loams, which have a reasonable representation of the three particle sizes.

Assessing texture. You can get some idea of the texture of your soil by giving it a squeeze test. Squeeze a handful of damp soil into a clump and give it a smack with the edge of your hand. See if it holds together or falls apart and, if so, how easily it falls apart. As you've probably guessed, a wet soil with lots of clay will stay clumped together, while a sandy soil will break apart easily, even when wet.

Readiness for Tillage

The same techniques you use to assess a soil's texture can be used to determine when a soil is too wet to be safely tilled. Simply dig down 4 or 5 inches with a shovel, take a handful of the soil, squeeze it into a ball, and see how easily it breaks apart. If it stays together, you'd best leave your tractor parked in the shed for another day or two.

Working a soil when it is too wet, especially with heavy equipment, will damage the soil's structure and cause compaction, both of which are undesirable, especially to an organic farmer. Compaction harms the soil's all-important aggregate structure, which means fewer and smaller pore spaces, resulting in less room for air and water. It also means there will be less room for the larger soil organisms, such as earthworms, mites, and springtails, to move around and fewer openings for plant roots to penetrate.

A field that is frequently wet or waterlogged may already have a compaction problem in the form of a dense layer 6 to 18 inches below the soil surface. Such a layer is often caused by the overuse of a moldboard plow. If this is the case, you need to change your tillage practices. A chisel plow or subsoiler will break up the compaction and improve drainage.

You can also try to shape damp soil into a thin sausage or ribbon shape. The greater the amount of clay, the longer the sausage. Sausages less than 1 inch long generally indicate that the soil has more sand particles than clay.

A soil's texture can be approximated by squeezing a handful into a clump and then trying to break the clump apart.

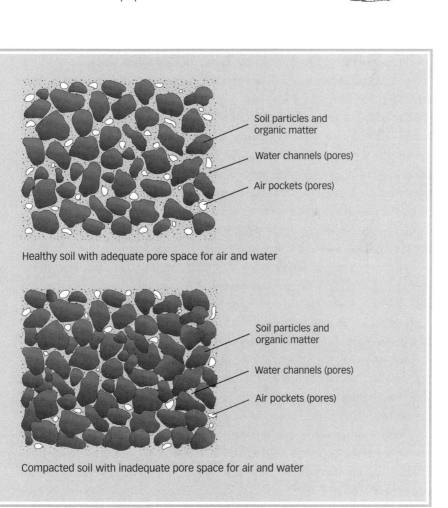

Healthy soil with adequate pore space for air and water

Soil particles and organic matter

Water channels (pores)

Air pockets (pores)

Compacted soil with inadequate pore space for air and water

Soil particles and organic matter

Water channels (pores)

Air pockets (pores)

Slow Roots Farm

Proprietor: Jacob Diaz

Location: Kingston, New York

Year Started: 2008; restarted under current name in 2010

Acreage: 6 acres, with 3 acres in production and 3 acres in cover crops

Crops: Mixed vegetables, strawberries

THREE DAYS AFTER GRADUATING FROM COLLEGE with a degree in environmental studies, Jacob Diaz came to live and work at our farm. That was in May 2003. He's been farming ever since. After his season with us, Jacob went on to plant and market his own vegetables on assorted lands in four different New York counties. He also put in a few years' milking cows on other peoples' farms. But all the while he was looking for a place to settle down.

He describes Esopus Creek Farm, where he landed in 2008, as a ghost farm — abandoned and overgrown. The landowner had leased his property to three different growers in the previous 2 years. None was able to make a go of it. But looking beyond the weeds and brambles, Jacob saw potential. He agreed to pay $6,000 a year to rent the land and have use of a small cabin and a tractor. It looked like a promising arrangement, and Jacob poured all his energy and a fair amount of his resources into building a farm operation that he could stay with for a long while. Then, in 2010, fate intervened: The man with whom Jacob had made his arrangement died, and his wife put the farm up for sale. Jacob assumed that his rental arrangement would no longer be valid and he would have to move.

Soon a buyer was found, and it looked as if the end was imminent. But once again, fate intervened, only this time she was more kindly. To Jacob's surprise, the new owners, a middle-aged husband and wife of independent means, liked what he was doing and invited him to stay on indefinitely and run his own farm business. The new owners and Jacob are currently in the process of negotiating a long-term lease arrangement.

Jacob has renamed his business Slow Roots Farm and is forging ahead, with the expectation that he will be staying put for quite a while. He has already invested some of his own money to erect an 8-foot-high deer fence around 6 acres of land. With help from a NRCS grant, he has built a 30- by 96-foot tunnel. The farm already had a functioning 20- by 45-foot greenhouse.

Today, Slow Roots Farm is growing as many as 250 varieties of about 50 different vegetables, along with a small planting of strawberries. Jacob admits it might be smart to trim those numbers down a bit.

The primary equipment at Slow Roots includes a 50-horsepower Kubota tractor, a 5-foot rototiller, and a 6-foot Perfecta field cultivator and bed preparer. To build fertility in his very well-drained, sandy soil, Jacob uses chicken manure, compost, and some organically approved mineral amendments. His main cover crops are winter rye, white clover, and buckwheat. Jacob is currently planting 3 acres of row crops and has 3 acres in cover crops, but there is room for expansion. At some point, he would like to bring animals onto the farm, possibly even a few milking cows.

Jacob has chosen Certified Naturally Grown status, rather than organic certification. He sells at two local farmers' markets and runs a 40-member CSA. He also does a little wholesaling on the side. It's a lot of work for one man to do, even with a little help from his friends.

Jacob Diaz thinks farming is a good life. He likes being engaged with nature in a positive way. He enjoys seeing the tangible results of a day's work in the fields. He believes healthy food, grown in a sustainable manner, has the power to transform the health of individuals, and even the health and well-being of entire communities.

Organic Matter

It would be hard to overstate the value of organic matter. It is the glue that holds soil together, and much more besides. Maintaining a good level of soil organic matter is one of the most important things an organic grower can do. Organic matter comes in a wide spectrum of forms, ranging from living and dead organisms to undecomposed vegetation, plant roots, manures, and mature and relatively stable humus — and all stages in between. Most of the various forms of organic matter are found in a soil's upper 6 to 12 inches.

Humus

Humus is the nearly stable end product of organic matter decomposition. It is lightweight, dark-colored, and fluffy and has a pleasant earthy smell. It can act like a sponge, being able to absorb 80 to 90 percent of its weight in water. Humus makes a soil more resilient — gives it more spring. It also has a gummy quality that helps bind other soil particles together into crumbs. These crumbs, or aggregates, give a soil good structure, or tilth, and help it resist erosion.

Humus is good at holding onto mineral nutrients, especially cations, that might otherwise leach out of the soil (more about cations later in this chapter). When attached to humus, nutrients remain available to plants on an as-needed basis. A good supply of humus also lessens the impact of soilborne diseases. In general, the more humus, the better.

What Organic Matter Does for Your Soil

- ► Helps soil retain moisture, which means more water for plants during dry periods
- ► Supplies plants with essential nutrients
- ► Prevents nutrients from leaching out of the soil
- ► Improves soil structure and creates pore spaces for water, air, and soil microbes
- ► Makes soil less prone to erosion by wind and water
- ► Helps control soilborne plant diseases by providing food and shelter for soil microbes and beneficial organisms

Prehumus

At the other end of the organic matter continuum are the as-yet-undecomposed plant remains and animal manures. These are just starting their journey to becoming humus, but this doesn't mean they are of little value in their more raw state. On the contrary. Undecomposed and partially decomposed organic matter (sometimes referred to as active carbon) provides necessary food for many of the organisms that exist in a healthy soil. When the amount of fresh organic material declines, so does the microbial population.

Organic farmers keep their soils' organic matter quotient high by avoiding excessive tillage (which results in the oxidation and loss of organic matter), by growing cover crops (which are later incorporated into the soil), and by applying composts, minerals, manures, bedding material from horse or dairy barns, and almost any type of uncontaminated vegetative matter. It's all grist for the mill.

Structure and Tilth

The terms "structure" and "tilth" have similar meanings when it comes to soil, but they are not totally synonymous. Both describe the physical condition of a soil — its mineral or solid component, along with its capacity for holding air, water, and organic matter. A soil with good structure or tilth will hold itself together nicely, yet be easy to work and plant. It will be well aggregated and highly porous. It will be better able to resist erosion, especially when subjected to extremes of weather (wind, rain, flooding) or human interference such as plowing, tilling, or cultivating.

The term structure is used by agriculturalists, but it might just as easily be used by an engineer or contractor when referring to a soil's ability to accommodate a road or a building.

Tilth is a little more farmer specific. It is more focused on how well a soil can support vegetation, how hospitable it is to plant roots and microorganisms, and how easily it enables them to move around and access nutrients and organic matter. A soil in good tilth will provide ample sites and surfaces for biochemical exchanges to occur. A soil in good tilth is a soil in good health. It is truly a living soil. Oregon's organic farming association named themselves Oregon Tilth for good reason.

The Web Soil Survey

To find out the name and characteristics of a soil and its suitability for vegetable growing, go to the USDA's Web Soil Survey (see Resources for website).

This site has soil maps and descriptions of soils for almost all the counties in the United States. Simply type in an address and up will pop an aerial photograph of the property you have specified and the nearby surrounding land. You can then request a soils map for your area of interest, as well as descriptive data for the soils on that map.

Soil pH

To soil scientists, pH is a measure of the concentration of hydrogen ions in solution, as in "percent hydrogen." To most farmers, it is, more simply, a measure of a soil's acidity or alkalinity. A pH of 7 is neutral — neither acidic nor alkaline. A reading below 7 tells us a soil is acidic — from slightly acidic to exceedingly acidic — depending on how low the number is. A pH reading above 7 indicates that a soil is alkaline (or basic).

A soil's pH is important because it largely determines how available various soil nutrients will be to growing plants. Many nutrients are less available to plants in highly acidic or alkaline soils.

The pH scale is logarithmic. This means that every whole number below 7 is 10 times more acidic than the number above it. Thus, a pH of 6 is 10 times more acidic than a pH of 7, and a pH of 5 is 10 times more acidic than a pH of 6, but 100 times (10 times 10) more acidic than a pH of 7. The same holds true for pH numbers above 7, only each number is 10 times more alkaline than the number immediately below it (a pH of 9.5 is 10 times more alkaline than a pH of 8.5 and 100 times more alkaline than a pH of 7.5). So you can see that, when it comes to pH, a few decimal points of difference can be quite significant.

The Ideal pH Range

Vegetable growers are generally happy when their soil is mildly acidic. For most vegetable crops growing on mineral soils, you want a pH between 6.3 and 6.8. Within this range, essential soil nutrients are more soluble and therefore more available to plants. This range is also good for many beneficial soil organisms.

Adjusting pH

When a soil's pH is too low or too high, problems arise. For example, though a soil may have an ample supply of phosphorus, if the pH is too low, the phosphorus will bind with aluminum and be unavailable to plants. At the same time, iron, manganese, and aluminum will become more soluble and reach undesirable levels. Adding more phosphorus is not going to help much. Calcium, magnesium, phosphorus, and molybdenum are most likely to be deficient in acidic soils.

When the pH is too high, phosphorus gets tied up with calcium and becomes deficient. Iron, manganese, copper, zinc, and boron will also be less available to plants.

Most of the soils in the eastern United States tend to be somewhat acidic and require periodic applications of limestone, or calcium carbonate, to bring them to peak productivity. Soils with high levels of organic matter tend to have more stable pH values and therefore require less frequent liming.

Applying Lime

Let's use our farm's gravelly silt loam as an example. Under normal cropping conditions, we expect the pH value of our soil to go down about 0.1 every 2 or 3 years. We can easily reverse this by applying an appropriate amount of agricultural limestone. The general rule of thumb for our silt loam is that 1,000 pounds of lime per acre will raise the pH by 0.1; therefore, if we wish to bring the soil's pH from 5.9 to 6.4, we would apply lime at a rate of 2.5 tons (5,000 pounds) per acre. That sounds like a lot of lime, but, fortunately, lime is one soil amendment that is relatively inexpensive, especially when purchased in bulk.

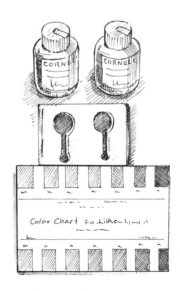

Soil pH test kit

The amount of limestone needed to raise the pH to a suitable level varies by soil type. In general, sandy loams require less limestone than silt loams, while clay loams require more. When you have your soil tested, you will be asked to identify it by name. If you don't know the name, you can find it on the Web Soil Survey. Liming recommendations will be based on this information, as well as on the crops you plan to grow.

Testing a soil's pH and applying lime, when needed, are two of the simplest yet most important things a grower can do. Soil samples can be taken to a local Cooperative Extension office for testing. Alternately, one can buy a simple, inexpensive soil pH test kit. Moderately priced pH test meters are also available through farm supply catalogs, but these don't always work with great precision. I'll have more to say about soil testing later in this chapter.

Calcitic versus dolomitic lime. When applying lime, a farmer often has a choice between using calcitic lime, which contains more calcium, and dolomitic lime, which contains some magnesium. Calcitic lime is generally about 40 percent calcium. Dolomitic lime is about 30 percent calcium and 10 percent magnesium.

It makes sense to spend a few extra dollars and have a full soil test, rather than one that simply tests pH. This will tell you a lot more about your soil, including its calcium and magnesium content, which will help you decide which kind of lime best suits your needs. Most growers choose calcitic lime, unless their soil has less than 200 pounds of magnesium per acre. If the soil is low in magnesium, dolomitic lime is a better choice.

If you don't have your own spreader, agricultural service companies will spread lime for you or possibly rent you a spreader that can be pulled behind a tractor. Avoid processed forms of lime, such as quicklime and hydrated lime. These can damage plants and soil organisms, and they are not permitted under the rules of the NOP.

When to apply lime. Late fall is a good time to apply lime — after your last harvest, but before the ground is frozen and while there is still a good chance of rain to aid infiltration. If your soil is already wet and you are concerned that heavy spreading equipment will cause compaction, you would be wise to wait until the ground is frozen and soil compaction is not an issue. Because it is so heavy, it is unlikely that lime spread in winter will blow away. The lime should infiltrate when the ground thaws in spring.

Plant Nutrients

In this section, we'll take a look at the nutrients that, when married with sunlight and water, enable plants to prosper and grow. They divide readily into three groups: The big three, other important nutrients, and micronutrients.

NPK – the big three. Though not the only nutrients vital for plant health, nitrogen, phosphorus, and potassium are the major nutrients that are more likely to be in short supply.

Synthetic chemical fertilizers that conventional farmers use are rated according to their respective percentages by weight of nitrogen, phosphorus, and potassium. A fertilizer blend rated 5:10:10, for example, will have 5 percent available nitrogen, 10 percent available phosphate, and 10 percent available potassium oxide. (Because phosphate and potassium oxide contain other elements, the amounts of elemental phosphorus and potassium in the blend are actually only 4.4 percent and 8.3 percent, respectively.)

Fertilizers for organic growers are made from nonsynthetic products and are rated similarly, but usually the numbers are much lower. Organic NPK blends of 4:2:4, or 2:4:2, are more typical. Unlike chemical fertilizers, organic blends are not likely to provide a quick fix. They release their nutrients more slowly and over a longer period of time; therefore, they do not harm soil organisms.

Other major nutrients. Plants also need calcium, magnesium, and sulfur. These are usually, though not always, available in sufficient quantities in a healthy soil.

Carbon, oxygen, and hydrogen are also critical to plant growth, but a farmer doesn't need to worry too much about providing them, especially if his or her soil has good structure and its proper quotient of air and water. Atmospheric carbon dioxide is the source for carbon. Oxygen is also obtained from the atmosphere or from air in pore spaces in the soil. And plants get the hydrogen they need from water within the soil.

Micronutrients. Also important for plant growth are nine micronutrients: boron, chlorine, cobalt, copper, iron, manganese, molybdenum, nickel, and zinc. Plants require these in much smaller quantities, and they are generally, and preferably, available through soil organic matter. Organic matter releases micronutrients to plants in a balanced manner, ensuring that they don't receive too much of any single one. Excessive amounts of micronutrients in a soil can become poisonous to plants. Except in clear and verified cases of deficiency, it is usually not necessary to add micronutrients to your soil.

Good Balance

Ideally, a fertile and well-managed soil will have an adequate supply of all plant nutrients. The primary sources of these nutrients will be the mineral component of the soil — the sand, silt, and clay particles — and the organic matter component, which is kept robust by the decomposition of crop residues, cover crops, and the periodic application of composts and possibly animal manures. Rotating crops and periodically leaving the land fallow will also contribute to soil health and fertility (I'll have more to say about these matters in chapters 10, 11, and 12).

Of course, the ideal does not always prevail, however much we might wish it did. But there are other remedies, aside from the normal good practices of cover cropping and using composts, for soils that may be deficient in some respect. You might use various commercially available and organically approved soil amendments to correct a nutrient imbalance, and we'll consider some of these later in this chapter. But before applying any commercial product, you should be sure it's what you need. The easiest way to find out is to have your soil tested (see page 216).

> ## Soil Nutrients
>
> **The big three:** nitrogen, phosphorus, potassium
>
> **Other major nutrients:** calcium, magnesium, sulfur, carbon, oxygen, hydrogen
>
> **Micronutrients:** boron, chlorine, cobalt, copper, iron, molybdenum, nickel, zinc

Cation Exchange Capacity

The major soil nutrients are classified as either cations or anions, depending on whether they carry positive or negative charges. Calcium, magnesium, and potassium are positively charged. They are the cations. Nitrogen, phosphorus, and sulfur carry negative charges. They are the important anions.

Cation exchange capacity (CEC) is a measure of a soil's ability to hold onto its positively charged ions (calcium, magnesium, and potassium) so they remain available to plants. In the world of electricity and, some might say, in the realm of human relations, opposites attract. For this reason, cations in the soil attach themselves to particles of humus (well-decomposed organic matter) or tiny particles of clay, both of which carry many negative charges.

This attraction between positively and negatively charged ions is important because it ensures that valuable cations are not leached out of the topsoil after a period of heavy rain or snow melt. CEC is just one of the many reasons organic matter is so important to an organic grower, especially a grower whose soil is high in sand and low in clay. Without a good quota of humus, essential nutrients may be lost.

> ## Cations and Anions
>
> **Major cations** (positively charged):
> - calcium
> - magnesium
> - potassium
>
> **Major anions** (negatively charged):
> - nitrogen
> - phosphorus
> - sulfur

The three major negatively charged ions (nitrogen, phosphorus, and sulfur) are mostly supplied to plants through organic matter. Of the three, nitrogen is most vulnerable to leaching. During cold, wet periods, a soil's nitrogen supply may be temporarily diminished because soil microbes are less active at lower temperatures. Phosphorus is usually present in soil as an insoluble compound, and is therefore less likely to be leached out. Sulfur is also less prone to leaching.

Soil Nutrient Deficiencies and Remedies

We've already talked about the various nutrients that plants depend on — from the big three to other major nutrients to the various micronutrients that are needed in very small quantities. Here, we'll take a closer look at primary nutrients and how to detect deficiencies in your soil and remedy them.

Nitrogen

Symptoms of nitrogen deficiency typically include yellowing plant leaves (especially the youngest ones), thin stems, and slow growth. On the other hand, a good supply of nitrogen, usually in the form of nitrate (NO_3), will cause rapid growth and healthy dark green leaves, along with increased yields of fruit and seed. But beware of too much nitrogen. When some plants, such as tomatoes, are given more nitrogen than they need, they will grow a large amount of foliage but not produce much fruit.

Nitrogen is the most volatile of the major nutrients and is difficult to measure with accuracy. Nitrogen levels fluctuate in response to changes in temperature and moisture; in general, a soil will have less available nitrogen when it is cold and wet. Growers tend to take the nitrate readings in a soil test with a few grains of salt. The best way to provide plants with a balanced supply of nitrogen is to maintain good levels of organic matter in the soil, preferably at different stages of decomposition (fresh green vegetation to mature compost).

The carbon-to-nitrogen (C:N) ratio. Too much organic material that is high in carbon but low in nitrogen (such as sawdust, wood shavings, and shredded tree limbs) can cause a short-term nitrogen deficit. Microorganisms will immediately go to work breaking down the high-carbon

Legumes and Nitrogen Fixation

Legumes, in conjunction with soil bacteria called rhizobia, are able to take atmospheric nitrogen and convert it into forms that other plants can use. This is known as "nitrogen fixing." Legumes include field crops such as peas and beans and forage or cover crops such as alfalfa, vetch, bell beans, and various clovers. These crops provide a valuable and highly appropriate source of nitrogen for organic and nonorganic growers alike.

When a suitable legume is grown and, at maturity, incorporated into the soil, all the nitrogen needs of a future nonlegume crop may be met. We'll look into this more deeply in chapter 10. Meanwhile, it's worth noting that the accuracy of the nitrogen values in a soil test will depend, to a large extent, on the thoroughness of the cover crop and manure application information provided by the grower.

material, but they need nitrogen to fuel their decomposing activity. They will take whatever available nitrogen they can find in the soil, depleting the supply that might otherwise be used by plants.

Vegetable growers need to keep an eye on this ratio between carbon and nitrogen. A C:N ratio of 20:1 to 30:1 should not pose a problem, but a large application of highly carbonaceous material that brings the C:N ratio up to 40:1 or 50:1 will most likely lead to a temporary, but serious, shortage of nitrogen. Fortunately, the problem is usually temporary. The microorganisms doing the decomposing will eventually die and release the nitrogen in their bodies back into the soil.

Fresh or composted animal manures are excellent sources of nitrogen. Fresh manure is less stable than composted manure and will release its nitrogen more rapidly. Note, however, that the NOP rules require that 3 to 4 months elapse before crops can be harvested from ground that has received fresh or uncomposted manure (see page 301 for more information).

Other sources of nitrogen include blood meal, fish meal, feather meal, cottonseed meal, and soybean meal. A sufficient amount of available nitrogen for most plants is 100 to 140 pounds per acre.

Phosphorus

Phosphorus is vital for getting seedlings and young plants off to a good start. It stimulates blooming and seed development and is essential to the genetic health of plants. Plant leaves turning purple, especially on the underside, as well as delayed growth and premature fruit drop, are often indications of phosphorus deficiency.

Phosphorus is usually present in soils in an insoluble form and is therefore not vulnerable to leaching. The activity of soil bacteria makes it available to plants. A phosphorus deficiency can often be traced to a pH that is outside the desirable range. As noted above, a pH that is too high or too low will cause phosphorus to bond with other elements in the soil in such a way that it becomes unavailable to plants. Agricultural limestone may be used to correct a pH that is too low, and elemental sulfur will bring down a pH that is too high.

Natural sources of phosphorus include rock phosphate, colloidal phosphate, bonemeal, and fish meal (before using any of these, check with your certifier; some brands may contain a disallowed substance). A soil test will indicate how much phosphorus your soil has in pounds per acre and whether more is needed. For most soils, an appropriate range for available phosphorus is 30 to 90 pounds per acre. If you have less than this, a soil test is likely to recommend that you add more.

Potassium

A potassium deficiency will cause poor root development, stunted growth, susceptibility to disease, and low yields. Curled and mottled leaves and leaves with scorched-looking, papery spots may be indications that your plants are not getting all of the potassium they need.

Potassium is water soluble; therefore, it is susceptible to leaching, especially in soils with a low CEC. The best way to hold onto potassium, aside from maintaining high levels of organic matter, is to keep as much plant cover in your fields as possible. Plants — be they crops destined for market, cover crops that will be turned back into the soil, or even weeds — will capture and hold on to available potassium. As cover crops, weeds, and crop residues decompose, the potassium they contain will be returned to the topsoil. Soil that is left bare for too long and subjected to heavy rain is likely to lose some of its available potassium to the subsoil.

Organically approved and readily available sources of potassium are greensand, granite dust, soybean meal, and langbeinite, which sells under the names Sul-Po-Mag or K-Mag. Langbeinite is also a good source of sulfur and magnesium. The typical range for potassium in most productive soils is 100 to 300 pounds per acre. A soil test should tell you how much potassium your soil has and how much more, if any, might be needed.

Magnesium

Magnesium deficiency is not common but can occur. The main symptom is yellowing lower leaves with green leaf veins. Eventually, affected leaves will turn brown and die. Magnesium deficiency will also reduce yields.

Magnesium helps plants metabolize phosphorus and assists in the uptake of other plant nutrients. Having plenty of organic matter in your soil will help maintain adequate levels of magnesium.

Two natural sources of magnesium are dolomitic lime (which is a mixture of magnesium and calcium carbonate) and langbeinite. A typical range for magnesium in a productive soil is 100 to 200 pounds per acre, with the higher range being more suitable for soils with high clay content.

Detecting Nutrient Deficiency Is Not So Easy

The physical appearance of a plant can suggest that a soil is deficient in a particular nutrient, but not always with full reliability. Some deficiencies cannot be detected by a simple visual inspection. Even an expert might not know for sure what is causing the problem. Stunted growth, low yield, and yellowing or curling leaves are not always the result of a nutrient deficiency; they might be caused by an insect or plant disease.

Calcium

Calcium deficiency leads to poor root growth and curling plant leaves. Blossom end rot in tomatoes may be caused by a deficiency of calcium or poor calcium uptake from uneven irrigation.

Calcium helps plants take in nitrogen and some other nutrients. Again, a generous amount of organic matter in your soil is the best insurance against a calcium deficit.

Either calcitic or dolomitic limestone can be used to correct a calcium deficiency. A desirable range for calcium is generally 2,000 to 4,000 pounds per acre. Most soils have adequate calcium when their pH is within the 6.3 to 6.8 range. In naturally acidic soils, the easiest way to maintain this range is by periodically adding agricultural limestone.

Carbon

Carbon is the most basic component in all living cells. It is essential for both plant and animal life. Plants contain more carbon than any other single element. Through the process of photosynthesis, they extract carbon from carbon dioxide in the atmosphere and from pore spaces in the soil. Most other forms of life get the carbon they need from plants or other photosynthesizing organisms such as algae.

If for some reason there is a limited supply of carbon in the soil, the plants' growth will inevitably slow down, though this is not likely to happen in the field. Contained environments, such as greenhouses, sometimes run short of carbon. This can usually be remedied by using a fan to bring in and circulate air from the outside.

Decaying organic material is a rich source of carbon, and soil organisms emit carbon as a by-product as they break down plant material. The best thing you can do to ensure that your plants have enough carbon is to keep the soil's quotient of organic matter high.

Sulfur

The symptoms of sulfur deficiency resemble those of nitrogen deficiency. Yellowing plant leaves may indicate the need for more sulfur. If plants are growing vigorously and have a healthy dark green color, they probably have all the sulfur they need. Plants get sulfur from organic matter and from pollutants in acid rain, and certain soil microbes are able to fix atmospheric sulfur.

Organic farmers might use Sul-Po-Mag and elemental sulfur if their soils are deficient in sulfur.

Soil Testing

The easiest way to learn about a soil's potential for growing specific vegetables and its supply of available nutrients is to take a sample and send it to a lab for testing. You might use a private lab or one that is associated with your state's land grant college (through your local Cooperative Extension office). Certainly, it is wise to use a lab that is familiar with soils in your region.

When you submit a soil sample, you will be asked to fill out a form. Depending on the lab you are working with, you'll need to provide certain information about the field or fields being tested, such as:

- Soil name

- Tillage depth

- Cash crops grown over the previous 3 years

- Crops planned for next 3 years

- Cover crops grown over the previous 3 years

- Manure applications (in tons per acre) over the previous 3 years

The more accurately you can answer these questions, the more you can rely on the recommendations given. For most diversified organic growers, the "crops grown" and "crops planned" are often listed simply

The Cooperative Extension System

The Cooperative Extension System is a nationwide network of offices that provide information and advice to farmers and other members of the public. Each Cooperative Extension office receives guidelines and guidance from the land grant college in its state. Nearly every county in the United States with some agricultural activity has a Cooperative Extension office. It's well worth paying the small annual fee to become a member.

Throughout most of its history, the Cooperative Extension has focused on conventional agriculture, and that's where the emphasis still lies. Increasingly, though, there is an awareness that organic farming is here to stay, and many Extension agents are becoming more conversant with organic methods, which is a very good thing.

Many Extension offices also administer a Master Gardener program. If you are planning to have a very small operation and are not intending to use any serious farm equipment, you may want to consider taking part because it can teach you a lot about growing vegetables and other plants.

Regardless of whether you are a conventional or organic grower, Cooperative Extension agents can provide plenty of useful information, such as how and where to get your soil tested.

as "mixed vegetables." But if you were planning to grow only corn or only garlic, you would definitely want to note this.

Analyzing the Results

A typical soil test will give you the pH of your soil and the percentage of organic matter. It will also tell you, usually in pounds per acre, how much phosphorus, potassium, calcium, and magnesium is present in forms available to plants. It should indicate if you have an excess, a deficiency, or a suitable amount of each of these important nutrients. If a soil is deficient in one or more nutrients, the recommendations section of the test will indicate how much should be added to correct the situation. If the pH is outside the optimum range, you will be given liming recommendations.

In addition, most labs will tell you the amounts of other nutrients, such as sulfur, aluminum, copper, zinc, manganese, and iron, that are present in your soil. These, too, are usually expressed in pounds per acre. The nitrogen content might be expressed as nitrate, but as I've already noted, soil nitrogen is volatile and difficult to measure accurately.

It's important to remember that soil tests attempt to measure the amounts of the various nutrients that are available to plant roots, which may be different from the actual amounts in the soil. Your soil's pH and temperature will affect how much of a given nutrient is accessible to your crop at a given time.

Results can vary by lab. The type of chemical extractant used to determine how much of a particular nutrient is in a soil sample varies by lab. The method employed will usually depend on such factors as the lab's equipment, the type of soil being tested, and the region the soil comes from. Results may vary from one lab to another. When interpreting results, it's helpful to know which extractant method was used. Your local Cooperative Extension agent or an agronomist at your state's land grant college should be able to shed light on these matters.

Extractant methods include Morgan, Modified Morgan, Mehlich 1 and Mehlich 3, Bray & Kurtz, and Olsen. You can capture the best impression of your soil by staying with the same lab, or labs that use the same extractant method, and comparing soil test results from the same fields over a good number of years.

How to Take a Soil Sample

1. Use a clean soil probe or dig down several inches with a clean shovel or trowel (stainless steel is best), and collect about ¼ cup of soil in a clean plastic bucket.
2. Take several samples from different spots within the same field.
3. Add all the samples to the bucket, and mix them together.
4. Remove any rocks or pieces of sod or vegetation.
5. Place about a cup of the mixture into a plastic bag.
6. Fill out any forms that are provided, and send them and the sample to the lab.

Agro-One Soils Laboratory
730 Warren Road Ithaca, NY 14850
Ph: 800.344.2697 ext. 2179
Fax: 607.257.1350
www.dairyone.com

Soil Analysis Report

Cornell University
College of Agriculture
and Life Sciences

Sample #: 70449170
Date Sampled: 10/8/2009
Date Received: 10/19/2009
Date Mailed: 10/26/2009

Crop, 3 Years Ago: Mixed Vegetables
Crop, 2 Years Ago: Mixed Vegetables
Crop, Last Year: Mixed Vegetables
Tillage Depth: 7 - 9 inches
Manure: No
County: Orange
Field / Location: 010
Soil Name: Mardin
Acres: 1.80
Statement ID: KEITH STEWART

Component	Mod. Morgan ppm	Mod. Morgan lbs/acre	Morgan Equiv. lbs/acre	Soil Test Levels				
				Very Low	Low	Medium	High	Very High
Phosphorus (P)	46	93	121	**************************				
Potassium (K)	318	635	643	**************************				
Calcium (Ca)	1,556	3,112		**************************				
Magnesium (Mg)	365	730		**************************				

Water		Calcium Chloride		No Till		Organic Matter (%)	Nitrate-N (ppm)	HWS Boron (lbs/acre)	Soluble Salts (mmhos/cm)
pH	Buffer pH	pH	Buffer pH	pH	Buffer pH				
6.8	6.4					4.7			

Other Nutrients, Mod. Morgan, lbs/acre								
Sodium (Na)	Aluminum (Al)	Sulfur (S)	Zinc (Zn)	Manganese (Mn)	Iron (Fe)	Copper (Cu)	Boron (B)	Molybdenum (Mo)
	29.8		2.2	11.4	3.9			

Soil Fertilizer Recommendations

		tons / acre	lbs / acre			lbs / 1000 sqft			
Year	Crop	Lime	N Range	P2O5	K2O	Lime	N Range	P2O5	K2O
1	Mixed Vegetables	0.0	120 - 140	20	0	0.0	2.8 - 3.2	0.5	0.0
2	Mixed Vegetables	0.0	120 - 140	20	0	0.0	2.8 - 3.2	0.5	0.0
3	Mixed Vegetables	0.0	120 - 140	20	0	0.0	2.8 - 3.2	0.5	0.0

Comments

Nutrient recommendations provided by Cornell University.
These are general comments. Always consult with your crop advisor for recommendations specific to your farm.
Yr1   Do not exceed 80 - 100 lbs/acre of N + K2O in the fertilizer band. If more K2O is required, plowdown or broadcast before planting.
Yr1   High soil P response to P unlikely in warm soil. Crops seeded or transplanted into cooler soils may respond to banded or starter P.
Yr1   K2O fertilizer recommendation above appropriate for coarse or moderately coarse soils normally used in vegetable crop production.
Yr1   Apply 100 lbs/acre of N at or near planting time.
Yr1   Broadcast 50% and apply the remainder in bands when setting plants.
Yr1   Seeding: the fertilizer band should be 2 inches below the seed and 2 inches to the side of the row.
Yr1   Sidedress 30 - 40 lbs/acre of N 3 to 4 weeks after thinning or setting the plants.
Yr1   Additional sidedressed N should not be required unless there was leaching from heavy rains.
Yr2   Do not exceed 80 - 100 lbs/acre of N + K2O in the fertilizer band. If more K2O is required, plowdown or broadcast before planting.
Yr2   High soil P response to P unlikely in warm soil. Crops seeded or transplanted into cooler soils may respond to banded or starter P.
Yr2   K2O fertilizer recommendation above appropriate for coarse or moderately coarse soils normally used in vegetable crop production.
Yr2   Apply 100 lbs/acre of N at or near planting time.
Yr2   Broadcast 50% and apply the remainder in bands when setting plants.
Yr2   Seeding: the fertilizer band should be 2 inches below the seed and 2 inches to the side of the row.
Yr2   Sidedress 30 - 40 lbs/acre of N 3 to 4 weeks after thinning or setting the plants.
Yr2   Additional sidedressed N should not be required unless there was leaching from heavy rains.
Yr3   Do not exceed 80 - 100 lbs/acre of N + K2O in the fertilizer band. If more K2O is required, plowdown or broadcast before planting.
Yr3   High soil P response to P unlikely in warm soil. Crops seeded or transplanted into cooler soils may respond to banded or starter P.
Yr3   K2O fertilizer recommendation above appropriate for coarse or moderately coarse soils normally used in vegetable crop production.

Visit our website www.dairyone.com/AgroOne for more interpretive information.   3R

Actual soil analysis report for our farm

Frequency of Testing

It's a good idea to have your soil tested every 3 years, preferably by the same lab and at approximately the same time each year. If you do this, you can confidently compare tests from different years and see how your soil is trending. When we have the soils on our farm tested, I like to see a gradual increase in organic matter and a steady supply of the major nutrients. And of course, I'm always interested in the pH.

Soil tests should be done at least a few months in advance of planting, so there is enough time for any recommended applications of lime or other amendments to take effect. We prefer to do ours in late fall, before the ground freezes.

Soil Tests Don't Give the Whole Picture

Most organic growers do not follow soil test recommendations blindly; they put them in the context of the bigger picture of soil health. But for the small amount they cost, soil tests are a good investment. They will give you accurate pH readings to supplement any pH testing you might do for yourself. They will keep you informed of major changes, for better or worse, that are occurring in your soil, and they will advise you on the possible need for nutrient amendments.

The overuse of fertilizers and animal manures is a waste of money and is damaging to the environmental. Fertilizer and manure that is unneeded and unused by plants will leach into groundwater and find its way into streams and rivers. The enormous dead zone in the Gulf of Mexico (it's the size of the state of New Jersey) is caused, in large part, by agricultural runoff in the form of unneeded synthetic fertilizer draining into the Mississippi River as it passes through America's heartland. The fertilizer leads to a proliferation of algae that exhausts the supply of oxygen needed by fish and other aquatic organisms — hence the dead zone.

For a more in-depth look at soil testing and interpretation, check out Cornell's website (see Resources).

Cornell Soil Health Assessment

Ecologically minded folks at the Cornell Nutrient Analysis Lab in Ithaca, New York, have developed a more holistic approach to soil testing than the lab tests described above, which focus on measuring available nutrients. Their Cornell Soil Health Assessment is more expensive than the standard tests, but in addition to performing a thorough nutrient analysis, it provides valuable information about a soil's overall health and physical condition.

The test measures and ranks such indicators as wet aggregate stability, available water capacity, surface and subsurface hardness, organic matter content, active carbon, potentially mineralizable nitrogen, and plant root health. Taken together, these indicators give a broad picture of a soil's essential biological and physical processes. If any one of them is found limiting, plant growth may be negatively impacted, especially under extreme weather conditions. Identifying a problem will enable you to develop appropriate management strategies. The professionals at Cornell can offer guidance in this regard.

Fortunately, you don't have to be from New York to avail yourself of this more comprehensive evaluation of your soil. Growers across North America and even from other parts of the world are sending in samples for analysis. The Cornell Soil Health Assessment is especially appropriate for organic farmers who want to maintain their soil in peak health and better understand the dynamic environment in which their crops grow. See Resources for the website that will tell you more about the Cornell Soil Health Assessment. A downloadable manual describes how the assessment works and how to interpret results.

Soil Organisms

We humans live on the surface of the earth, and that is what we know best. We are familiar with the life we see around us — the multitude of animals, birds, insects, and plants that inhabit our terrestrial sphere. Many of us have learned to recognize, identify, and relate to these other forms of life. Some people study them in great detail and have gained a better understanding of the endless interactions and chains of interdependence that exist in the nonhuman realm.

Few people, though, devote much time or thought to the world under their feet — the fungi, bacteria, worms, beetles, and various other organisms that compose the soil's biotic community. Yet the scope and ecological complexity of this netherworld and its inhabitants rivals the scope and complexity of the surface world we are more familiar with.

Soil organisms are the planet's great recyclers. They are in the business of breaking down all forms of spent life into its elemental parts. In an almost godlike fashion, they transform death, decay, and excess into new living creations. Without them, the cycles of life and death on this planet would grind to a halt.

The sheer numbers are staggering. In the top 6 inches of an acre of healthy soil, there are literally trillions of diverse, living organisms with a combined weight of as much as 10 tons. In terms of biomass, the life contained in such a sample can easily surpass the life that is on top of it. Every organic farmer worth his or her salt understands how important it is to keep the soil microbes and larger organisms well nourished and in a balanced state.

Microorganisms

The smallest soil organisms are fungi, bacteria, actinomycetes, protozoa, and algae, of which there are many distinct species. These organisms are generally not visible to the naked eye and may accurately be termed microorganisms. Their primary function is to break down dead plant matter and convert it into soluble forms that living plants can use. They are especially good at starting the decomposition process.

Soil microbes have been described as a sort of external digestive system for plants — similar to the bacteria and fungi in our stomachs that help us digest the food we eat. Microscopic soil organisms help plants obtain and digest minerals and nutrients. Instead of residing within the plants themselves, these fungi and bacteria live in the soil and attach themselves to plant roots. The relationships between these organisms and plant roots are numerous, symbiotic, and often not well understood; however, we do know that without soil organisms many nutrients would not be available to plants.

Rhizobia

Certain bacteria, called rhizobia, are able to convert atmospheric nitrogen into a form that plants can use. They do this in collaboration with the roots of leguminous crops such as peas, beans, and clover. This process is called "nitrogen fixation" and can provide a significant amount of the nitrogen needs of an existing or subsequent crop. For this reason, organic farmers include legumes in their crop rotation plans whenever they can.

When you grow legumes, it's a good policy to inoculate the seed with rhizobia bacteria before planting, just in case the bacteria are in short supply or absent from your soil (see page 235 for more information).

Mycorrhizae

Mycorrhizae are fungi that live in the soil and play a vital role in plant health and productivity. Like rhizobia, they too enter into a symbiotic relationship with plant roots. They are able to convert insoluble nutrients (e.g., phosphorus) into biological forms that plants can use. In return, mycorrhizae receive carbohydrates from their host plants. They also produce antibiotics that protect plant roots from pathogens and, incidentally, are often used in human antibiotic medicines.

Larger Soil Organisms

These include nematodes, mites, centipedes, millipedes, springtails, water bears, termites, various beetles, slugs and snails, and, of course, the well-known and much-revered (at least by organic farmers and gardeners) earthworm. Some of these, such as nematodes (which look like tiny worms), slugs, snails, and earthworms, consume mostly plant material. Many others are secondary consumers or what we might term predators. They dine on very small organisms, such as bacteria and fungi, as well as larger members of the soil community. It's a dynamic and competitive world down there, and sooner or later just about everything and everyone gets eaten.

Some fairly large animals spend a good part of their lives underground. Along with moles, rabbits, woodchucks, gophers, and prairie dogs also burrow into the soil and import organic material from above. Their holes allow water to penetrate more deeply.

Beneficial or Not

Most soil organisms are immensely beneficial to farmers but a few are not, though even they are likely to play a role in the wider ecology of living things that may often transcend our human interests. The larger animals mentioned above are not exactly welcome on most farms, since they like to eat what many farmers grow and may dig burrows that a horse or cow can slip into and break a leg.

Certain nematodes feed on plant roots and can transmit diseases in the process. Some fungi are responsible for such plant diseases as downy mildew and late blight, which many growers are unhappily familiar with. Certain bacterial diseases are also an ever-present threat. We'll look into these matters more deeply in chapter 22.

The organic farmer's task is to create conditions that are favorable for beneficial organisms but less so for those that undermine his or her efforts. The best way to do this is to maintain a balanced supply of plant nutrients and healthy levels of organic matter at varying stages of decomposition.

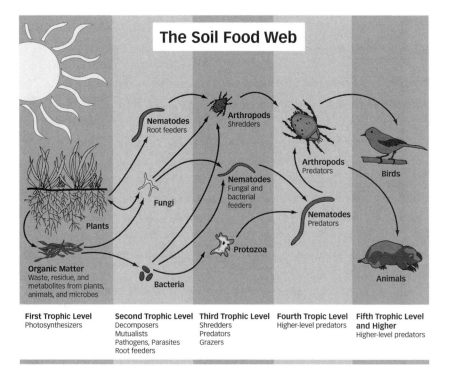

This diagram shows the movement of energy (captured initially by plants, through photosynthesis) through the bodies of a variety of living organisms. Note the changing functions and hierarchy at the different trophic levels.

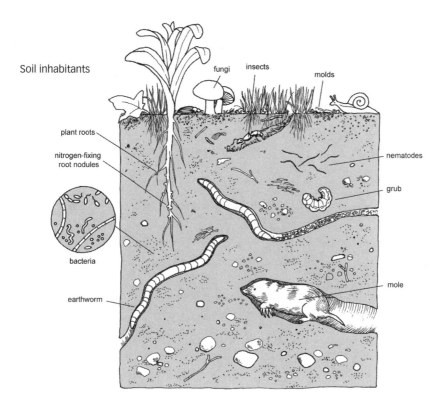

Soil inhabitants

fungi insects molds

plant roots

nitrogen-fixing root nodules

bacteria

earthworm

nematodes

grub

mole

The Noble Earthworm

Any discussion of life in the soil would be incomplete without paying special homage to the noble earthworm. Earthworms not only consume and process an enormous amount of organic material, they also create minitunnels to travel underground. These byways provide important channels for air and water to enter the soil.

There are more than two thousand species of earthworms on our planet. They come in many different types and sizes — from mere millimeters to several feet long. Some, such as nightcrawlers, ingest plant remains, along with minerals and bacteria, on the soil's surface; others consume organic matter already incorporated into the soil. An earthworm can consume its weight in organic matter every day.

Worm castings are essentially excrement left behind after the worm has digested what it needs. It is valuable excrement, however, since it provides plants with a super-rich source of nitrogen and other nutrients such as calcium, potassium, and magnesium.

Nutrient Density

The term "nutrient density" is getting plenty of use these days. It refers to the measurable quantity of nutrients (usually calcium, magnesium, selenium, iron, copper, manganese, and zinc) in a particular food. USDA food-testing data show a steady decline in the nutrient content of food since the mid-twentieth century, presumably from varietal selection, monoculture, the widespread use of synthetic pesticides and fertilizer, and the resulting impoverishment of our soils.

Many organic growers believe that our nation's preoccupation with ever-higher yields and greater profits (usually for the agribusiness giants and seldom for farmers) may have served only to dilute the mineral content of the food we eat. It often appears that an increase in harvest volume amounts to little more than an increase in water and carbohydrates.

It has long been known that the amount and diversity of nutrients can vary markedly from one type of vegetable to another. More recently, we are seeing that nutrient levels can vary within the same vegetable, depending on how and where it was grown.

In other words, broccoli may be good for you, but not all broccoli is equally good. Depending on the health of the soil in which the broccoli was grown and that soil's natural supply of minerals and nutrients, some heads of broccoli are likely to be more nutrient dense (and therefore better for you) than others. Different varieties of the same vegetable may also have different nutrient loads. Those varieties that are bred to grow faster and bigger are often the least well endowed.

The health of the biotic community in the soil and its ability to help plants digest minerals and nutrients may be a factor in achieving maximum nutrient density. Lab tests to determine the nutrient content of a food are expensive. For this reason, some farmers use an instrument called a "refractometer" to conduct a simple field test. A refractometer measures the density of a plant's juices or sap, known as the Brix level. Some growers believe that higher Brix levels are found in plants that are more nutrient rich.

Some earthworms can live for 10 to 15 years, and many live for 5 or 6. All earthworms are hermaphrodites — they have both male and female parts, but they still get together to reproduce. A solitary worm cannot have offspring. Depending on the soil's porosity and moisture content, certain species of worms can tunnel their way 5 or 6 feet into the ground, opening up deep channels for plant roots, air, and water. In doing this, they bring subsoil closer to the surface and mix it with organic matter.

A healthy soil can have as many as a million worms per acre. Soils that are continuously drenched with agricultural chemicals, especially synthetic forms of nitrogen, usually have very few earthworms. Excessive tillage also reduces the earthworm population. The goal should be to have as many worms in your soil as possible.

The importance of balance. Soil fertility depends on the presence of minerals, nutrients, organic matter, and a broad spectrum of living organisms — all in balanced proportions. Too much of one nutrient might block or interfere with the uptake of another. Too much carbon, for example, will cause a temporary nitrogen deficit as soil microbes use available nitrogen to fuel their decomposing activity. A pH that is too high or too low will immobilize phosphorus and other nutrients. Potassium and magnesium, when present in excess, will interfere with each other's availability.

Good agricultural practices will enhance your soil's fertility. These include crop rotation, allowing land to lie fallow, and using composts, cover crops, animal manures, and possibly mineral amendments.

Good practices also require that a farmer avoid excessive tillage, which is detrimental to soil aggregates and organic matter, and refrain from using heavy equipment on wet soils, which results in compaction and structural damage.

CHAPTER 9 RECAP

Review of Terms

aggregates. Small clumps made up of minerals, organic matter, air, and water that give a soil good structure and the ability to resist erosion

anions. Negatively charged ions necessary for plant growth; the major anions are nitrogen, phosphorus, and sulfur

cation exchange capacity. The degree to which a soil can hold onto and exchange cations; it is highly dependent on the presence of organic matter and clay particles

cations. Positively charged ions necessary for plant growth; the major cations are calcium, magnesium, and potassium

compaction. Occurs when excessive pressure/weight is applied to a soil, especially a wet soil; air is expelled and soil particles are pressed closely together, reducing pore space

loam. A soil with roughly equal amounts of sand, silt, and clay

mineral. Solid elements derived from rocks, many of which are necessary for plant growth

mycorrhizae. Soil fungi that help make insoluble nutrients available to plant roots and receive carbohydrate from plants in return

nutrient. Solid, liquid, or gaseous elements that are necessary for plant growth

nutrient density. A measure of the various individual nutrients present in a food, usually a vegetable, fruit, or grain

pH. A measure of a soil's acid concentration; critical for nutrient availability and uptake

rhizobia. Soil bacteria that function symbiotically with legumes to convert atmospheric nitrogen into forms usable by plants

structure. Describes a soil's physical condition, stability, and ability to hold air, water, and organic matter

texture. Describes a soil's relative proportions of sand, silt, and clay, which will affect its ability to retain nutrients, water, and air

tilth. Similar to soil structure but with focus on the soil's ability to support plant and microbial life

Web Soil Survey. A USDA website that maps and describes soils for most areas of the United States

Chapter 10

....

Cover Crops and
Green Manures

THE MORE I FARM, THE MORE I APPRECIATE cover crops
and green manures. They don't go to market with us. Our customers never see them. They don't bring in any cash. But these
soil-improving crops play a vital role in the overall health and productivity of our farm.

Cover and green manure crops are seldom grown for human consumption. Neither are they likely to be fed to animals on a vegetable
farm — unless the farmer uses horses or oxen in place of tractors, in
which case their aboveground parts may come in handy as fodder. The
most obvious purpose of a cover crop, as the name implies, is to keep the
soil covered while it's not being used for other agricultural purposes. But
a green manure can do the same thing.

So what exactly is the difference between a cover crop and a green
manure? By definition, a green manure is generally incorporated into
the soil while still green and before it has set seed. A cover crop might
be incorporated at any stage. Beyond this, there often is not much difference. Sometimes the terms are used almost interchangeably. Both help
replenish and maintain healthy soil, and both can perform several overlapping and even identical functions.

Which of the two terms a farmer chooses to use will often come down
to his or her primary goal. With cover crops, the emphasis is usually on
protecting the soil from erosion. With a green manure, the emphasis is
on adding organic matter and often nitrogen to the soil, though these
things might also be asked of a cover crop. In the following discussion,
when in doubt I will use the term "cover crop." It has broader connota-
tions and is often used as a stand-in for green manure anyway.

Eight Reasons to Plant Cover Crops and Green Manures

Most organic farmers are strong believers in the value of cover crops and green manures. There are many good reasons to incorporate them into a rotation plan.

1. Prevent Erosion

Depending on where you live on our planet, it can take thousands, tens of thousands, or even millions of years for a mature soil to form. Unfortunately, regardless of how long it took to form, any soil, if abused, can be lost in a very short space of time. And lose soils we do. Erosion rates in the United States have declined over the past 25 years, but they are still alarmingly high. According to the USDA, on farmland alone, we are still losing about 1.7 billion tons of topsoil to erosion every year. Cover crops can make a major dent in this rate of loss.

Rill and sheet erosion. All forms of vegetation, including cover crops and green manures, hold the soil together and help protect it from erosion. The danger is greatest during the winter months and spring thaw, when the land might otherwise be bare. Sloping fields are especially vulnerable. Periods of heavy rain and snowmelt can easily sweep topsoil downhill. You know this is happening when you see narrow channels, known as rills, on your sloping ground.

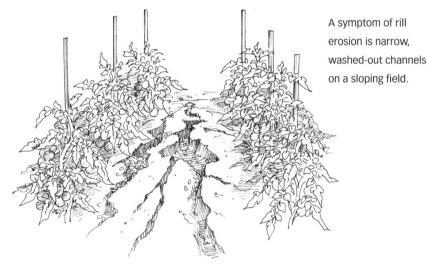

A symptom of rill erosion is narrow, washed-out channels on a sloping field.

Sheet erosion occurs when a thin layer of soil is swept downhill across a much wider area. With both rill and sheet erosion, the eroded material might be deposited at the bottom of a field or find its way into a pond or stream. Often, its ultimate destination is the ocean.

Wind erosion. Strong winds can blow away topsoil when land is left bare and dry. This is one reason farmers who have not planted cover crops are usually happy to see a nice blanket of snow on their land during winter. Wind erosion can occur during a dry summer as well.

When preventing erosion is your main concern, you'll be thinking along the lines of a cover crop more than a green manure. That said, if a green manure is already in place, it could do a very adequate job of preventing soil erosion.

2. Add Organic Matter to the Soil

At some point, cover and green manure crops are usually turned back into the soil with a tillage implement such as a plow, disc, spader, or rototiller. On a smaller scale, one might use a shovel or fork. Alternately, they might be mowed and left as a mulch on the soil surface.

Whichever method is used, sooner or later these crops contribute organic matter to the soil. When they do, they provide food for the trillions of organisms that inhabit the vast, terrestrial underworld. As the processes of consumption and decomposition take place, organic matter undergoes various stages of transformation before it finally becomes humus.

Importance of humus. I've already championed humus in chapter 9. Now I'd like to reiterate why it's so good to have humus in your soil:

- Its gummy quality helps bind soil particles into crumbs or aggregates, making them less vulnerable to erosion.

- Its lightness and aggregate structure promote pore spaces, which make a soil more hospitable to both plant roots and microbial life.

- It has a spongelike capacity to absorb and retain moisture in the upper levels of the soil where plant roots are most active.

- It holds on to positively charged mineral nutrients, such as calcium, magnesium, and potassium, in forms that plant roots can use. The mechanism by which this occurs is known as cation exchange capacity, or CEC (for more information on CEC, see chapter 9).

Humus, but Not Humus Alone

Mature humus has many attributes, but it is not a major source of food for soil organisms. Organic matter in its various pre-humus states — sometimes referred to as active organic matter — does a better job of feeding the biotic community. This is why a healthy soil will have organic matter at different stages of decomposition, as well as a good supply of humus. After all, it's important to keep the little guys well fed.

Soils that are poor in humus are compromised in several ways: They are more erodible, less favorable to root and microbial activity, less able to hold onto reserves of moisture, and more likely to suffer the loss of mineral nutrients as excess water passes through them. All these conditions are common to soils that don't get their fair share of organic matter, the building block of humus.

Benefits of roots. The roots of cover crops and green manures also contribute organic matter and improve soil structure. As they reach out in search of water and nutrients, roots open up spaces in the soil. When the plants die and their roots decompose and are consumed by soil organisms, the spaces or channels left behind allow still more air and water to penetrate and, at the same time, facilitate further microbial movement and activity.

Reducing amendments. By adding to the soil's supply of organic matter, a green manure reduces, and might even eliminate, the need for other soil amendments such as compost, animal manure, or organic fertilizer blends. Put another way, a well-chosen green manure can reduce or eliminate input costs.

3. Suppress Weeds (Smother Crops)

Some cover crops and green manures are very good at suppressing weeds. The faster-growing ones can outpace weeds at their own game. They monopolize the supply of nutrients and water in the soil as well as the sunlight coming from above, all three of which are needed by weeds. This weed-suppressing feature is especially helpful to a farmer in early and midsummer, on land that is waiting to receive a late-season vegetable crop. A fast-growing and short-lived cover crop, such as buckwheat, if seeded in early summer, will keep weeds at bay until the time is right to plant such crops as kale or carrots for fall harvest. Cover and green manure crops used for this purpose are sometimes called "smother crops."

4. Inhibit Weeds (Allelopathy)

Certain cover and green manure crops attack weeds on two fronts. In addition to mounting stiff competition for light, moisture, and nutrients, they produce toxic compounds that inhibit the germination of nearby seeds and slow the growth of adjacent plants of different species. This effect is known as "allelopathy." But note the use of the words "inhibit germination" and "slow growth" — allelopathic crops are not likely to stop these processes altogether. They are not herbicides. They slow down the weeds. That's all.

Some commonly used cover and green manure crops that possess allelopathic capabilities are winter rye, oats, and sorghum sudangrass. The toxic substances that these crops produce remain in effect while they are growing and for a few weeks after they are turned into the soil. They are most effective when chopped up into small pieces by a rotary or flail mower before incorporation. Allelopathic crops can also have an inhibiting effect on weeds when cut and used as mulch.

A word of warning: Allelopathy affects vegetables just as much as weeds — probably more so. Direct-seeding a cash crop soon after incorporating or plowing down an allelopathic crop is not a good idea. Allow at least several days to pass, or several weeks if the soil is cool and low in moisture; both conditions will slow decomposition. You can allow less time if the cash crop is transplanted rather than seeded.

5. Capture and Recycle Nutrients (Catch Crops)

Most farmers know that leaving ground bare for more than a short period is a good way to squander agricultural capital. You need some form of plant cover to ensure that the soil and its nutrients are not lost to erosion. But there's another way to lose valuable nutrients, and that is through leaching.

Prevent leaching. When the main intention is to prevent nutrient leaching, the cover or green manure crops used are often referred to as "catch crops." They capture and hold on to nutrients that might otherwise leach down below the root zone of most plants. When they die and decompose, the nutrients they have amassed are returned to the upper soil layers.

To be fair, weeds are capable of doing the same thing when they are incorporated into the soil. But there are two problems with weeds in this regard: One, they do it at their own variable pace rather than on a schedule that you might want; and two, they always seem to have procreation on their minds. Neither of these traits endears them to farmers.

Safeguard the water supply. By capturing nutrients that might otherwise leach out of the soil, cover or catch crops help safeguard the water supply. Excess nitrogen, phosphorus, and some other plant nutrients, when they enter surface streams, ponds, lakes, or aquifers, have negative consequences. Regardless of whether they originate from organic or petrochemical sources, they can become pollutants. Cover or catch crops are especially needed when uncomposted manure or petrochemical fertilizers high in nitrogen are used.

Using catch crops. When a field is double-cropped in the same season, a catch crop might be sown between the two vegetable plantings — for example, an early-season planting of snow peas, followed by a mid-summer catch crop of buckwheat, followed by a fall planting of lettuces.

Or a catch crop might be planted in fall, after a main-season vegetable crop is harvested — for example, potatoes harvested in September, followed by winter rye. The rye, as long as it is planted early enough, will do a good job of catching and holding on to nutrients that could easily be lost in the winter months.

Of course, one could argue, with some justification, that the rye is also functioning as both cover crop and green manure, since it will undoubtedly protect the field from erosion in winter and provide organic matter when it is incorporated into the soil in spring. So you see the interchangeability of these three terms. It all depends on your primary goal.

6. Break Up Compacted Soil

Deep-rooted cover crops are capable of penetrating a plow pan, a compacted layer left by the repeated use of a moldboard plow. The roots of some cover crops can even reach deep into the subsoil, well below a plow pan, to break up brittle and dense layers of low porosity. Such layers, when they occur, are referred to as hardpans and fragipans. They usually comprise cemented alluvial silica. The minerals and nutrients obtained at these lower levels rise up through the cover crop's vascular systems and later become available for other plants to use.

But don't expect instant gratification. It might take more than a couple of years of growing soil-improving crops before compaction at deeper levels is lessened.

7. Add Nitrogen to the Soil

Nitrogen is an essential but volatile plant nutrient that is sometimes in short supply in an otherwise fertile soil. Leguminous plants (think peas, beans, clover, vetch, soybeans, and alfalfa) are valuable to a farmer because they help import atmospheric nitrogen into the soil. They do this by entering into a symbiotic relationship with soil bacteria called rhizobia.

As already noted, when they meet up with an appropriate legume, rhizobia are able to extract nitrogen from pore spaces in the soil and convert it into a mineralized form usable by plants. The legume's role is to play host to the rhizobia bacteria, which attach to its roots. There is a catch, however. In exchange for providing extra nitrogen, the rhizobia siphon off some of the carbohydrates (up to 20 percent) that the legume has acquired through photosynthesis. It's a give-and-take arrangement.

Inoculating Legumes

Before planting a legume, inoculate the seeds with rhizobia bacteria, just in case there are none, or very few, present in your soil. The bacteria are safe, easy to use, and not expensive. It's important to note that there are different species of rhizobia, just as there are different species of legumes, and not all are compatible with each other and able to enter into this special nitrogen-enriching arrangement, at least not in the most fruitful way.

A poor match between legume and rhizobium might produce some nitrogen, but nowhere near as much as a good match. Most seed companies sell legume inoculants and will tell you which one is right for the crop you are planting. After a few years in a soil with no host legumes, rhizobia populations decline dramatically. To be on the safe side, it's smart to always inoculate.

Selection and storage. At the beginning of each season, we usually buy three different 1- or 2-ounce packets of inoculant: one to be used with alfalfa and clover, one to be used with field peas and vetch, and one for regular garden-variety peas and beans. The regular peas and beans are cash crops, not cover crops or green manures, but they are still legumes and therefore capable of acquiring atmospheric nitrogen, albeit in lesser amounts, as long as they are paired up with the right rhizobia.

Rhizobia bacteria are living organisms and probably won't survive in a plastic bag for more than a year, so it's wise to always use new inoculants and store them in a refrigerator or other cool place. Once you've opened the bag, fold it over and seal it with a paper clip or tape between uses.

Testing for nitrogen fixation. You know nitrogen fixation has occurred when you can find little white nodules on the roots of your legume plants. Dig up a plant gently to avoid separating the nodules from the roots, then slice into a nodule with a sharp, thin-bladed knife. If you detect a pink or red color, you can be sure that the rhizobia bacteria, which inhabit the nodule, are actively fixing nitrogen. A black or dark-colored interior indicates that the bacteria are inactive or dormant.

Managing Nitrogen Deficits

Immediately after a nonlegume cover crop is turned into the soil, there is a high probability that existing nitrogen will be tied up by soil microbes and become temporarily unavailable to a subsequent vegetable crop. This is referred to as a short-term "nitrogen deficit" or is sometimes called "robbing." When the microbes die, the nitrogen they have "robbed" is released and becomes available to plants again, but this may not happen as quickly as a farmer would like.

The process will move along more rapidly in a warm, moist soil, and more slowly when the ground is wet and partially anaerobic conditions exist (see page 213 for more about this phenomenon).

Dry and fibrous versus moist and succulent. A nitrogen deficit is more likely to occur when a dry, fibrous crop, such as late-stage winter rye or sorghum sudangrass, is turned into the soil. Such crops have a high C:N ratio — more than 30 parts carbon to 1 part nitrogen — and it takes a lot longer for soil organisms to break them down in what is essentially a case of in-field composting. After incorporation, therefore, it is wise to allow at least a few weeks to pass before planting a cash crop, especially a direct-seeded one. The lack of available nitrogen will not affect seed germination, but it is likely to result in weak and stunted plants with yellow leaves.

How to Inoculate Legume Seed

Always inoculate seed immediately before planting.

1. Put the amount of seed you plan to plant in a plastic bag or bucket.
2. Dissolve a teaspoon of unrefined sugar or molasses in a half pint of warm water.
3. Slightly moisten the seed by adding a small amount (about ½ teaspoon per pound of seed) of the water/sugar mix to it. Stir with a wooden ladle or stick. Be careful not to use too much water to moisten the seeds — it will make them clump together, and they will be difficult to sow.
4. Sprinkle in the recommended amount of inoculant (instructions on the inoculant package will tell you how much to use per pound of seed). The goal is to coat most or all of the seeds lightly with inoculant.

A more succulent, herbaceous green manure crop, such as buckwheat, after just 6 or 8 weeks of growth, will contain a higher percentage of nitrogen relative to carbon than mature rye or sorghum sudangrass. Because it is light and has high moisture content, it will break down much more rapidly, meaning any nitrogen deficit will be short-lived. In this case, the interval between plow-down and the planting of a new crop can be very short.

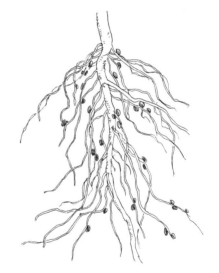

Legume roots with nitrogen-containing nodules

Remember, more mineralized nitrogen in the soil is beneficial to the leguminous crop that is hosting the rhizobia bacteria, and even more so to nonlegumes (presumably vegetables) that follow. More available nitrogen means there will be less need to import nitrogen in other forms such as animal manure. If your main purpose is to add nitrogen to the soil, think green manure.

8. Attract Beneficial Insects

Attracting beneficial insects and disrupting pest and disease cycles are not usually the primary reasons to plant soil-improving crops, but they are valuable added benefits. Many cover crops and green manures, especially the ones that flower, provide good habitat for predatory, beneficial insects. True, some of these crops might also harbor insect pests, but this isn't altogether a bad thing.

Diversity works. If you want to keep the predators around, you need a source of prey to feed them. Think ladybugs and aphids. The key is balance. When you have a healthy population of predatory beneficials and other life forms (such as small mammals, birds, arthropods, and a full complement of microorganisms), you are far less likely to have a major problem with insect pests and plant diseases.

A healthy, diverse system will have a greater ability to maintain or restore equilibrium. Industrial monoculture, on the other hand, lacks ecological balance and for this reason, requires ever-increasing inputs of pesticides and fertilizer that only make a bad situation worse. Plant, insect, and microbial diversity are an antidote for monoculture and the destructive use of agricultural chemicals.

Attract bees and other beneficials. When in flower, cover and green manure crops provide pollen and nectar for bees as well as other beneficial insects that use this food source when their preferred prey are in short supply. In general, a variety of flowering plants, including many wild species that occupy hedgerows and idle land, will greatly lessen the chance of a serious pest or disease outbreak in your vegetable crops. But not all wild plants are desirable. You should avoid or remove those that harbor diseases that could be transmitted to crop plants (for more information, see chapters 11 and 12).

Eight Cover and Green Manure Crops

It's well beyond the scope of this book and the knowledge of its author to discuss each of the many cover and green manure crops currently being used in North America and the numerous niches they fill. Lucky for me, that's not really necessary. Over the past few decades, there's been a growing recognition of the value of these crops in sustainable agrarian systems. Accordingly, farmers, university researchers, Extension specialists, and others have conducted many trials and studies. The results are

Intercropping

Some farmers use a practice known as intercropping to maintain diversity and disrupt pest and disease cycles. This involves the use of alternating strips of cover or green manure and cash crops. The cover strips provide erosion control, shelter for beneficials, and nitrogen fixation (in the case of a legume) and, at some point in the future, will replenish organic matter.

available for all to see, in various publications and research reports, and often online. See Resources for some good books and websites.

Once you settle into farming and fully appreciate the benefits that a healthy soil confers, you'll want to avail yourself of as much of this information as you can. There's plenty to learn. The goal is to constantly feed your soil, to keep it alive, productive, and in a balanced state. One of the principal ways to achieve this is to identify and use the right cover crop, or combination of crops, to meet your needs — in your soil, your climate zone, and your timeframe.

In this section, I'll take a more detailed look at eight cover and green manure crops that have been used repeatedly on our farm. Five are nonlegumes and three of them are legumes.

Depending on what you're looking for — short-term summer cover and weed suppression, heavy biomass production and allelopathy, deep-rooted winter erosion control, rapid cool-weather growth followed by winter kill, or an effective living mulch — you'll find it among these first five nonlegume cover and green manure crops: buckwheat, sorghum sudangrass, winter rye, oats, and annual ryegrass. Each has its own distinct attributes.

The last three crops I will discuss — red clover, white clover, and hairy vetch — are legumes. We already know that legumes contribute a net gain of nitrogen to the soil. For this reason, they are usually viewed more as soil-enriching green manures than as cover crops. The red and white clovers discussed on pages 250–254 are both perennials. They take longer to establish than buckwheat, oats, rye, or sorghum sudangrass, so they are seldom used as short-term cover and weed-suppressing crops.

To gain maximum benefit, it's best to leave them in place for at least a year. (If you're looking for nitrogen fixation, quick turnaround time, and some weed suppression, field peas or bell beans would be a better choice.) The last legume I've chosen to discuss here is hairy vetch. It's a winter annual that can fix a lot of nitrogen. But first, to the non-legumes.

Buckwheat

Buckwheat (*Fagopyrum esculentum*) is a fast-growing, broad-leaved annual that thrives in the summer heat and does a great job of smothering weeds and scavenging for phosphorus. It can blossom in 5 or 6 weeks and set viable seed a week or two later. Because buckwheat plants are tender, green, and succulent, they can usually be turned into the soil without needing to be mowed. Another of buckwheat's attributes is that it breaks down quickly once incorporated. This means you can transplant vegetables into ground that had living buckwheat on it only a week or two before.

Buckwheat is normally used as a midsummer cover crop, catch crop, and weed suppressor. It does best with adequate rainfall, though it is

Five Questions to Ask Before Choosing a Cover or Green Manure Crop

Your answers to the following questions will help you narrow your crop options and make a sound choice.

1. What are your primary, secondary, and tertiary goals — erosion control; organic matter production; nitrogen fixation; weed suppression; nutrient recycling and management?
2. How long will the cover or green manure crop be in the ground and during what period of the year?
3. What are the high and low temperatures and average rainfall expected during this period?
4. Will the cover or green manure crop be incorporated into the soil or mowed and used as a mulch?
5. How soon after incorporation or mowing will a cash crop be planted, and what will this cash crop be?

Buckwheat

AT A GLANCE

Seeding rates: 60 pounds per acre for drilled seed; 80 to 100 pounds per acre for broadcast seed

Range: Widely adapted. Can be grown in most parts of North America that have adequate rainfall and at least 2 months of frost-free growing days. Will not tolerate frost.

Major attributes: Fast growth; weed suppression; attractive to beneficials; rapid decomposition

moderately tolerant of drought. It does not produce a large amount of biomass.

Bees like buckwheat. While in flower, buckwheat provides habitat and an excellent source of nectar for honeybees and other beneficial insects. An acre of buckwheat can attract and nourish thousands of bees. It's a delight to see them gently probing the white blossoms, but a wise farmer will not let the bees partake for very long. Once flowers appear, buckwheat seeds are not far behind.

Two crops for the price of one? If you let the seeds develop to maturity, you will be inviting a second round of volunteer buckwheat, not long after you've tilled in the first one. If you have no plans for a subsequent cash crop, this might be a reasonable option. After all, there's no charge for the second batch of seed. But to most farmers, buckwheat that has gone to seed becomes a weed — with

Buckwheat

unintended and unwanted consequences. Most of us adopt a tough-love attitude and till in the little white flowers before they are able to develop viable seed, even though it means disturbing the bees and shortening their stay at the dinner table.

Buckwheat Planting Sequence: Before Fall Greens and Root Crops

1. Sow buckwheat in a clean seedbed in early June, or after all danger of frost, and preferably before rain.
2. Incorporate into soil with a rototiller 7 to 10 days after flowers appear and before seeds set. (If we plan to plant within a few days of incorporation, we might mow the crop before tilling to hasten decomposition.)
3. Plant lettuce and brassica transplants as early as a few days after incorporation, or direct-seed crops such as carrots and turnips a week or 10 days after incorporation.

We frequently use buckwheat as a short-term summer cover on fields that will be planted with lettuces, carrots, turnips, and brassicas in July and August.

Sorghum Sudangrass

If you're looking for a fast-growing cover crop that will compete well with weeds and give you major biomass production during the hot days of summer, sorghum sudangrass (*Sorghum bicolor* × *S. bicolor* var. *Sudanese*) might be a good choice. Also called Sudex or Sudax, it is a cross between sorghum — a forage or grain crop that looks a lot like corn — and sudangrass, and it can grow more than 10 feet tall in a couple of months, depending on the variety you select and whether you choose to mow it.

Sorghum sudangrass grows best under hot and moderately moist conditions. It can tolerate some drought, but excessive dryness will throw it into dormancy. Cool weather will slow the plant's growth, and a frost will kill it.

Mowing makes sense. If mowed two or three times and allowed to regrow, an acre of sorghum sudangrass can produce more than 15,000 pounds of dry matter or biomass. That's about as good as it gets in organic matter production. For maximum biomass, mow the crop when it reaches a height of about 4 feet.

Mowing sorghum sudangrass also stimulates root growth, both in terms of root mass and root penetration and depth. The roots of this crop are renowned for their ability to penetrate and break up compacted subsoil. Yet another good reason to mow it is to knock out any annual weeds that might have gotten a foothold in the shade the cover crop provided, but be sure to leave at least 6 inches of stubble. Mowing too low can kill the plant.

Sorghum sudangrass should definitely be mowed before incorporation, especially if you have allowed the plants to reach a significant height. The more it has been chopped up, the sooner it will decompose.

Sorghum Sudangrass Planting Sequence

We have used sorghum sudangrass on fields that will be planted to garlic in fall and have been satisfied with the results. On our southern New York farm, the sequence looks like this:

1. Sow sorghum sudangrass into freshly tilled ground in late May or early June, or after the soil has warmed to at least 65°F (18°C) and, ideally, just before a nice rainfall.
2. Mow crop when it reaches a height of 4 feet — usually about a month after seeding.
3. Mow again when crop reaches 4 feet for the second time — about the end of July.
4. Mow for the final time at end of August. This time, we mow as close to the ground as possible.
5. Incorporate the crop into soil with a rototiller or disc about a week after final mowing (early September). Sometimes, we chisel-plow before tilling or using the disc, to open up the ground.
6. Till field lightly in mid-October, in preparation for garlic planting.
7. Plant garlic in late October or early November. At this point, there is still some evidence of the sorghum sudangrass, but there's not enough to create a problem for the garlic, which will be running off its own store of energy through winter and early spring.

Even when mowed before incorporation, sorghum sudangrass can take a couple of months to break down.

Beware of the nitrogen deficit. Planting a cash crop soon after plowing down sorghum sudangrass will not yield good results. The plant's tough consistency and high C:N ratio (high carbon content) makes decomposition a slow process and a short-term soil nitrogen deficit almost a certainty. This condition can be somewhat ameliorated by applying manure or some other natural soil amendment that is high in nitrogen.

Allelopathy. Sorghum sudangrass contains compounds called cyanogenic glucosides, which have an allelopathic effect on many annual weeds. These

Sorghum sudangrass

compounds are also known to disrupt some pest and disease cycles in the soil, especially those associated with certain species of harmful nematodes, but not all varieties of sorghum sudangrass are equally effective in this respect. Some have more cyanogenic glucosides than others.

Sorghum Sudangrass

AT A GLANCE

Seeding rates: 40 pounds per acre for drilled seed; 50 pounds per acre for broadcast seed

Range: Sorghum sudangrass is widely adapted. It can be grown throughout most of North America where soil temperatures reach at least 65°F (18°C) for 2 months or more and rainfall is not an extreme limiting factor. If irrigated, it will, of course, grow in hot, dry regions. Will not tolerate frost.

Major attributes: Rapid growth; major biomass production; subsoil penetration; weed suppression by both smothering and allelopathy

A farmer who is planning to use sorghum sudangrass as fodder for livestock should choose a variety that is low in these compounds because they can convert to prussic acid in an animal's stomach and become toxic. A vegetable grower who is interested in the plants' allelopathic qualities would choose a variety that is high in cyanogenic glucosides. Your cover crop seed dealer should be able to advise you on these matters.

Winter Rye

Winter, or cereal, rye (*Secale cereale*) is a hardy annual cover crop. It should not be confused with either annual or perennial ryegrass, both of which can also be used as cover crops. In temperate regions with cold winters, winter rye is usually seeded from late summer to midfall, after summer crops are harvested. It grows rapidly in cool, late-season weather and provides excellent erosion control over winter.

When spring arrives, winter rye resumes its growth and can reach a height of 5 feet by late May, at which point it is prudent to mow the crop and turn it in, if you haven't already done so. If you give the plant enough time to develop mature seeds, you can have rye coming back at you, as an unwanted and difficult-to-eradicate weed, for years to come. Some growers harvest the mowed rye for use as a mulch.

Far-reaching roots. Winter rye has a fibrous and far-reaching root system, which enables it to scavenge nutrients at deeper levels and recycle them into the growing zone that most vegetables occupy. This is a major plus. Though it cannot capture atmospheric nitrogen, as leguminous crops can, rye is adept at finding and storing nitrogen that is already in the soil before it leaches down below the root zone. The crop's aggressive growth makes it a good cool-season weed suppressor and high biomass producer (up to 10,000 pounds per acre).

Allelopathy. Like sorghum sudangrass, winter rye contains allelopathic compounds while it is both growing and decomposing. It will inhibit the germination and growth of many

Winter rye

Winter Rye

Seeding rates: 60 to 120 pounds per acre for drilled seed; 90 to 160 pounds per acre for broadcast seed. Use higher rates (200–300 lb/acre) when seeding later in fall. When seeding with a legume, 50 to 60 pounds per acre is adequate.

Range and zones: Widely adapted throughout North America but performs best in more temperate zones with cool or cold winters. Hardy to Zone 3.

Major attributes: Cold hardy; rapid growth in cool weather; erosion control; good biomass production; nitrogen scavenger; allelopathic

other plants, including weeds and vegetables; therefore, always wait 3 or 4 weeks after incorporating rye before direct-seeding any cash crop (1 to 2 weeks should be enough if the cash crop is being transplanted). Mowing before incorporation will hasten breakdown.

Winter rye is often seeded with a legume such as hairy vetch or clover. The legume will counteract rye's tendency to monopolize soil nitrogen in spring.

Oats

Oats (*Avena sativa*) prefer cool weather and moist conditions. Their rapid growth and allelopathic qualities enable them to suppress and inhibit weeds. Their fibrous root system does a good job of holding soil together and preventing erosion.

In most of North America, oats can be seeded in April or May for early-season cover, or in late August or September for winter cover. Oats planted in spring can produce a not-too-shabby 6,000 to 8,000 pounds of dry matter per acre. For fall-seeded oats, 2,000 to 4,000 pounds per acre is more typical.

Winter kill. Because they are less hardy than rye, oats will die in winter in Zones 1 through 6 and much of Zone 7. This is known as "winter kill." It is not necessarily a bad thing. We often seed oats in fall in fields that will be planted to vegetables in April or May the following season. Because the oats are already dead, they are easy to incorporate into the soil and will break down rapidly. Sometimes we skip the incorporation phase altogether and let the dead oats act as a mulch, which we transplant right into. With winter rye — our other major winter cover — this would not be possible because it survives very cold weather.

Combine with a legume. In early fall, many growers combine oats with a legume, such as field peas, hairy vetch, bell beans (a small-seeded fava bean), or clover (white, red, or crimson). They grow nicely together and decompose quickly once incorporated. The oats act as a nurse crop, providing shade for the legume. The main function of the oats is to provide erosion control during fall and winter; the function of the legume is to supply nitrogen to the soil for a spring cash crop. The combination provides microorganisms with a balanced diet.

Spring-seeded oats may also be combined with fast-growing annual legumes such as bell beans, which, within as little as 2 months, can fix more than 100 pounds of nitrogen per acre, or with field peas, which can fix an impressive

Oats

Oats

Seeding rates: 100 pounds per acre for drilled seed; 150 pounds per acre for broadcast seed. When seeded with a legume, the rates can be cut in half.

Range and zones: Widely adapted. Can be grown in most parts of North America where rainfall is adequate. Not suited to arid regions. Winter kills in Zones 1 through 7.

Major attributes: Rapid cool-weather growth; erosion control; good biomass production; weed suppression by both smothering and allelopathy

180 pounds of nitrogen per acre. In this case, the oats act as both a biomass producer and nurse crop for the legume.

Annual Ryegrass

Annual ryegrass (*Lolium multiflorum*) is a fast-growing, versatile, and widely adapted grass that does an excellent job of suppressing weeds. It performs best on good ground, but it can also tolerate poorly drained, rocky soils with low fertility. Annual ryegrass is a palatable forage crop for most livestock. It prefers cool weather but should not be planted too late in fall. Seed the crop at least 40 days before a killing frost.

Annual ryegrass has a shallow but extensive root system, which aids water infiltration, improves

Annual ryegrass

soil structure, and prevents erosion. It captures and holds on to excess nitrogen in the soil and scavenges for other nutrients. When mowed over the course of a season, annual ryegrass can produce 4,000 to 8,000 pounds of biomass or dry matter per acre, but it should not be mowed lower than 4 inches.

Annual Ryegrass

AT A GLANCE

Seeding rates: 10 to 20 pounds per acre when drilled; 20 to 30 pounds per acre when broadcast. If combined with a legume, reduce rate by about 30 percent.

Range and zones: Widely adapted throughout North America but prefers lower temperatures. Hardy to Zone 6.

Major attributes: Rapid growth; weed suppression; erosion control; good biomass production; soil builder

Permanent Beds and Living Mulches

The benefits of permanent beds with strips of living mulch between them
are numerous:

- ▸ They make it easy to restrict both foot and tractor traffic to the path-
  way or living mulch area. This greatly reduces soil compaction in the
  planting zone.
- ▸ Living mulches provide habitat for beneficial insects during both the
  growing season and winter. If they contain flowering plants, they will
  provide pollen for bees and other beneficials.
- ▸ During winter and heavy rainstorms, living mulches provide excellent
  erosion control.
- ▸ When harvesting vegetables, living mulch pathways provide a clean,
  grassy surface on which to kneel or set down harvested crops.

Annual ryegrass seeded in pathways between beds of vegetables

As a living mulch. Annual ryegrass is often used as a "living mulch."
It may be seeded into an already established crop, a practice known as
"overseeding." When used in this way, it is sometimes combined with a
low-growing legume such as white clover. Once a living mulch is established, it no longer makes sense to cultivate your cash crop.

Annual ryegrass can also function as a living mulch in a permanent
bed system, but in this case the plant is seeded between beds, rather than
within the rows of vegetables themselves. We sometimes use it in this
manner between beds of perennial herbs or, less permanently, between
rows of tomatoes, peppers, and other crops that will be in the ground
for at least 3 or 4 months. The pathways are usually mowed on a regular
basis to reduce competition with the adjacent crop plants.

A couple of words of caution: In areas of low rainfall, or when water for irrigation is in short supply, annual ryegrass will compete aggressively with crop plants for whatever moisture is available in the soil. If left to go to seed, annual ryegrass will reseed itself, and unless a long-term living mulch is desired, it can become a weed problem. Vintners and orchardists often allow the crop to reseed itself.

Red Clover

Red clover (*Trifolium pratense*) is a short-lived perennial legume. It prefers cool, moist conditions and grows well in humid areas of the northern United States and Canada, where it is usually seeded in spring. In the humid Southeast, the crop may be used as a fall-planted winter annual.

Like most other clovers, red clover grows slowly in the beginning, and therefore is not an effective weed suppressor at that time. Once established, however, the plant puts on impressive growth and will prevail over most competition. For early weed suppression, sow red clover with a quick-to-establish nonlegume such as buckwheat, sorghum sudangrass, or oats, all of which can function as a nurse crop.

If cut a few times over the course of a full growing season, red clover can produce as much as 7,000 pounds of biomass or dry material per acre and 100 or more pounds of nitrogen. It has an extensive lateral root system, which holds soil together, and a strong taproot that can penetrate compacted layers and scavenge for phosphorus and other nutrients. The taproot can reach down several feet. The nutrients captured at these depths become available to shallower-rooted crops after the clover dies.

There are two distinct types of red clover: mammoth red and medium red. If both are cut only once, mammoth red clover will grow taller than medium red (up to 3 feet) and produce more biomass, but when each is cut multiple times, medium red clover will come out the winner, both in terms of biomass and nitrogen production. This is because medium red clover grows back more rapidly than mammoth red after each cutting. Which is best for you will depend on how long you expect your crop to be in the ground and how often you plan to mow it.

Within these two types, there are numerous strains or varieties. Many of them are locally adapted and will show varying levels of cold tolerance and disease resistance depending on where they are grown. With this in mind, it's always wise to buy seed from a supplier who markets primarily in the region where you live.

Because it is tolerant of shade, red clover can be overseeded into an existing vegetable crop. After the vegetable crop is harvested, the clover keeps growing as a cover crop and soil improver. Red clover is attractive to many beneficial insects, but it is also vulnerable to a number of plant diseases and, for this reason, should not be used in the same fields repeatedly.

In general, red and most other clovers are used as green manures when the intention is to add nitrogen to the soil, rather than simply the largest amount of organic matter possible. Because they take longer to establish than buckwheat, oats, rye, or sorghum sudangrass, they are seldom used as short-term cover and weed-suppressing crops. To gain maximum benefit, it's best to leave the perennial clovers in place for at least a year.

Red clover

Red Clover

AT A GLANCE

Seeding rates: 8 to 10 pounds per acre when drilled; 10 to 15 pounds per acre when broadcast. When seeded with a nonlegume crop, 6 to 8 pounds per acre is adequate.

Range and zones: Temperate and humid regions of North America for spring seeding; fall seeding in the humid South. More widely adapted than most other clovers. Hardy to Zone 4.

Major attributes: Nitrogen fixation; biomass production; erosion control; subsoil penetration; phosphorus scavenger

Estimating Biomass Yield

It is quite easy to arrive at a rough estimate of the amount of dry matter or biomass that your cover and green manure crops are producing. Let's say you have a field of winter rye that was planted the previous fall. Now it's late spring, the crop is 4 feet high, and you're ready to incorporate it into the soil. Here's what you need to do to determine how much biomass you've netted:

1. On a dry day, just before plow-down or mowing, identify a section of the field that has average growth, both in crop density and height.
2. Measure out a 4-foot-square plot (2' by 2'). Delineate the plot with pieces of 1×2 or 2×4, or strips of metal.
3. Use hedge clippers or pruning shears to cut all the crop plants within the plot area. Cut as low to the ground as possible — down to ¼ inch if you can. Place the cuttings in an open container.
4. Leave the container and cuttings to dry in the sun for a few days, or dry them in an oven at 140°F (60°C) for about 36 hours. When the stems break easily in your hand, they are dry enough.
5. Weigh the dried cuttings.
6. Apply the following formula:

(43,560 ÷ plot size) × weight in pounds = pounds per acre of biomass

For example, let's say you come up with a weight of 0.75 pound of dried winter rye from your 2- by 2-foot plot.

Knowing that there are 43,560 square feet in an acre, and 4 square feet in your sample plot, the calculation goes thus:

(43,560 ÷ 4) × 0.75 = 8,167.5

Your estimated biomass yield from the winter rye is 8,167.5 (I would round it off to 8,000) pounds per acre. If you have the time and would like more accuracy, take several samples from different parts of the field. Treat each one separately, then take the average of the results.

If you are dealing with a crop that will get mowed a few times before being incorporated, such as sorghum sudangrass, you'll need to calculate biomass before each mowing, then add the subtotals.

White Clover

White clover (*Trifolium repens*) comes in a few varieties: Dutch White, New Zealand White, and Ladino. It is a hardy, shallow-rooted, low-growing perennial that does well in temperate zones with adequate moisture. Like red clover, its tiny seeds make a slow start, but after 2 or 3 months, this plant will take full control.

Living mulch. The shorter varieties of white clover (Dutch White and New Zealand White) grow 6 to 12 inches tall and make an excellent living mulch. They are often seeded between rows of vegetables or herbs, or used as an understory cover in orchards and plantings of berry bushes. Both respond well to mowing and can handle heavy foot and vehicular traffic, all the while helping to protect the soil from erosion and convert atmospheric nitrogen into a plant-friendly mineral form.

We often seed New Zealand white clover between rows of tomatoes, peppers, and parsley. (I prefer the New Zealand variety because it has done well for us and, moreover, comes from my country of origin.) In the first several weeks after seeding, it might seem that fast-growing annual weeds are getting the best of the clover. All the while, though, the weeds are providing valuable shade and enabling the clover to gain a foothold. After a few mowings (to a height of 3 or 4 inches), the weeds generally run out of steam, and the clover comes into its own.

Overseed clover into existing cash crops. We also overseed this legume (and red clover, too) into plantings of winter squash and pumpkins. We always cultivate these cucurbits until they have grown too big for a tractor to enter the field without doing major damage.

Immediately after the last cultivation and just before rain (if we can arrange it), we broadcast the clover seed using an over-the-shoulder broadcast spreader. Because covering the clover in an acre of vining squash plants is not practical, our hope is that the tiny seeds will slip into openings in the loose, freshly cultivated soil and find a happy medium in which to germinate. Rain within a day or two of seeding definitely helps. We judge ourselves successful in this

White clover

White Clover

AT A GLANCE

Seeding rates: 6 to 8 pounds per acre when drilled; 8 to 10 pounds per acre when broadcast; 10 to 14 pounds per acre when overseeded into existing rows of an established vegetable crop; 4 to 6 pounds per acre when sown in combination with other cover crops

Range and zones: Widely adapted in the eastern half of North America and along the Pacific coast. Not suited to arid regions. Hardy to Zone 4.

Major attributes: Nitrogen fixation; living mulch; erosion control; can handle heavy traffic

---

undertaking when we encounter a budding field of clover in fall as we go about harvesting our butternut and acorn squash.

The Ladino variety of white clover grows taller and produces more biomass and more nitrogen than the other two. It is commonly used as a forage crop and soil builder but is less well suited to the role of a living mulch. It is also less heat tolerant and less durable than the shorter varieties.

Once established, white clover can be quite aggressive. It spreads by using creeping stems, known as stolons. When used in a permanent bed system, it will gradually encroach into the beds and compete with cash crop plants, especially for moisture. Tilling annually to redefine the beds will keep the clover in check.

Hairy Vetch

Hairy vetch (*Vicia villosa*) is an aggressive, vining legume and a cold-tolerant winter annual that is more likely to be employed as a nitrogen fixer and green manure than as a cover crop. It can tolerate a wide range of soils.

Stems of hairy vetch can reach more than 10 feet long and will remain largely prostrate unless they have some support. Accordingly, vetch is often seeded with a nurse crop such as oats or winter rye. The faster-growing grain crop gives the legume something to climb on. If seeded alone, hairy vetch will create a thick, tangled mass.

Spring weed suppression and biomass production. Hairy vetch is best planted in late summer or early fall, 35 to 45 days before the first frost. Don't expect a lot of growth before winter sets in, but when spring comes along, the vetch will take off and do its work. It has a smothering effect on weeds. It's possible for it to produce up to 7,000 pounds of biomass per acre, along with as much as 250 pounds of nitrogen. These are big numbers, reachable only under excellent growing conditions.

In more average soils and conditions, the yields may be cut in half, but they are still impressive. Hairy vetch

Hairy vetch

can also be seeded in early spring and incorporated into the soil in early or midsummer, but it will produce less biomass and nitrogen.

Vetch and rye as mulch. In late spring or early summer, when a fall-seeded vetch/rye combination has reached maturity and the vetch is setting flower, some growers choose not to turn the plants into the soil. Instead, they use a flail mower to create a bed of finely chopped mulch, then transplant a summer crop such as tomatoes into it.

Hairy Vetch

AT A GLANCE

Seeding rates: 20 to 25 pounds per acre when drilled; 35 to 40 pounds per acre when broadcast or overseeded. Reduce by about 30 percent when sowing in combination with oats or winter rye.

Range and zones: The eastern half of the United States and parts of Canada. Can be grown in drier regions if irrigated. Hardy to Zone 4.

Major attributes: High biomass and nitrogen production; weed control in spring; the most cold-tolerant of winter annual legumes

The mowing needs to be low and thorough to ensure that both the vetch and rye are killed; otherwise, they will compete with the vegetable crop. The layer of mulch also needs to be fairly dense to ensure that weeds cannot come through it. The residue of the rye and, to a lesser extent, vetch, have allelopathic qualities that can inhibit weed seed germination. In this example, the vetch and rye function as a weed-suppressing and moisture-conserving mulch in a no-till context.

Estimating Approximate Nitrogen Yield

To estimate the amount of nitrogen in a legume cover crop, use the same steps you would use to arrive at the biomass or dry weight of the crop before it is mowed or turned into the soil (see box on page 252). On average, about 3.5 percent of the aboveground portion of an annual legume in its green stage consists of nitrogen.

So, if you have a crop of red clover with a total biomass of 3,250 pounds per acre, you could expect a nitrogen gain of 113.75 pounds (0.035 × 3,250). But don't expect an immediate burst of nitrogen — the bulk of it will not be available to a cash crop until the red clover is fully decomposed.

If you are estimating the amount of nitrogen to be gained from a more woody, perennial legume, assume a nitrogen content of only 2.0 or 2.5 percent.

Don't bet the farm on your estimates of nitrogen content in leguminous crops. They are approximations only. Nitrogen content will vary depending on the legume in question, the effectiveness of the nitrogen-fixing bacteria in the soil, and the plant's stage of growth when you incorporate its tissue into the soil (legumes contain the most nitrogen just before flowering).

To get a more reliable figure, send a sample of your crop to a soil testing lab for analysis. Just be sure to answer the questions on the soil test questionnaire, especially those that relate to your cover cropping, as accurately as possible. Knowing how much nitrogen will be available to a subsequent cash crop, and over what period of time, is valuable information. It will tell you whether additional nitrogen is needed and how much.

Hairy vetch can be overseeded into existing vegetables, such as broccoli or kale, in late summer, as long as there are about 40 frost-free days remaining. During establishment, the vetch will not tolerate much foot traffic. And of course, once it is sown, all hoeing and cultivation must stop.

The right crops for you. The eight cover and green manure crops discussed above are just a sampling. Six of the eight were grown on our farm in 2010. In future years, the lineup might be a little different. There are many others to choose from, each with its own set of benefits and sometimes drawbacks.

Deciding which crop or combination of crops to use in a specific situation is not always easy. It will depend on your primary goal, along with a few other variables, such as the time of year, the length of time the crop will be in the ground, and where you are located. The salient point is that cover and green manure crops can do much good for those farmers and gardeners who are motivated to use them.

Watching the Weather

The climate that prevails in your neck of the woods — how hot it gets; how cold it gets; the average amount of sunlight — will have a major bearing on the cover and green manure crops you can grow and the length of time you can grow them. Plant hardiness zones, frost-free-day maps, and length of growing season are three useful tools to help you determine how crops will fare in the open field on your land. Here, we'll take a quick look at each of them.

Plant Hardiness Zones

The USDA's Plant Hardiness Zone Map shows the cold-weather extremes that occur across all geographic locations in North America. There are a total of 11 zones in the United States and Canada. Zone 1 has the lowest average annual minimum temperature and Zone 11 the highest.

For example: Zone 4, which contains Burlington, Vermont, has an average annual minimum temperature of –20 to –30°F (–29 to –34°C); whereas Zone 8, in which lies Seattle, Washington, on the other side of the continent, has an average annual minimum temperature between 10 and 20°F (–12 and –7°C).

When a cover crop or green manure, such as hairy vetch, is described as being hardy to Zone 4, this means it can experience temperatures as low as –30°F (–34°C) and not die. If it's colder than –30°F, all bets are off, and it might well get colder in Zone 4 every once in a while, because the zone map temperatures are *average annual* minimums, not multi-year lows.

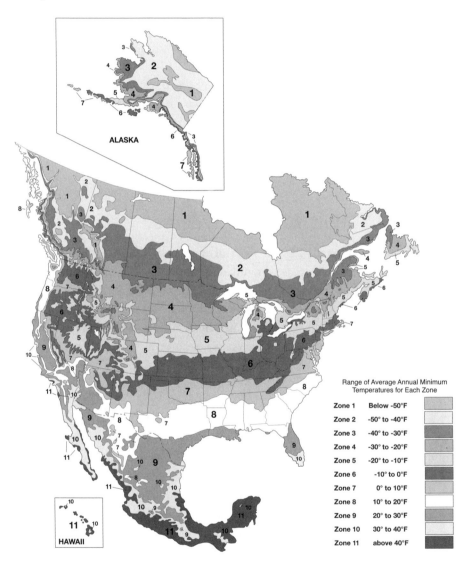

Range of Average Annual Minimum Temperatures for Each Zone	
Zone 1	Below -50°F
Zone 2	-50° to -40°F
Zone 3	-40° to -30°F
Zone 4	-30° to -20°F
Zone 5	-20° to -10°F
Zone 6	-10° to 0°F
Zone 7	0° to 10°F
Zone 8	10° to 20°F
Zone 9	20° to 30°F
Zone 10	30° to 40°F
Zone 11	above 40°F

Plant Hardiness Zone Map

Plant Hardiness Zones 2 through 10 are further divided into "sub-zones" *a* and *b*, which narrows the temperature range by half and therefore gives a more accurate reading. For example:

Average Annual Low Temperature

Zone 6: −10 to 0°F (−23 to −18°C)

Zone 6a: −10 to −5°F (−23 to −21°C)

Zone 6b: −5 to 0°F (−21 to −18°C)

If you don't have a Plant Hardiness Zone Map on hand, you can find one by going to the website of the United States National Arboretum and clicking on USDA Plant Hardiness Zone Map. The map allows the user to enlarge any area for greater detail. Be sure to use the most up-to-date zone map. Climate change in recent years, presumably from all the fossil fuel we like to burn, has caused many zone lines to shift. A map for all of North America is also available online, but at the time of the writing of this book, the hardiness zones for Canada had not been recently updated.

Limitations of plant hardiness zones. Because they don't take into account hot weather extremes, plant hardiness zones are not always a good indicator of which plants will flourish in what area. Some regions have mild winters and mild summers; others have cold winters and hot summers. You may live in an area where winter temperatures don't usually drop much below freezing, but if you seldom experience temperatures above 80°F (27°C) in summer, you're not likely to do well with such crops as okra and watermelons — or tomatoes and peppers, for that matter.

Simply put, the Plant Hardiness Zone Map tells you, on average, how cold it will get in winter. This is useful information for a farmer or gardener to have, especially when dealing with perennial plants and trees, but it is by no means the whole story.

Frost-Free Days

Another useful measure of climate, and the types of plants that should perform well in your locale, is the number of frost-free growing days that can be expected where you live. This information can be found at the website of the National Climate Data Center (NCDC), which is part of the National Oceanic and Atmospheric Administration (NOAA). Once you get to the site, just search for "Freeze/Frost Maps."

There you'll find three maps: the first gives the average number of days without frost for all regions of the nation; the second gives the average date of the last spring frost; and the third gives the average date of the first fall frost. Every vegetable grower should be familiar with this information, at least for the area in which he or she lives and farms.

State Cooperative Extension offices can also readily provide data on frost-free days. What they offer might be more useful if they include data on local microclimates — such as deep valleys or areas close to large bodies of water — where frost dates might deviate from the averages. (Note: Technically speaking, there is a difference between a frost and a freeze. Frosts occur at 32°F [0°C]; freezes, which are sometimes referred to as killing frosts, occur at 28°F [−2°C] or below.)

Growing Season

Sometimes the terms "frost-free period" and "growing season" are used interchangeably. This is misleading. To my way of thinking, the growing season starts well before the last frost of spring and continues well after the first frost of fall.

We routinely plant peas, onions, lettuces, brassicas, and other crops a few weeks before the last spring frost, and our fall-planted garlic has always put on 8 or 10 inches of growth in spring before the last frost occurs. At the other end of the season, root crops, brassicas, and many other greens do famously well long after the first frost of fall. In fact, many of them taste better once they've endured some nights in the 20s (−6 to −2°C) and even below.

Of course, there are also the heat-loving and cold-sensitive crops that will survive only when temperatures are above freezing. For these, the number of frost-free days is very relevant.

Cover Crop Seeding Methods

There are different ways to sow cover crop and green manure seeds. Spreading seed by hand will work for very small plots. But if you're dealing with areas larger than a few thousand square feet, you'll find that an over-the-shoulder broadcast seeder will come in very handy. For much larger areas, it may be worth investing in a tractor-mounted broadcast seeder or a tractor-drawn seed drill. Here, I'll discuss the pros and cons of these different approaches.

Broadcast Seeders

Most small farmers soon purchase an over-the-shoulder broadcast spreader/seeder. These hand-crank, rotary devices can be used to spread granulated or pelletized fertilizer, as well as cover crop seed. The typical model can handle up to about 20 pounds of seed at one time. All models require that the operator be willing to do a fair amount of walking, unless the area being seeded is small.

Getting an even spread of seed, at the rate you want, will take a little practice. Use just a small amount of seed to begin and see how it goes, or experiment with outdated seed on a clean surface such as a large concrete floor. For larger acreages, you may use a tractor-mounted three-point-hitch broadcast seeder/spreader.

Covering seed. Once seed is spread with a broadcast seeder, in most cases it will need to be covered. Small plots may be raked by hand. For larger areas, farmers normally use a tractor with a disc or spring-tine weeder. The trick is to not cover the seed too deeply. If you do, some of it may not germinate or germination may take place over a longer period of time, which can be a serious negative if you're hoping to get a jump on the weeds.

For most small seeds, such as many of the clovers, a soil covering of ¼ to ½ inch is adequate. Larger seeds, such as buckwheat, oats, or sorghum sudangrass, will germinate satisfactorily with anywhere from ½ to 1½ inches of soil on top of them.

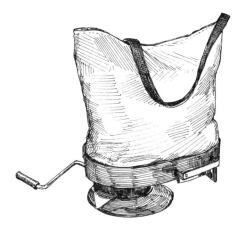

Over-the-shoulder broadcast seeder

Three-point-hitch broadcast seeder

Frost seeding. Very small seed, such as white clover, is sometimes left uncovered in the expectation that it will soon find its way into cracks in the soil. If the ground freezes and thaws, or receives a heavy rain, it will help this process along. Seeding into ground that is experiencing repeated freezing and thawing is known as "frost seeding."

Seed Drills

If you're planning to plant cover crops on more than a couple of acres of land at one time, you might choose to invest in a seed drill that can be pulled behind a tractor. The main advantage of a seed drill, whether it be a tractor-pulled implement or a simple hand-push model, is that the seeds can be placed at a more precise spacing and depth, which should result in more uniform germination and, ultimately, a more uniform crop.

Seed Germination Test

Before sowing any cover crop or vegetable seed that is more than 1 or 2 years old, it's a good idea to conduct a simple germination test. Take a sampling of seeds (anywhere from 20 to 100) and spread them out evenly on one half of a damp, but not saturated, paper towel (firm brown hand towels work better than super-absorbent towels used in the kitchen). Fold the other half of the towel onto the seeds and then loosely roll it up and place it in a plastic bag. Use a magic marker to write the date, number, and type of seeds you are testing on the plastic bag and leave the bag partially open. Now, place the entire package in a warm place (around 75°F [24°C] is ideal) and away from direct sun or artificial light.

After a few days, check the seeds daily and be sure to keep the paper towel moist by adding a sprinkle of water when necessary. Note the date on which the first seeds begin to germinate. If they are viable, under the above conditions most cover crop and vegetable seeds should germinate in 10 to 14 days, and often much sooner. At this point, or before, count the number of seeds that have germinated and appear to be turning into plants. Don't count any that are moldy or rotting. If you started with 50 seeds and end up with 40 that have germinated, you have a germination rate of 80 percent ($[40 \div 50] \times 100$).

Because of their greater accuracy, seed drills will give good results with less seed than a broadcast seeder would require. Another nice thing about using a drill is that the seeds are sown at the desired depth. No after-seeding covering is necessary. If you don't have a tractor-operated seed drill, you might be able to rent one from your local USDA Soil and Water Conservation District office.

Seeding by Hand

For small areas, broadcasting seed by hand is a reasonable option. When seeding pathways or narrow strips between planted beds, this is usually the best way to go. Follow up with a light raking or hoeing.

Seeding Tips

The old truism—if a job is worth doing, it's worth doing well—applies to growing cover and green manure crops as much as it does to most other tasks in life. Here are a few pointers to ensure that your cover crops get off to a good start.

Don't skimp on seed. Most cover crop seed is not expensive. It's better to put down more than is necessary rather than not enough. Old seed won't germinate as well; if that's what you're using, increase the rate by 50 to 100 percent. Better yet, when dealing with seed that is more than a year or two old, take a small number of them and do a germination test with a damp paper towel (see box on page 262). This will give you a germination rate, which will help you decide how much seed to use.

Use a clean seedbed. Sow into a clean seedbed if at all possible. When you seed into a field that already has weeds in it, even small ones, your crop will be fighting an uphill battle and quite possibly a losing one. It's no fun to look out on a poorly established cover crop with numerous patches of weeds or, worse yet, weeds intermingled throughout. By the time the crop is ready to be mowed or incorporated into the soil, many of the weeds, especially the annuals, will probably have gone to seed. This just means more weed problems down the road.

When a cover crop doesn't take well and has weeds spread throughout, many farmers will turn it into the soil prematurely before the weeds get a chance to set seed. If you are intercropping or overseeding a cover into an already established cash crop, be sure to hoe or cultivate immediately beforehand.

Watch for rain. Unfortunately, praying for rain seldom works. A better policy is to watch the weather and seed your cover crop just before you expect a nice rain event. If there is no rain in the forecast, you might consider sprinkler irrigation at the end of the day or in the evening, when there is likely to be less evaporation. The point here is that almost all seeds germinate best when they have some moisture around them to help bring them out of dormancy. Soil-improving crops are no exception.

Rotate cover crops. Don't grow the same cover crop repeatedly in the same field. Rotate from one year to the next, just as you do cash crops. Constant covers of clover and other legumes can result in a buildup of pathogens and diseases.

It is also unwise to use a cover crop that is in the same family as the cash crop you intend to follow with. For example, don't use yellow mustard or some other brassica cover crop in spring and summer before planting a brassica cash crop such as kale, collards, or broccoli in fall. Pests and soilborne diseases common to the cover could transfer over to your cash crop (see chapter 11 for more information on crop rotation).

Sow more than one cover crop. I've already noted that it's often wise to seed more than one cover or green manure crop at the same time. There are many possible combinations; they often include a legume and a high-biomass-producing nonlegume. One example is winter rye and hairy vetch seeded together in fall. The vetch brings extra nitrogen to the table, and the rye provides erosion control over winter and abundant organic matter the following spring. Because the rye grows fast in the cool late-season weather, it stays ahead of the weeds. It also functions as a nurse crop, creating a favorable environment for the shade-tolerant vetch and providing something the vetch can climb on.

Another nurse-crop combination is sorghum sudangrass and red clover, with the sorghum sudangrass providing a hospitable environment for the clover. Buckwheat might also be added to this mix, but both the buckwheat and sorghum sudangrass will need to be mowed at least a couple of times to prevent them from choking out the clover altogether or going to seed. This combination might be seeded in early summer and not plowed in until the following spring.

Bell beans and oats are another good combination for nitrogen fixation and biomass production. These two vigorous and rapid growers can be seeded together in early spring and turned into the soil a short couple of months later. They also perform well when planted in early fall.

Review of Terms

allelopathy. The ability that some plants have to produce toxins that inhibit germination and slow the growth of competing vegetation

carbon:nitrogen or C:N ratio. The amount of carbon in organic material relative to the amount of nitrogen. C:N ratios higher than 25:1 (25 parts carbon to 1 part nitrogen) are likely to cause temporary immobilization of nitrogen. Most animal manures have a C:N ratio of about 15:1. Wood shavings or sawdust could be as high as 400:1. Most other plant material falls somewhere in between.

catch crop. A soil-improving crop whose major purpose is to capture nutrients that might otherwise leach out of the soil

cover crop. A soil-improving crop whose principal purpose is to prevent erosion

frost seeding. Seeding into frozen or near-frozen ground, with the expectation that the seeds will work their way into the soil as its surface freezes at night and thaws during the day

green manure. A soil-improving crop whose main purpose is to increase soil organic matter content and often to add nitrogen

inoculation. Treating legume seed with various species of rhizobia bacteria to ensure their presence in the soil for the purpose of fixing nitrogen

intercropping. The use of alternating strips of cover crop and cash crop to prevent erosion and provide habitat for beneficial insects

living mulch. A cover or green manure crop maintained between rows of vegetables and herbs or in vineyards and orchards; usually mowed periodically

nitrogen deficit. The temporary unavailability of nitrogen after a high-carbon cover crop is incorporated into the soil

nitrogen fixation. The ability of leguminous plants to form a symbiotic relationship with soil bacteria called rhizobia, which results in the conversion of atmospheric nitrogen into a mineralized form usable by plants

nurse crop. A cover or green manure crop whose purpose is to provide shade or support for another crop, usually a legume

overseeding. Also referred to as undersowing and interseeding, the seeding of a cover or green manure crop into an existing cash crop — literally, over the top of it

rhizobia. Soil bacteria that associate with the roots of legumes and convert atmospheric nitrogen into a plant-ready form

smother crop. A cover crop whose major purpose is to shade out and suppress weeds

Fogline Farm

Proprietors: Johnny Wilson, Caleb Barron, Jeffrey Caspary

Location: Soquel, California

Year Started: 2009

Acreage: 38 acres, with 3 acres in row crops and the balance in apples, grapes, and pasture

Crops and Products: Mixed vegetables, herbs, livestock, and value-added products

L IKE MANY YOUNG PEOPLE, Johnny Wilson got bitten by the farming bug in his activist days, when he glimpsed the dark side of America's industrial food system. After college, he traveled and worked on farms in Central America. On returning to the United States, he enrolled in the University of California, Santa Cruz, Apprenticeship in Ecological Horticulture and honed his growing skills. He then moved East and spent a season interning at Stone Barns Center for Food and Agriculture, a nonprofit farm and educational center, in Westchester County, New York.

At Stone Barns, Johnny worked with managers Jack Algiere and Craig Haney and experienced a successful, well-funded, integrated vegetable and livestock farm in action. He returned to California with the goal of starting his own farm and putting his accumulated knowledge to work. Like many other young would-be farmers, the main obstacle he faced was the high cost of land.

Four Men Linked Together

Bruce Manildi's family has owned a 38-acre ranch in the Santa Cruz mountains for more than one hundred years. Over the course of its history, the Manildi Ranch, most of which is on relatively level ground, has produced vegetables, grapes, and apples, but it's been quite a while since the land generated any serious income. If sold to a developer, it would be worth over $3,000,000, but Bruce Manildi, now in his 70s, doesn't want to cash in that way. He wants to see the place remain as open space and a farm.

In 2008, the California Land Link program brought together Johnny Wilson and Bruce Manildi and helped forge a possible answer to their respective aspirations. With two partners, Caleb Barron and Jeffrey

Caspary, Johnny now has a 5-year lease on a portion of the Manildi Ranch and hopes to extend it well beyond that.

The three partners are now in their fourth year of operation. Together they grow 3 acres of mixed vegetables, raise 50 heritage-breed hogs, and manage 400 layer hens and 1,000 meat birds. The hogs are butchered off-farm at a USDA slaughterhouse. The broilers are processed on-farm by the three partners — anywhere from 100 to 150 each week. It's messy work, but it helps pay the bills — each 4-pound bird sells for around $24.

Fogline Farm is represented at five year-round farmers' markets and operates a small but growing CSA. The three partners take turns doing the farmers' markets, most of which are half-day affairs and within a 30-minute drive of where they live. The farm takes on five or six interns each year. It will be certified organic in 2012.

Fogline Farm's lease agreement with Bruce Manildi grants the partners use of the ranch's existing equipment. This includes a 25-horsepower Kubota L245H tractor, an 80-horsepower Lamborghini crawler tractor, and a 35-horsepower Massey Ferguson. A chisel plow on the Lamborghini is used for primary tillage, and a 60-inch rototiller on the Kubota prepares beds for planting. The farm also has a disc harrow and an assortment of mowing equipment for use in the vineyard and apple orchard.

For a late-season cover crop, the partners prefer a mix of bell beans, field peas, hairy vetch, and winter rye. This combination is seeded in October and incorporated into the soil in March or April. The most commonly used summer cover is a mix of sorghum sudangrass, barley, and cowpeas. The farm makes 15 to 20 yards of its own compost each year.

Johnny Wilson and his partners are intent on integrating the vegetable, herbal, and livestock arms of their business as much as possible. In addition to selling vegetables, eggs, and meat, they are developing value-added products that both complement and use the fresh food they produce. They rent space in a nearby commercial kitchen and make their own sausages, pâtés, and bacon using herbs and sometimes vegetables, such as fennel, grown on the farm. They produce smoked dry peppers and make their own roasted red pepper jam and hot sauces. They are working toward a future in which most of the food they produce is not only sold fresh but also represented in value-added products.

The three partners have taken out loans and are working hard to make Fogline Farm a success. So far, they are not putting much money in the bank, but they believe they are headed in the right direction.

Chapter 11

....

Crop Rotation

C ROP ROTATION AND DIVERSITY OF CROPS are fundamental tenets of organic farming, and the two go hand in hand. After all, it's impossible to rotate crops if you've only got one. When you have many crops, your options for moving them around increase exponentially.

As I have mentioned elsewhere in this book, diversity is pivotally important to organic growers. For starters, diversity spreads risk in the field, and farming is a notoriously risky business. If all your eggs are in one basket and something bad happens to that basket, you're left with a lot of broken eggs or no eggs at all. The year we got late blight and lost over 2,000 tomato plants, I got a deeper appreciation of diversity. That year, more than most, I was extremely glad to have our many other crops to keep us going.

Unless you plan to sell wholesale to a processing plant or supermarket chain, having a variety of vegetables and herbs to offer is almost always a good thing. It insulates you against the vagaries of the marketplace as well as the uncertainties of the field. And why sell just one item to a customer when you could sell a half dozen or more?

Last, but not least, a variety of crops enables you to rotate them to different locations on your farm, which is the primary subject of this chapter.

Six Reasons to Rotate Crops

Improve soil health, maintain and enhance nutrient levels, control plant diseases and insect pests, put a brake on weeds, add nitrogen to the soil — these are all compelling reasons to practice crop rotation. Let's take a closer look at each of them.

1. Enhance Soil Health and Fertility

Most organic growers strive to keep plant cover on their land for as much of the year as possible so that root development occurs throughout the growing season. Rotating between cover and cash crops, between plants with shallow and deep root systems, and between early and late-season crops will generally be beneficial to a soil's structure, health, and fertility.

Cover crops and green manures. These should be included in any rotation plan. They contribute organic matter, which improves a soil's tilth, aggregate structure, and water-holding capacity, and they capture and recycle nutrients that might otherwise leach beyond the reach of most vegetables. Leguminous cover crops enrich the soil by capturing atmospheric nitrogen. Cover crops are also used to suppress weeds (more on this coming up) and protect the soil from erosion.

Alternating between shallow and deep-rooted crops. Different vegetables and cover crops establish themselves in different ways. Some have shallow but dense and fibrous root systems that stay close to the soil surface. The roots of peas, lettuces, and onions, for example, don't go very deep but do have good lateral spread. Buckwheat has shallow but densely fibrous roots that are good for adding organic matter and stimulating microbial activity in the topsoil.

Deep-rooted plants confer different benefits. Some of the deeper-rooted vegetables are winter squash, tomatoes, and parsnips. Among cover crops, sweet and red clovers and sorghum sudangrass form strong taproots that can penetrate as many as 5 feet below the surface in their search for water and nutrients. In the process, these deep-rooted plants relieve soil compaction and open channels for the roots of future cash crops and the movement of soil organisms. They also recycle nutrients from subsoil to topsoil, which is where most vegetable roots reside.

Late-season crops. A majority of vegetable crops do the bulk of their growing and root development in spring and summer. Late-season cover crops such as rye and wheat can do a lot of active growing in the cooler months. They are especially useful for capturing and holding on to soil nitrogen that might otherwise be lost to leaching in winter.

2. Vary Nutrient Demand

To avoid soil nutrient imbalances, it is better not to plant the same crop, or a member of the same crop family, in the same location from one year to the next. Each vegetable you grow places different nutrient demands on the soil. Some have high nitrogen needs; others might require more phosphorus, potassium, or calcium, or more of a particular micronutrient. The differences are usually greatest with respect to

Nitrogen Needs of Crops

Heavy feeders (more than 100 pounds per acre): broccoli, brussels sprouts, cabbage, cauliflower, celery, leeks, peppers, potatoes, spinach, sweet corn

Moderate feeders (75–100 pounds per acre): beets, chard, collards, eggplant, kale, lettuce, onions, parsley, pumpkins, summer and winter squash, tomatoes

Light feeders (50–75 pounds per acre): Beans, carrots, most herbs, peas, radishes, scallions, turnips

nitrogen. For example, potatoes, broccoli, and cabbage are heavy feeders. They require significantly more nitrogen to prosper than do peas, beans, and carrots; therefore, a rotation plan that places broccoli immediately after potatoes is ill-advised. The broccoli is likely to have insufficient nitrogen to satisfy its needs; however, a rotation that allows broccoli to follow beans or carrots (crops with relatively low nitrogen needs) would make good sense.

If one has no choice but to plant two high-nitrogen-demanding crops in a row, a nitrogen-rich soil amendment, such as composted chicken manure or soybean or fish meal, might be applied in between plantings.

3. Control Plant Diseases

Growing the same crop in the same spot every year makes it easy for some plant diseases to establish themselves and become a recurrent threat. Moving the crop to a different location might solve the problem. This is especially true of root diseases caused by parasitic fungi and nematodes, whose populations can build up in the soil when the same crop is grown repeatedly. But avoiding disease is not always as simple as moving the crop to a different field or location. Here are some other factors to consider.

Identification. What is the scientific name of the disease and its strain?

Not all diseases can be controlled by crop rotation. And some strains may be more virulent and persistent than others.

Winter survival. Can the disease in question survive winter in the soil or on plant debris? Does it die at a certain temperature? Depending on the answers to these questions, it might be necessary to burn, bury deeply, or remove plant debris.

Persistence without host. How long can the disease persist without its host crop? Some diseases, most notably fusarium wilts, can survive in the soil for 5 years or more, even in the absence of a host plant. When dealing with such diseases, a long rotation away from the host plant is necessary. For most diseases, however, a rotation that allows 4 years to pass before repeating the same crop is adequate.

Other hosts. Are there weeds or wild plants that can act as hosts to the disease in the absence of the susceptible crop? If there are such plants on your farm, it will be necessary to control or remove them as part of your strategy.

Dispersal strategy. Is the disease soilborne or wind-borne? If the disease has spores that can be transported on the wind and it can survive your farm's winters, it is likely to find its host crop elsewhere on the farm, regardless of rotation, as soon as weather conditions are suitable for spore dispersal.

Soilborne diseases are more readily controlled by rotating susceptible crops; however, it should be noted that infected soil can be transported around the farm on tractor tires, equipment, and even the soles of your boots. Care must be taken to avoid having this happen.

Biofumigants. Some crops, most notably members of the brassica family, contain secondary compounds known as glucosinolates, which have natural biofumigant properties. When worked into a rotation, these plants, especially when grown as cover crops, can help suppress soilborne diseases and control some nematodes.

When dealing with diseases, the first step is always positive identification. If you're not sure what ailment you're faced with, the vegetable specialist at your local Cooperative Extension office should be able to help. Once you know your adversary, there are numerous publications and websites that will provide information on the disease's life cycle, hosts, and dispersal strategies. I'll take a closer look at plant diseases in chapter 22.

4. Control Insect Pests

Planting the same crop or crop family in the same location repeatedly can be a bonanza for certain insect pests and trouble for you. But as with rotating crops to avoid plant diseases, there are a number of variables to take into account.

Timed plantings. At what stage in the life cycle of the insect is it most likely to harm your crop? Many insects do most of their damage in the immature or larval stage. Depending on when this stage occurs, it may be best to plant early or delay planting.

Flea beetles, a common pest of brassicas, do their worst damage on our farm in May and June, especially on small and tender seedlings. We use row cover to protect early brassica plantings, but by July we can transplant seedlings of kale, broccoli, collards, and mustard greens without having to worry too much. By using row cover on the early brassicas, we reduce the likelihood of larger populations of subsequent generations of this pest.

Pest mobility. Is the pest able to fly or move about easily? In their adult stage, many pests can move freely from one corner of your farm to the other. This usually occurs soon after the insect emerges from overwintering. For rotation to work effectively, your crop needs to do most of its growing while the pest is in a relatively immobile stage or after the pest has determined that there is nothing to eat on your farm and moved on.

The Colorado potato beetle, a serious pest of potatoes, has limited mobility after it emerges from overwintering. It prefers to walk to its host plant, and it's a slow walker. Rotating potatoes some distance away from a previous planting is a good way to reduce or delay infestations of this pest.

Overwintering. Is the pest an overwintering resident, or does it fly in from warmer climes? Rotating crops will have little effect on insects that fly in from elsewhere unless the rotation takes into account the life cycle stages at which the pest does the least and the most amount of harm. In the Northeast, potato leafhoppers, imported cabbageworms, and cabbage loopers fly in from points south. It's hard to rotate away from them.

Presence of beneficials. Most beneficial insects are predators or parasitoids. If you have them on your farm, they will either eat or parasitize the insects that are causing you a problem. Beneficials often do not emerge until the subject of their interest is present in sufficient numbers to satisfy them. A good rotation plan might include delaying the planting of a vulnerable crop until the beneficials are in town. Hedgerows and strips of fallow land with diverse plant species will provide habitat for over-wintering beneficials.

Other host plants. Some insect pests can subsist on wild plants while waiting for you to provide the crop of their choice. Flea beetles, for example, can make a living off wild mustards until you plant spring brassicas for them. If possible, you may want to eradicate these other hosts.

As is the case with plant diseases, the management of insect pests might call for more than simply moving crops to different locations. Factors such as pest life cycle, overwintering habits, dispersal strategies, and the presence of other host plants must be taken into account. I'll take a close look at insect pests in chapter 22.

5. Reduce Weed Pressure

Most organic farmers spend a lot of their time combating weeds. Any benefit they can receive from crop rotation is a major plus. The idea is to make life more difficult for the weeds and prevent them from reproducing. Here are 10 ways to accomplish this.

Rotate between early- and late-season crops. Each species of weed has its own timetable for setting seed for the next generation. Chickweed, for example, is a cool-season annual that produces seed in spring, dies off in the summer heat, then germinates in fall. Lamb's quarters and pigweed are summer annuals that germinate in late spring and summer; they produce seed from mid- through late summer. You can disrupt the reproductive cycles of these and other weeds by planting cash crops that have different maturity dates.

Rotate between crops that are more and less vulnerable to weeds. Crops that have narrow leaves or take a long time to mature (e.g., onions, leeks, and shallots) are less able to compete with weeds than fast-growing and broad-leafed greens, such as lettuces and members of the brassica family. It is far easier to control weeds and prevent seed formation in the fast-growing, broad-leafed crops; therefore, they are a good choice to precede the slow-growing crops that are more vulnerable to weeds.

Rotate between crops grown in plastic mulch and those grown in bare soil. Black plastic mulch is an effective weed blocker. Many growers use it for heat-loving plants such as tomatoes, basil, peppers, and summer squash. Often, they mulch the aisles between the strips of plastic with straw, weed-free hay, or aged wood chips.

Another approach is to seed a low-growing clover or grass in the access aisles. Regularly mowing the aisles favors the clover or grass, which functions as a living mulch, but exhausts any weeds that may gain a foothold. A subsequent bare-soil crop, such as a leafy green or root crop, will experience less weed pressure when planted after a rotation that uses plastic and living mulch.

An alternative to plastic mulch is a woven weed barrier, which is every bit as effective at blocking weeds. This product is much more expensive but can be reused for several years. Chapter 20 delves more deeply into the subject of weed management.

Rotate between crops grown in organic mulch and bare soil. A 2- to 3-inch layer of organic mulch, such as straw or a wood shavings, will also block weeds, though not as effectively as plastic mulch or a woven weed barrier. Some weeds will inevitably find their way through the mulch, but these can be pulled by hand with relative ease.

The great thing about using an organic mulch in a rotation plan is that, after it has served as a weed barrier, it gradually decomposes and contributes a large amount of organic matter to the soil. The decomposition process can be sped up if the mulch is incorporated into the soil with a chisel plow or disc after the crop plant is harvested. Because our farm is close to a couple of horse farms, we've used bedding from horse stalls to mulch our fall-planted garlic for many years. For more details on how this works, see chapter 14.

Rotate between short- and long-season crops. The longer a crop is in the ground, the harder it is to stay ahead of the weeds and prevent them from going to seed. Fast-growing plants such as lettuces and spinach, when transplanted, may spend as few as 6 weeks in the field. Two or three light cultivations should be enough to keep the weeds in check. Celeriac and brussels sprouts, on the other hand, usually remain in the field for at least 4 months. They present a greater weed-control challenge. It's a good idea to rotate between these two extremes.

Rotate between cash crops and cover crops. Fallow periods enable you to establish a fast-growing cover crop, such as buckwheat or sorghum sudangrass, which will outcompete and suppress most weeds. These summer covers might be followed by a fall planting of garlic or fast-growing greens.

Rotate between annual and perennial crops. On diversified vegetable farms, most cash crops are annuals. The perennial crops are usually cover crops or green manures such as perennial ryegrass or any of several different kinds of clover. When allowed to grow for a year or more and mowed at regular intervals, these perennial (or biennial) crops can effectively suppress annual weeds. Almost any cash crop will benefit when it is preceded by a perennial or biennial cover.

Rotate between transplanted and direct-seeded crops. Direct-seeded crops are more vulnerable to weeds than transplanted crops. When vegetable transplants are set into a clean bed, they should have a good head start over subsequent weeds that have yet to germinate. On the other hand, crops that must be direct-seeded, especially those with small seeds, such as carrots, can be engulfed in fast-germinating annual weeds before reaching a few inches in height.

It is always better to plant carrots, for example, after a crop that has received good weed control. Competition from weeds is less severe with such direct-seeded crops as beans and peas, which have larger seeds and will germinate and grow more rapidly. Still, they require a clean seedbed.

Rotate between cash crops and bare fallow periods. Leaving your soil fallow and bare for a period of time will, of course, give full rein to the weeds. But this can work to your advantage. Allowing, and even encouraging, weeds to germinate before planting a cash crop gives you an opportunity to destroy them. This approach is often referred to as the "stale seedbed" technique. The ground to be planted is usually prepared several weeks in advance. Weeds are allowed to germinate and grow, but when 1 or 2 inches high are dispatched using shallow cultivation or a flamer.

Many small-seeded annual weeds (pigweed, lamb's quarters, galinsoga) need light to germinate. The goal is to kill the weeds that are already growing but not create conditions that will help others to germinate. A flame weeder, because it does not disturb the soil, works best.

Another strategy, known as the "false seedbed" technique, is an even more aggressive approach to eliminating weeds. This technique requires a longer bare fallow period in which weeds are repeatedly encouraged to germinate and repeatedly destroyed, usually by cultivation.

With the false seedbed technique, the cultivation can be deeper, to bring additional seeds to the surface where they will readily germinate. The intention is to deplete the soil's supply of weed seed, or the seed bank, over a period of time. Some farmers even go so far as to use old row cover to warm the soil, or sprinkler-irrigate after each cultivation, to speed the germination of a new round of weeds.

Rotate between crops according to ease of cultivation. Some crops are more readily cultivated than others. For example, fast-growing brassicas and lettuces are easier to cultivate with a wheel hoe or tractor and cultivator than are sprawling vine crops such as pumpkins or zucchini. Because of this, more weeds are likely to survive and go to seed in the vine crops. Direct-seeding a weed-sensitive crop, such as carrots, after a vine crop that has not received thorough weeding would be asking for trouble. It's better to follow the vine crop with lettuces or brassicas, which lend themselves to easier weed control, or perhaps tomatoes and peppers in a plastic mulch.

6. Capture Nitrogen

Legumes, both as cash crops (peas and beans) and cover crops (clovers, vetch, field peas, or bell beans), are fundamental to any rotation plan on a diversified farm. This is because of the legume's ability, in conjunction with soil bacteria, to convert atmospheric nitrogen into a form both they

Crop Rotation Basics

- ▸ Rotate between vegetable families (to vary nutrient demand and reduce pest and disease problems).
- ▸ Rotate between shallow- and deep-rooted crops.
- ▸ Rotate between nitrogen-fixing crops and crops with high nitrogen needs.
- ▸ Rotate between cash crops and cover crops and green manures.
- ▸ Rotate between early- and late-season crops.
- ▸ Rotate between crops that compete well with weeds and those that compete poorly.
- ▸ Rotate between transplanted and direct-seeded crops.
- ▸ Rotate between crops grown in plastic or organic mulch and crops grown in bare soil.
- ▸ Avoid bare fallow periods except when practicing stale seedbed or false seedbed techniques.
- ▸ Keep cover on land during winter months.
- ▸ Plant perennial crops on steep slopes.
- ▸ Plant catch crops on fallow land.

and other plants can use. A legume crop is especially valuable when it both precedes and follows any crop that has high nitrogen needs, such as broccoli, cabbage, and potatoes.

Not all legumes are equally good at capturing atmospheric nitrogen. The amount captured will differ depending on the following variables:

- Species of legume planted

- Length of time it is allowed to grow

- Amount of biomass it produces

- Whether some portion is harvested for mulch or animal fodder

- Type of soil in which it is growing

- Prevailing temperature and moisture conditions

Legumes grown in nitrogen-rich soils will naturally fix less atmospheric nitrogen because they will have less need of this nutrient. Because some legumes, especially the perennial clovers and hairy vetch, are costly to plant and slow to establish, it makes most sense to use them only when they are needed and can be left to grow for a year or more.

Legumes typically capture 50 to 200 pounds or more of nitrogen per acre. Peas and beans grown for harvest come in at the lower end of this range; clovers, bell beans, and vetches are at the higher end. This doesn't mean, however, that the nitrogen contribution of cash-crop peas and beans is not worth having.

How to Create a Rotation Plan

In this section, we'll take a step-by-step approach to the development of a multiyear rotation plan. There are plenty of details to observe, so start out with a clear head and allow yourself a block of time. The steps will be easier to follow and make more sense if you have a real-life situation in front of you.

Step 1. Develop a Crop List and Determine Spacing

First, you need to know what crops you are going to grow, the quantity of each crop (number of transplants, or row feet for direct-seeded crops such as carrots and beans), and how much space each crop will require.

Sample Crop List and Space Requirements

Values are for field plantings in spring/summer (before June 31) and summer/fall (after June 31).

Crop	No. of Plants or Row Ft. Spr./Sum.	Space Needed in Bed Ft. Spr./Sum.	No. of Plants or Row Ft. Sum./Fall	Space Needed in Bed Ft. Sum./Fall	Rows Per Bed	Inches Between Plants w/in Row
Broccoli	0	n/a	1,500 Pl.	750 BF	3	18
Carrots	1,680 RF	560 BF	3,000 RF	1,000 BF	3	1–2
Kale	0	n/a	4,500 Pl.	1,500 BF	3	12
Lettuce	5,000 Pl.	1,250 BF	7,400 Pl.	1,850 BF	4	12
Tomatoes	1,400 Pl.	2,800 BF	0	n/a	1	24

To calculate space requirements, you'll need to nail down a few things: The width of your planting beds; how many rows of each individual crop you will be placing in a bed; and how much space you'll leave between plants within each row. For a closer look at space requirements, see chapter 3.

Also relevant is the season in which each crop will be planted and grown. It will be helpful to put this data in a spreadsheet format. Shown above is an example of what it might look like for a small sampling of vegetables for spring/summer and summer/fall plantings.

Step 2. Draw Rotation Units on Your Field Map(s)

Using the field map you created in chapter 3 (see page 41), divide the field into separate sections or rotation units. Where possible, it is best to make each section or unit approximately the same size — a quarter-acre, half-acre, or whatever size you think will work best for you. In the example shown on the next page, the rotation units in fields A, C, D, and E range from 16,562 square feet to 16,800 square feet, giving them an average size of 16,673.5 square feet, which is close to 0.4 acre. Field B, because of its elongated and irregular shape, has three rotation units of 18,958 square feet or 0.44 acre each.

Step 3. Establish Planting Beds

Each rotation unit can now be divided into planting beds. The number and length of the beds will vary depending on the shape of the rotation unit. An elongated rectangle will have fewer and longer beds than a rotation unit that is a square of the same area.

The width of the beds will normally depend on the size of your tractor and tillage equipment. In the example given on page 281, for the sake of simplicity, the beds are 7 feet on center, which is quite large. Beds of this size will give you 4 to 5 feet of planting area and 2 to 3 feet for tractor wheel tracks and access paths. For more on bed width, see chapter 3. Again, where possible, try to create rotation units and beds that can contain roughly the same acreage of any given vegetable or grouping of vegetables. In general, the greater the diversity of crops on your farm, the more rotation units you should have.

Multiple fields. If you have several fields, make a map showing the location and dimensions of each. The individual fields can then be divided into rotation units of roughly equal size. Depending on the size of your farm, the number of fields, and the number of rotation units, you might need several sheets of paper. Make a half-dozen copies of the map or maps you have developed.

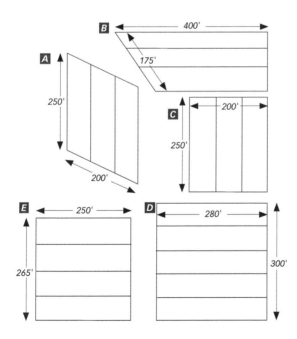

Field map showing field dimensions and rotation units · Each unit will accommodate several planting beds.

Step 4. Account for Special Considerations

Now that you have your field maps and rotation units, and have divided each unit into planting beds, note any special conditions that are relevant to a field or individual unit. Such considerations might include:

Soil. Is it different in some way from soils elsewhere on your land? For example, is it more or less deep, shallow, well drained, poorly drained, sandy, or rocky than other soils you have?

Aspect. Is the field or management unit level, south facing, or north facing? South-facing or level fields are better for both early spring crops and heat-loving crops.

Slope. What is the degree of slope? Steep slopes are usually more difficult, and more dangerous, to work with a tractor — and more prone to erosion when tilled.

Is any part of a field at the bottom of a slope that might receive run-off or be prone to flooding? For a more detailed discussion of slope, see chapter 2.

Proximity to other features. Is any part of the field close to a clean water source that will make for easy irrigation? Is it close to a wooded area or hedgerow that might give cover to deer, woodchucks, or other freeloading herbivores?

Fence. Does the field have a fence that can stop the aforementioned herbivores?

Spreadsheet Version of Fields and Rotation Units with Number of Beds

Field Name	Field Dimensions (in feet)	Field Size (sq. feet/acres)	No. of Rotation Units	Average Size of Rotation Units (sq. feet)	No. of Beds per Rotation Unit
A	200 × 250	50,000/1.15	3	16,666	9
B	162.5 × 350	56,875/1.31	3	18, 958	6, 7, or 8
C	200 × 250	50,000/1.15	3	16,666	9
D	280 × 300	84,000/1.93	5	16,800	8
E	250 × 265	66,250/1.52	4	16,562	9
TOTAL		307,125/7.06	18		

NOTE: Because Field B is not rectangular, the numbers given in the Field Dimensions column are averages of the different lengths and widths:

(400 + 300) ÷ 2 = 350

(150 + 175) ÷ 2 = 162.5

MAP OF FIELDS, ROTATION UNITS, AND PHYSICAL AND CULTURAL FEATURES

Farm Size: 13.75 acres

Tillable Fields: 6.78 acres

Farm Map showing fields, rotation units, number of beds per rotation unit, and physical and cultural features

Variable Field and Rotation Units

There is no rule that requires every crop to be rotated through every rotation unit that you have created, or that every field and rotation unit be equal in size and shape. Because of the special considerations already outlined on page 280, and probably others, certain crops may be ill suited to certain parts of your farm. Fields with steep slopes and shallow soils, for example, will be better suited to long-lived perennial herbs such as sage, thyme, or oregano, rather than short-season annuals. These fields should also be tilled less frequently.

Low-lying land that is prone to wetness would be a good choice for perennial water-loving plants such as mint or lovage. Unfenced areas that are frequented by deer and woodchucks are a bad place to put tasty greens that might get eaten by the four-legged competition before you can get them to market; however, these areas could make a fine home for onions, garlic, or shallots, which, so far, only humans have developed a taste for.

Step 5. Separate Crops into Families

Because vegetables within the same family are often vulnerable to the same pests and diseases and place similar nutrient demands on the soil, most growers develop a rotation plan that separates the taxonomic families. The objective is to avoid having members of the same family follow each other from one year to the next. This approach also makes it easier to introduce fallow periods and appropriate cover crops into your rotation.

Vegetable Families

Alliaceae (**alliums**): chives, garlic, leeks, onions, scallions, shallots
Chenopodiaceae: beets, chard, spinach
Compositae: chicory, endive, escarole, Jerusalem artichokes,
    lettuce, radicchio
Cruciferae (**brassicas**): arugula, brussels sprouts, cabbage,
    cauliflower, collards, kale, kohlrabi, mache, mustard greens,
    radish, rutabaga, and also many Asian greens such as bok choy,
    mibuna, mispoona, mizuna, tatsoi
Cucurbitaceae (**cucurbits**): cucumbers, melons, pumpkins,
    summer and winter squash, zucchini

Lamiaceae: basil, marjoram, mint, oregano, rosemary, sage, thyme
Leguminosae (legumes): beans, lentils, peas, soybeans
Solanaceae (nightshades): eggplant, okra, peppers, potatoes, tomatoes
Umbelliferae: carrots, celeriac, celery, fennel, parsley, parsnip (also anise, cilantro, dill, lovage)

Step 6. Pencil In a Rotation Plan

Your rotation plan can be penciled right into the rotation units on your field map or maps. Of course, this is easier said than done because there may be several variables to consider, such as those mentioned above — the acreage needed; the planting date; the nature of the soil; the slope and aspect of the field. If you are planting a lot of one crop, it might require several rotation units. A small planting might require less than a full rotation unit.

You would probably choose to plant carrots or daikon radish on a deep, rock-free soil — if you have this — rather than in a shallower soil with plenty of stones and rocks. You might choose to group crops that are grown in either plastic or organic mulch, or crops that lend themselves to similar succession plantings.

You might even group certain crops to facilitate harvest or some other management strategy. If you have the option, you should choose south-facing or nearly level ground for your earliest crops. Lettuces and other leafy greens that are appealing to woodchucks and deer might be placed in a field or rotation unit that is protected with a fence or a patrolling dog.

A Spreadsheet for Storing Data

In addition to rotation maps, many growers use a spreadsheet program to describe the location of their crops by field, rotation unit, and planting bed. A spreadsheet will enable you to easily note such data as the planting and harvest dates of each crop, the amount of labor required to produce the crop, the yield, problems encountered, and even the gross sales. Not all of this information is directly relevant to a crop rotation plan, but it's good stuff to keep track of nonetheless.

CROP ROTATION PLAN FOR
SPRING/SUMMER, YEAR 1

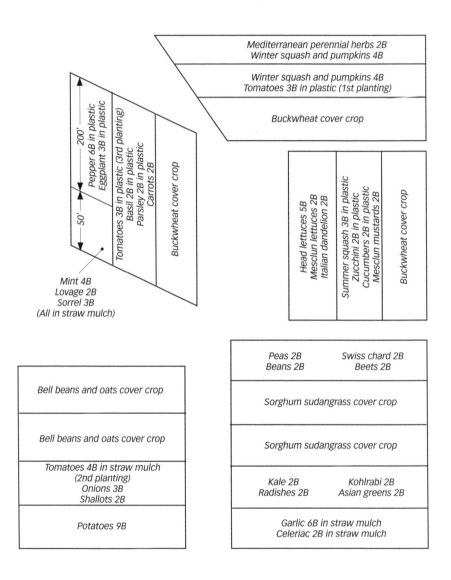

Mediterranean perennial herbs 2B
Winter squash and pumpkins 4B

Winter squash and pumpkins 4B
Tomatoes 3B in plastic (1st planting)

Buckwheat cover crop

200'

50'

Pepper 6B in plastic
Eggplant 3B in plastic

Tomatoes 3B in plastic (3rd planting)
Basil 2B in plastic
Parsley 2B in plastic
Carrots 2B

Buckwheat cover crop

Mint 4B
Lovage 2B
Sorrel 3B
(All in straw mulch)

Head lettuces 5B
Mesclun lettuces 2B
Italian dandelion 2B

Summer squash 3B in plastic
Zucchini 2B in plastic
Cucumbers 2B in plastic
Mesclun mustards 2B

Buckwheat cover crop

Bell beans and oats cover crop

Bell beans and oats cover crop

Tomatoes 4B in straw mulch
(2nd planting)
Onions 3B
Shallots 2B

Potatoes 9B

| Peas 2B | Swiss chard 2B |
| Beans 2B | Beets 2B |

Sorghum sudangrass cover crop

Sorghum sudangrass cover crop

| Kale 2B | Kohlrabi 2B |
| Radishes 2B | Asian greens 2B |

Garlic 6B in straw mulch
Celeriac 2B in straw mulch

CROP ROTATION PLAN FOR
SUMMER/FALL, YEAR 1

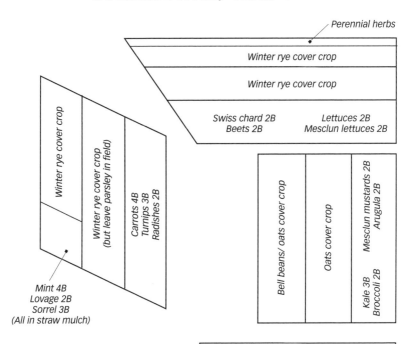

Perennial herbs

Winter rye cover crop

Winter rye cover crop

Swiss chard 2B
Beets 2B

Lettuces 2B
Mesclun lettuces 2B

Winter rye cover crop

Winter rye cover crop
(but leave parsley in field)

Carrots 4B
Turnips 3B
Radishes 2B

Mint 4B
Lovage 2B
Sorrel 3B
(All in straw mulch)

Bell beans/ oats cover crop

Oats cover crop

Mesclun mustards 2B
Arugula 2B

Kale 3B
Broccoli 2B

Lettuces 5B
Escarole 2B
Italian dandelion 2B

Broccoli 1B
Kale 3B
Cabbage 2B
Asian greens 3B

Winter rye/Hairy vetch cover crop

Winter rye/Hairy vetch cover crop

Oats cover crop

Winter rye/Hariy vetch cover crop 6B
Scallions 2B

Garlic for next year (plant in October) 6B

Winter rye/Hairy vetch cover crop

Winter rye/Hairy vetch cover crop

Step 7. Create Future Plans

The next step, which is a bit more challenging, is to create rotation plans for the next 3 or 4 years, or even farther into the future. This is why you need multiple copies of the field map or maps.

You will need to take into account many other variables, such as those enumerated above in "Six Reasons to Rotate Crops" and "Twelve Crop Rotation Basics." Think about dodging pests and diseases, alternating deep- and shallow-rooted plants and those with different nutrient demands, incorporating cover crops and green manures, and controlling weeds. It's enough to keep you awake at night.

You also face the challenge of predicting market demand a few years down the road — for everything you plan to grow. No easy task! The truth is: If you're growing many different crops, it will often be virtually impossible to make allowances for every factor to the extent that you might wish. Just do the best you can, and try not to let yourself get tied up in a knot.

Other Considerations for Rotating Crops

Don't read too much into the placement of crops in the illustrations on pages 284, 285, and 287. These maps are merely examples of how different vegetable families might be moved around on a hypothetical farm with a hypothetical crop list. They take into account some, but by no means all, of the factors that a grower might want to consider.

Every farm will have its own set of variables, and some will present greater crop rotation challenges than others. It's relatively easy to create a rotation plan for a farm that has all level ground, is of one soil type and depth, and is surrounded by an 8-foot-high deer fence. Far more challenging, from a crop rotation perspective, will be a farm that has fields of different aspect and relief, with different soils, that are separated by hedgerows and patches of woods.

Unforeseeables. Many experienced growers feel it is unrealistic to have every feature of a rotation plan set in stone several years in advance. Life has a tendency to present us with unforeseeable and changing circumstances from time to time. When this happens, you may need to modify the plan. Here are some examples of such unforeseeable events:

- Market demand can shift, or your market may become flooded with a particular crop.

CROP ROTATION PLAN FOR
SPRING/SUMMER, YEAR 2

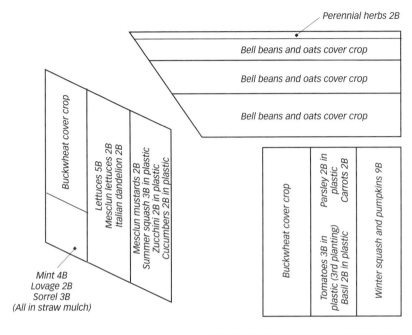

- You might be unable to find the variety of potato or garlic seed stock you need, or the seed you receive might be unusable.

- Other farmers might do a better job of growing a certain crop and undersell you.

- Extreme weather might preclude the planting of a favorite crop.

- A pest or disease might become such a problem that you decide to eliminate a family of vegetables from your crop plan for a period of time.

If any of these things should happen, it is important that you make adequate changes to your plan to take into account the overriding economic imperative. This imperative requires that you be flexible enough to meet the challenges that come your way so that you can make enough money to stay in business and continue with your chosen vocation.

Rome was not built in a day. A good crop rotation plan that reaches into the future is not something you are going to develop overnight. The important thing is to be aware of all the factors so that you can make informed choices. With time, you will develop a better sense of what is going on in your fields, which rotational sequences work best, and which don't work so well. The overall health of your land and the quality of your vegetables and herbs will tell you when you're doing a good job.

See Resources for some good books on crop rotation.

Six Reasons to Rotate Crops

1. To enhance soil health and fertility
2. To help prevent soil nutrient imbalances
3. To help control plant diseases
4. To help control insect pests
5. To reduce weed pressure
6. To capture atmospheric nitrogen

How to Create Your Own Rotation Plan

1. Develop a crop list and determine space requirements.
2. Draw rotation units on your field map(s).
3. Divide rotation units into planting beds.
4. Account for special considerations.
5. Separate crops into their respective families.
6. Pencil in a rotation plan.
7. Create future plans.

Chapter 12

Building Fertility

I N CHAPTER 9, WE DISCUSSED THE IMPORTANCE of organic matter in building soil fertility. In chapter 10, we considered the key role that cover crops and green manures play toward the same end. Now let's take a look at compost, animal manures, and some of the other amendments that can be used in a soil fertility program.

Compost

Composting is the ultimate form of recycling. It is the highly efficient dismantling of spent life into forms living plants can use. This is something that nature has been doing, with help from the sun, the rain, and a vast assortment of soil organisms, since the first plants and microbes appeared on the earth. Humans have found various ways to speed up the composting process, but it always boils down to the same thing — the biological decomposition of organic materials such as plant debris and animal manures into a rich, humus-like substance that improves soil fertility and tilth, conserves moisture, and suppresses plant diseases.

Whenever you harvest and remove vegetables from your fields, you are exporting soil nutrients. The more you harvest, the more you export. You can't export nutrients for very long without depleting the supply and running into a real problem. So you have to find ways to return nutrients and minerals to the soil or put back what you have taken away. Applying compost to your land is one of the best ways to do this.

Types of Composting

Composting is not a small subject. Entire books have been written about it, and views vary on exactly how it should be done. Here, we'll focus on the approaches that are most commonly used on small farms: sheet composting, windrow composting, and compost heaps. The method or methods you choose will depend to a large extent on the equipment and materials at your disposal.

Sheet Composting

This is the simplest way to make compost, but it is also the slowest. It involves spreading undecomposed organic matter, such as leaves, grass clippings, uprooted weeds, crop remains, vegetable scraps, straw, and wood shavings, across your fields in a thin layer. The material can be left to break down on the surface or be incorporated into the soil with a roto-tiller, disc, or plow.

Weed seed. You can sheet-compost living weeds, but it is critical that you mow them as close to the ground as possible before they develop seeds. A thin layer of slowly decomposing material on the surface of a field will not heat up as does a well-managed compost pile. This means that any weed seed will remain viable and, at some point, is likely to germinate and create more work for you. A properly managed compost pile, on the other hand, should get hot enough to kill most weed seed.

Nitrogen and carbon. Fresh, green material such as grass clippings and immature weeds, both of which are relatively high in nitrogen, can break down in a few weeks of warm weather, especially if they get some rain. In contrast, high-carbon material, such as wood shavings, wood chips, and sawdust, will take much longer to decompose and might, depending on the quantity applied, cause a temporary shortage of nitrogen in the soil.

As previously noted, this shortage happens because soil organisms commandeer and monopolize whatever nitrogen they can find to help them break down the highly carbonaceous matter. For more on this subject, see chapters 9 and 10.

Unless you are planning to leave a field fallow through the growing season, it makes sense to sheet-compost in fall, to allow enough time for the material to decompose and for the supply of nitrogen to reconstitute.

Windrows

Many small farmers who own a tractor with a front-end loader use the windrow method of composting. It involves laying out the material to be composted in long rows — usually 6 to 10 feet wide, about 3 to 4 feet high, and whatever is a workable length. There should be enough space on either side of the windrows for the tractor to move freely so that it can turn and aerate the material, redistribute moisture, and mix the warmer and cooler portions of the pile.

The material that is composted might be solely plant based or it might contain the manure of herbivorous animals. As with sheet composting, too much carbonaceous matter will slow down the process. Ingredients with an average C:N ratio of about 30:1 are ideal. This might include some plant matter or manure that is high in nitrogen and other material that is much higher in carbon. In general, the more variety in the compost pile the better. Once finished, a mature compost should have a C:N ratio of about 25:1 to 35:1.

Average C:N Ratio of Commonly Composted Materials

Material	C:N Ratio	Material	C:N Ratio
Soybean meal	5:1	Horse manure	30:1
Chicken manure	10:1	Weeds	30:1
Alfalfa meal	15:1	Fruit scraps	35:1
Grass clippings	15:1	Leaves	60:1
Sheep manure	15:1	Straw	75:1
Cow manure	20:1	Sawdust	350:1
Vegetable scraps	25:1	Hardwood chips	600:1

Compost Calculators

An online compost calculator (see Resources) will provide you with an estimate of the C:N ratio of your finished compost based on the type and amount of each ingredient you put into it. All you have to do is enter the numbers. Take, for example, a compost made from the following ingredients:

- ▶ 100 cubic feet of fresh chicken manure
- ▶ 150 cubic feet of dry oat straw
- ▶ 75 cubic feet of fresh vegetable waste
- ▶ 100 cubic feet of fresh weeds
- ▶ 75 cubic feet of sawdust

According to the Green Welcome Wagon website, the finished compost will have a C:N ratio of 26:1.

Most farmers use a tractor bucket to turn their compost windrows, though a pitchfork and human muscle will do the job, albeit a lot more slowly. You should turn the piles at least five times in the first couple of weeks, then less frequently for a few weeks thereafter. The goal is to have the windrows maintain an internal temperature of at least 131°F (55°C) during their most active phase. A long, probing compost thermometer will tell you whether you're successful in this regard.

Compost windrow turners. If you want to get really serious about composting, you might look into getting a compost windrow turner. These machines straddle windrows and turn the material with revolving paddles, injecting air, redistributing heat, and generally speeding up the composting process. They work best when the windrow is on a hard and level surface.

Compost windrow turners are large, highly specialized, and expensive (at least $15,000). You need a large tractor to operate one. They don't make much sense unless you have a lot of material to work with; however, as your business grows, you may wish to rent a compost turner, if there is one nearby, or purchase one with several other farmers. I know only one farmer who has a turner. To justify the cost of the machine, he not only applies the compost he makes in his own fields, he also markets it throughout the state.

Compost windrows · Allow enough space between the windrows for a tractor with a front-end loader to turn and aerate the composting material.

Compost Heaps

Small heaps or piles are an easy way to compost kitchen scraps, weeds, and unsold vegetables that cannot otherwise be used. On our farm, these heaps are usually about 5 feet wide, 3 feet high, and whatever length is convenient. It's best to start with a layer of existing compost (preferably with worms and a cross section of other organisms in it) at the base, then add layers of fresh material as they become available. The more diverse the plant matter and vegetable waste in the pile, the better.

Shredded material, if it is available, will break down faster because it provides more surface area for microorganisms to work on, but not many farmers go to the trouble of shredding. It's also a good idea to throw on a few shovels of soil from time to time.

Turn the pile with a fork or a tractor bucket every few weeks to speed up the composting process and cut down on visits from raccoons, rats, opossums, and other scavenging critters. Once your pile is 3 feet high, don't add any more fresh material; start a new one.

Multiple piles. Because we generate a fair amount of vegetable waste on our farm, we like to have at least three piles going at the same time, each at a different stage of decomposition. The piles are positioned so that we can reach them with a tractor that has a front-end loader. This enables us to turn the decomposing material easily and remove it once it has become finished compost.

Contained piles. If you don't like the look of sprawling piles of vegetable waste near your house, you can build a three-sided, U-shaped containment structure out of concrete blocks, old pallets, or any other rigid material. The composting pile will look neater but will be more difficult to access and turn with a tractor bucket.

The Indore Method

The Indore method of composting was developed by the English agronomist Sir Albert Howard, who is considered by many to be the father of modern organic farming. Developed at the Indore research farm in India in the early twentieth century, it was the first widely publicized, scientific approach to composting. It is essentially a layering system that has held up well over time.

With the Indore method, the compost piles range from 3 to 5 feet high and 5 to 10 feet wide, with variable lengths. Piles are started on ground from which the sod has been removed. Several layers (usually

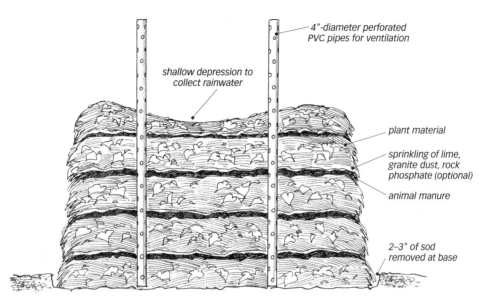

4"-diameter perforated
PVC pipes for ventilation

shallow depression to
collect rainwater

plant material

sprinkling of lime,
granite dust, rock
phosphate (optional)

animal manure

2–3" of sod
removed at base

This approximation of an Indore compost pile shows the deliberate layering
of diverse material.

6 to 8 inches thick) of plant material — preferably a combination of dry
and green matter, such as wood shavings, sawdust, leaves, straw, grass
clippings, and weeds — are sandwiched between shallower layers (about
2 inches thick) of manure or other nitrogen-rich material and a thin layer
(about ⅛ inch thick) of topsoil.

You can add a sprinkling of limestone, granite dust, and rock phosphate to the soil layers, and insert two or more vertical ventilator pipes
made from 4-inch-diameter perforated PVC or concrete reinforcement
wire during the early stage of layering. Form a shallow depression at the
top of the pile to collect rain. If rainfall is insufficient, you should add
water during the layering process.

A pile structured in this way should quickly reach temperatures of
131°F (55°C) or higher. Once the multiple layers of diverse ingredients
have reached the desired height, the ventilation pipes are removed,
enabling the farmer to turn the pile with a tractor bucket. Finished compost should be ready in 3 months.

You need to pay close attention to detail in the Indore method; therefore, it is a little more demanding of a farmer, but it is a very effective system. It distributes the different materials well, introduces soil organisms into the pile, and creates a good balance between carbon and nitrogen.

Sources of Organic Material for Compost

Organic material is all around us. It's just a matter of moving it to a suitable spot on your property so that you can turn it into compost. Besides what you generate on your farm, there are numerous other possible sources. Here are some of them.

Restaurant leftovers. You can supply nearby restaurants with clean garbage cans and pick them up when they are full, but you have to make sure that meat scraps are not included with the vegetable waste. If you can find them, vegetarian and vegan restaurants are the best choices.

Other peoples' kitchen waste. CSA members, farmers' market customers, food coops, or other entities that receive your produce may be persuaded to return their food waste to you, especially if you make it easy for them by providing containers. Tell them it's a meaningful way to participate in growing the food they eat and far more environmentally sound than sending their scraps to a landfill.

Local municipal waste. Local municipalities might be willing to deliver to you leaves, wood chips, and other compostable materials for which they have no use and place to store. Many towns hire contractors to trim and chip the trees growing under power lines. These companies can be a great source for wood chips.

Nearby livestock farms. If you live near a farm that has any type of livestock, you might have a ready source of manure. Chicken houses, and horse and other animal farms, usually have far more manure than their owners can deal with, and they are very happy for you to take it off their hands. Generally, it's just a matter of picking up the stuff yourself or paying someone to truck it for you. The manure does not have to come from animals that are raised organically; however, your certification provider might require that you provide a letter stating that there are no prohibited substances in the manure you are using (see the next section on the NOP rules for using manure).

Other possibilities. High schools, college cafeterias, and many institutions that feed large numbers of people throw away huge amounts of vegetable waste every day. This is all potential compost for your fields. The challenge is to show these organizations how they can separate the material you need from all the other waste products that you don't want (such as tinfoil; meat scraps; plastic knives, forks, and cups). It's not an easy task, I'll admit, but it needs to be addressed sooner or later if we want to build a truly sustainable society. Try to convince these food-providing organizations that giving you their compostable material will reduce the flow of waste to the landfill, benefit their community, and win them some good, green PR.

Composting ⟩ My Story

OVER THE YEARS, we've relied heavily on wood shavings, straw, and animal manure to make the bulk of our compost because these materials have been readily available to us at either low cost or no cost at all. We've been satisfied with the standard windrow method and use a tractor with a front-end loader for mixing, turning, and aerating the piles.

When possible, we've combined nitrogen-rich chicken, cow, or sheep manure with horse bedding that contains plenty of straw and wood shavings, along with a peppering of horse turds. We usually allow several months for the composting process to take place.

Using a tractor and a manure spreader, we try to spread about 40 tons (approximately 60 cubic yards) of finished compost per acre on each of our fields every 3 or 4 years. This amounts to a layer about ½ inch thick.

We usually do the spreading a month or more before planting a crop for market. Sometimes, we spread compost in the early fall and incorporate it into the soil before planting a winter cover crop. Compost that is spread in fall can be less mature — it will continue to break down during winter and early spring. Our kitchen scraps and market leftovers, as noted above, are composted in smaller piles located closer to the farm buildings.

Blue Heron Farm

Owners: Robin Ostfeld and Lou Johns

Location: Lodi, New York

Year Started: 1987

Acreage: 17 acres, with 13 acres in production and 4 acres fallow

Crops: Mixed vegetables and herbs

L OU JOHNS AND ROBIN OSTFELD OF BLUE HERON FARM grow an impressive range of vegetables and herbs, as well as flowers and strawberries. They have been certified organic since 1987. In addition to their great diversity of crops (just about everything you can think of from asparagus to zucchini), Lou and Robin have opted for a diverse marketing plan. Their current outlets include a farmers' market, two food co-ops, a winter CSA, and several restaurants.

Crop rotation, cover crops, fallow periods, and composted chicken manure are integral parts of Blue Heron's program to build soil health and fertility. The farm hires a crew of four or five full-time workers and two or three part-timers. It also uses a fair amount of machinery. Here's a list of the main elements:

- John Deere 5303 4WD 63-horsepower diesel tractor
- Case 1240 4WD 65-horsepower tractor
- Farmall 350 tractor with belly-mounted cultivators
- Ford 1710 high-clearance offset tractor
- Five-tine field cultivator
- Lilliston cultivator
- Beet-knife cultivator on the Farmall 350
- 70-inch Kuhn rototiller
- Mechanical transplanter
- New Idea manure spreader

As their farm business has grown, Robin and Lou have built four greenhouses and two high tunnels. They've also converted portions of one of their barns into refrigerated and heated storage space. Selling root crops and other cold-weather vegetables through a winter CSA is now a significant part of their business.

At Blue Heron Farm, transplanting is done with a pair of Mechanical transplanters that are offset on a toolbar. Rows 1 and 3 are planted with the tractor moving in one direction and rows 2 and 4 with the tractor going the opposite way. Rows are spaced 16 inches on center. Two-row plantings, for crops such as summer squash, are spaced 36 inches apart and usually planted by hand in furrows created by cultivator shanks.

Direct-seeded crops are planted with a gang of belly-mounted Planet Juniors on the three-point-hitch of the Ford 1710. All between-row cultivating is done with tractor-mounted cultivators. In-row weeding is done with hand hoes and fingers.

Permanent Bed System

In their early years of farming in Lodi, Lou and Robin felt their crops were not as productive as they should be. They noticed that their shallow silt and clay loam soils were easily compacted by both tractor and human traffic. They found themselves planting and harvesting under sometimes wet and muddy conditions. After 4 years of less-than-satisfactory crops, they decided to leave behind full-width tillage (tillage across an entire field) and convert to a permanent bed system with perennial, grassy paths that would absorb all the tractor and foot traffic.

Once this decision was made, the first order of business was to widen the wheel tracks on all the tractors, to enable them to straddle rather than ride on top of the 70-inch swath left by their rototiller. In the beginning, they allowed the 18-inch-wide rear tires of one of their tractors to define the path width. Now, they have moved to much wider 50-inch paths. There are a few reasons for this: The wider paths are more forgiving of any encroachment when the rototiller isn't kept perfectly straight; they make it easier to maintain a healthy vegetative cover (typically a mix of native grasses and weeds, overseeded with hard red and creeping fescue and Dutch and Alsike clovers); they facilitate harvesting; and they provide habitat for beneficial insects. But, Lou cautions, 50-inch-wide paths call for a fair bit of mowing, especially in spring and early summer. This is done with a walk-behind DR field and brush mower that has a 26-inch deck. Lou has installed shielding on the sides of the rotovator (both in front and rear) to prevent tilled soil from being sloughed into the pathways.

According to Lou Johns, the healthier crops, better yields, and less insect and disease pressure that Blue Heron now experiences are in large part attributable to the permanent bed system established on the farm many years ago.

Animal Manure

The NOP has created its own definition of compost containing animal manure and how that compost should be made. Certified organic growers must adhere to strict rules when it comes to using such compost or using straight animal manure in their fields.

Raw manure and the 90/120-day rule. Raw and uncomposted animal manures can be applied to organically managed fields but with the following restrictions: You must wait 90 days before harvesting any edible crop whose harvested portions are aboveground and have no contact with the soil or soil particles (for example, corn or trellised tomatoes or peppers). If your crops are root crops or have an edible portion that is in contact with the soil or soil particles, 120 days must elapse between manure application and harvest. Such crops would include lettuces and other greens, which might have soil splashed on them during a heavy rain. (Note: NOP rules are periodically revised; always be sure you are following the latest version.)

Definition of compost. When it comes to compost that contains animal manure, the NOP's definition of compost presents something of a challenge to many organic growers. The rule states:

> Producers using an in-vessel or static aerated pile system must maintain the composting materials at a temperature between 131°F and 170°F [55 and 77°C] for 3 days. Producers using a windrow system [which is much more common among small farmers] must maintain the composting materials at a temperature between 131°F and 170°F [55 and 77°C] for 15 days, during which time the materials must be turned a minimum of five times.

Any material containing animal manure that does not meet the minimum temperature threshold cannot be called compost, which means it must comply with the 90/120 days-before-harvest rule mentioned above. This is the case even if the pile has been sitting for a year or more and looks and smells just like any other compost.

The NOP's requirement is based on the belief that subjecting composting material to temperatures of at least 131°F (55°C) for a specific number of days will destroy harmful bacteria that might be present in the manure and could be pathogenic to humans. This might be true, but farmers and gardeners have been composting with animal manures for centuries without having to meet these strict temperature requirements.

A not entirely satisfactory solution. The NOP rule requires that the temperature of the compost pile be monitored daily and that the farmer keep a log of these readings for the certification inspector to see.

Many growers have found that, even when using a tractor bucket to turn their compost pile, it can be difficult to maintain the high temperatures over the entire 15-day period. When they are not successful, or do not have the time to spend on this process, they treat their compost as though it were raw manure and follow the 90/120-day rule. This solves the problem but, under a normal cropping rotation, limits the number of days in the year when the material, which cannot bear the name of "compost," can be applied to a field.

I never said farming was easy.

Other Soil Amendments

Compost, cover crops, fallow periods, and — on occasion — raw or composted animal manures (if available) are the best ways to keep a soil fertile and in good tilth. Farmers who use these materials and practices should seldom have a need for additional amendments. If, for whatever reason, you find your soil deficient in a specific nutrient — and this should always be confirmed by soil testing, not just guesswork — there are a number of other products to reach for. Here are some of them, listed according to the nutrients or minerals they contain.

A word of caution: If you are a certified organic grower, be sure to check with your certifying agency before using any individual product. Inadvertently using a disallowed product will only complicate your life.

Nitrogen: Alfalfa meal, cottonseed meal, feather meal, fish meal
and fish emulsion, hydrolyzed fish, blood meal and dried blood
Phosphorus: Rock phosphate, colloidal phosphate, bonemeal, blood
meal and dried blood, cottonseed meal
Potassium: Greensand, sulfate of potash, Sul-Po-Mag, soybean
meal, granite meal and granite dust, potassium magnesium
sulfate (langbeinite), sulfate of potash
Calcium: Calcitic limestone, calcium sulfate (gypsum), aragonite

Magnesium: Magnesium sulfate, dolomitic limestone

Sulfur: Elemental sulfur, magnesium sulfate (Epsom salts), calcium sulfate, potassium magnesium sulfate (langbeinite), sulfate of potash

Micronutrients and trace minerals: Azomite, kelp meal, seaweed

There are a number of commercially available organic fertilizer mixes. They come in granulated, pelletized, and liquid forms and usually contain the big three — nitrogen, phosphorus and potassium — in various ratios. They tend to be expensive if you use them over large areas.

THE CROPS WE GROW

Chapter 13

··

Our Most Profitable Crops

THIS BOOK IS NOT INTENDED TO BE a comprehensive guide to growing all the crops that can be grown in North America. That would be a mammoth task and is hardly necessary anyway, as information of this sort is readily available from many sources (see Resources).

Instead, in this chapter I'll examine the major crops (excluding herbs and garlic, which have their own chapters) that are grown on our farm. I'll tell you why we grow them, when we grow them, how we grow them, and in what quantities we grow them. You'll notice that tomatoes get the most attention. That's because they're a big item for us.

A quick disclaimer before I begin: There are many ways to climb a mountain and more than one mountain to climb. As already noted, what works on one farm might not be ideal for another, for a variety of reasons, foremost among them being differences in climate, soils, equipment, resources, labor, and markets. The focus should be on playing the hand you're dealt, not mimicking too closely what others do. And remember: In the farming game, you can be dealt a different hand every year.

Bush Beans

Why. Bush beans are easy to seed and easy to weed. They grow quickly. They capture atmospheric nitrogen. They sell well. The only negative is that they take quite a bit of time to pick. Some varieties produce better than others; experiment with a few different ones to determine which work best in your region. We like 'Provider'—it's not a particularly fancy bean, but it produces well, tastes good, and germinates in cooler soil. 'Jade' is another good one, but it likes warmer weather.

When. Plant after all danger of frost has passed and the soil temperature is at least 60°F (16°C). In our case, that is usually mid- to late May. We normally lay some old row cover, weighted down with rocks, over the first couple of plantings to hold on to soil warmth at night. We continue seeding beans every 10 days or so until mid-July. In our region, after July 20, there's seldom enough time for beans to mature before cooler weather sets in. Beans will grow sluggishly in cool weather and cannot stand a frost.

We normally seed about 200 row feet of beans at a time. The multiple succession plantings ensure that we have plenty of good-tasting beans to harvest throughout summer. If the picking gets slim or the beans get tough and dry, we take a pass and wait for the next planting to produce. We could probably sell more beans if we grew them, but we don't want to devote too much time to picking this crop in midsummer, when there are a lot of other things to do.

How. We direct-seed beans into a freshly tilled bed using an Earthway walk-behind seeder. We sow two rows per bed — 36 inches between rows, 2 to 3 inches between seeds, with 1 inch of soil on top.

If the soil is dry and no rain is expected, we lay drip irrigation tape on top of the newly seeded rows and let it run until there is a continuous line of water over the top of the beans. This promotes even and rapid germination, which results in less weed competition. When weeds do appear, they are easily controlled with wheel hoes. We always inoculate bean seed with rhizobia bacteria before planting to ensure nitrogen fixation.

Broccoli

Why. Well-grown, organic broccoli is in high demand at our markets. It is nutritious, tastes good, and fetches a premium price. We seldom come home with unsold broccoli, but getting good-sized heads is not easy. I recommend starting out with a modest trial planting.

When. In our first years of farming, we planted broccoli in spring, for an early-summer crop, and again in early summer for a fall crop. In some years we did okay with the first seeding, but often, hot weather in June caused the heads to flower before they reached a decent size. That was disappointing and unremunerative. Now we plant only a fall crop.

Beware of what most seed catalogs or seed packets give for the days to maturity. These numbers can be misleading. They are usually 55 to 85 days (depending on the variety grown), but you need to know whether the numbers given are from the date of direct seeding in the field or from transplanting. Read the catalog description carefully to find out. Also, the "days-to-harvest" numbers given in seed catalogs apply to plants grown under optimal conditions and maximum day length. Broccoli destined to be harvested in October or November experiences shortening days through most of its growing life; therefore, it needs more time in the field.

We seed most of our broccoli in the second half of May and the first half of June, transplant through most of July, and harvest from mid-September, through October, and into November. In other words, we allow as many as 120 days, and sometimes more, from seeding to harvest. If we waited until July to seed broccoli, then transplanted in late August or September, we'd get nice, leafy plants but very diminutive heads by the time cold weather set in and growth slowed down dramatically.

There are several varieties of broccoli to choose from. We've had reasonable results with 'Fiesta', 'Belstar', and 'Waltham'; consult local farmers to find out which do best in your area. Select for either summer or fall harvest, and plant two or three varieties with different maturity dates, so you'll have an extended harvest period. We transplant about two thousand head of broccoli each year.

How. We start broccoli in our greenhouse in 72- or 98-cell flats, one or two seeds in each cell, then thin them down to one seed per cell soon after germination. We give the seedlings a foliar application of fish emulsion and kelp every other week. When they are big enough and have well-developed root systems (usually after about 6 weeks), we harden them off outside, then transplant them to the field with a water-wheel transplanter — three rows per bed, 18 inches between rows, and 18 inches between plants.

For broccoli heads to size up, they need good soil fertility, plenty of moisture, and adequate spacing. We've learned the hard way to plant only into well-rested and well-composted ground and to irrigate during dry spells. We use wheel hoes and hand hoes to control weeds. Weeding is especially important in the first month following transplanting. After a month or so, the broad broccoli leaves do a good job of shading out weeds.

Pests. The imported cabbage worm and cabbage looper are two major pests of broccoli and, indeed, of all members of the brassica family. We control these pests by spraying the plants with *Bacillus thuringiensis (Bt)*, variety *kurstaki*, using a backpack sprayer. We start spraying a few weeks after transplanting, or as soon as we see the adult insects flying about. The adult version of the cabbage worm is the common white butterfly; the adult cabbage looper is a 1½-inch, brownish, night-flying moth.

Neither the butterfly nor the moth do any actual damage; it is the larval or caterpillar stage of the pests that does the eating. We continue to spray twice a month for the first 6 weeks and then about every 10 days as the broccoli heads begin to form. We want to make sure no worms get into the developing heads. We delay spraying if rain is imminent. The biological insecticides Entrust or Monterey Garden Spray (both formulations of the spinosad organisms) are other organic options for controlling these pests.

Cabbage

Why. The cabbage is a common and unglamorous vegetable, but it has good nutritional value and stores well. Also, cabbage that is grown in a healthy, organically managed soil tastes much better than the bland stuff you get in supermarkets. It won't take long for your customers to catch on to this. We've found that cabbage is an increasingly popular crop.

When. We seed and transplant cabbage at about the same time as broccoli, or a little later. Some growers do an early-season crop, but we've found that cabbages grow better, taste better, and sell better in the cool days of autumn. Cabbage varieties vary widely: Some are suited to early-spring planting; others to mid-season planting and fall harvest. Some varieties store better than others, and these usually take longer to mature. Grow a few different types so that you are covered for early- and late-fall harvest. You might also pick one variety for long-term storage and winter sales. Varieties we've used over the last several years include 'Early Jersey Wakefield', 'Storage #4', 'Primax', 'Impala', 'Golden Acre', and 'Farao'. We transplant about 1,500 cabbages each year.

How. We use the same seeding, transplanting, and spacing methods with cabbage as we use with broccoli. And like broccoli, cabbage needs good fertility and plenty of water to form good heads. It also falls victim to the imported cabbage worm and cabbage looper, as the names imply. In our region, spraying with Bt is essential.

Carrots

Why. Carrots are one of those staple crops that almost every vegetable eater wants sooner or later. They are a good late-season storage crop and a good seller. They like a loose, fertile sandy loam but will do okay in any decent soil with enough organic matter to hold onto its moisture.

When. We prefer to seed most of our carrots from late June to mid-July, for harvest from mid-September through December. Many growers plant carrots in early spring — as soon as the ground can be worked — with the expectation of a summer harvest, but early plantings have not worked out so well for us. They take longer to germinate in the cooler soil and generally don't size up as well, or taste as sweet, as those seeded in midsummer. Our favorite carrot varieties are 'Danvers', 'Red Cored Chantenay', and 'Nantes Fancy'. We seed at least 3,500 row feet of carrots each season.

How. Their one drawback is that carrots must be direct-seeded. Because they start out as tiny seedlings and grow slowly for their first few weeks, a fair amount of hands-and-knees weeding can be necessary during their early growth. This has to be factored into the farm plan. Unweeded carrots will not prosper.

We seed carrots with a Planet Junior walk-behind seeder. We put in three rows per bed, 18 inches between rows, and we often do a couple of passes with the seeder, making each row 2 or 3 inches wide. After emergence and during early weeding, we thin the carrots to a spacing of 1 to 2 inches between plants.

Carrots can take anywhere from 1 to 3 weeks to germinate, depending on soil temperature and moisture. The longer they take, the more weeding we need to do. We've found that using drip irrigation lines to dampen the soil over the top of seeded carrots considerably speeds up germination. It also reduces crusting of the soil surface, which can impede germination.

Another trick to minimize weeding, is to go over the seeded rows with a flame weeder a day or two before you expect germination to occur. Here, good timing is essential: If you flame-weed too early, you won't make much of an impression on the weeds; if you do it too late, you'll desiccate the emerging carrots and have to plant again. In midsummer, assuming warm soil and moisture from drip lines, we generally come in with a backpack flame weeder on day 6 and make a quick pass down each row. It's very gratifying (and time-saving) to see the little carrots emerge a day or two later, with no weeds around them at all.

When it comes time to harvest our carrots, we loosen the soil with a garden fork, then simply lift them out. We remove the tops and wash the carrots before selling them.

Celeriac (Also Known as Celery Root)

Why. Celeriac is an interesting late-season storage crop. It forms a bulbous root about the size of a softball and has the texture of a carrot, but with a celery-like flavor. Fedco's seed catalog calls it "the frog prince of root vegetables." I like that description. It did take us a while to educate our customers about how to use this crop, but now that many of them know what to do with it, they look forward to their celeriac each year.

When. Celeriac is a slow grower. We start it in our greenhouse in the first half of March and move it to the field a couple of months later. We don't usually harvest it until mid-September. In our region, this crop holds well in the field until mid- to late November, but then it should be harvested, cleaned, and stored in a root cellar. We grow about 1,200 celeriac each year. We've done well with the varieties 'Mars' and 'Brilliant'.

How. We start celeriac in 128-cell flats and transplant them to the field with a water-wheel transplanter. We put in three rows per bed, 18 inches between rows and 12 inches between plants within the row. We often mulch our celeriac with aged and composted horse bedding from a nearby horse farm. This keeps the soil cooler and helps retain moisture. Celeriac, like celery, definitely needs a continuous supply of moisture, especially during its bulbing stage. The mulch also reduces weed pressure, but over the course of a long summer, a fair amount of hand weeding is always necessary. In the years that we don't mulch this crop, we use wheel hoes and hand hoes to take care of the weeds, and we irrigate more heavily during dry spells.

Italian Dandelion

Why. Italian dandelion, which is really a chicory, grows quickly and easily throughout much of the season. It is a good seller at our New York City markets. When small, Italian dandelion can be eaten raw in a salad with other ingredients; otherwise, it is best steamed or stir-fried with garlic and olive oil. This crop is high in nutrients, but its somewhat bitter flavor, albeit followed by an after-sweetness, may not endear it to all customers. Tell people to give it a try, or maybe a couple of tries. It is, for many, an acquired taste. I started out not caring for it but now am a big fan.

When. We seed Italian dandelion in our greenhouse every 3 weeks from early March until late May, then every week through June. This gives us a continuous harvest from June through October, though production drops off a bit in the summer months. The plant prefers cooler weather and is at its peak in June and again in September and October. Demand is greatest in fall. This plant can handle light frosts but not heavy ones. 'Clio' and 'Catalogna Special' are our favorite varieties. We transplant about three thousand dandelions each season.

How. We seed Italian dandelion in 128-cell flats — two seeds per cell, then thin them down to one after germination. We transplant them 5 or 6 weeks later with our water-wheel transplanter — three rows per bed, 18 inches between rows, and 12 inches between plants. The crop has moderate fertility and water needs. Because it grows rapidly, we can easily control weeds with wheel and hand hoes.

Kale and Other Brassicas

The genus *Brassica* contains many agricultural crops, including kale, collards, mustards, cabbage, cauliflower, arugula, broccoli, broccoli raab, brussels sprouts, turnips, and a variety of Asian greens. These crops are also referred to as cruciferous vegetables. We've already looked at cabbage and broccoli. Now let's consider some of the open-leaf, nonheading brassicas.

Why. Brassicas are relatively easy to grow in cool and even cold weather. They are increasingly in demand for their high nutritional value and interesting flavors. They can reach maturity in the field in 40 to 60 days and withstand temperatures well below freezing. In fact, their flavor improves after frost, and the hardiest of the kales and collard greens keep selling at our markets right into December.

Varieties. We grow six different kales — 'Winterbor', 'Lacinato', 'Rainbow Lacinato', 'Red Russian', 'White Russian', and 'Siberian'. The first two are the most popular, but we normally sell at least 90 percent of each variety we bring to market. We transplant about 9,000 kales each season. We plant 2,000 Champion collard greens. We also do about 1,000 plants each of 'Red Giant' mustard and 'Green Wave' mustard, as well as 600 of mizuna, 1,200 of tatsoi, and 500 of mibuna.

When. The open-leaf brassicas can be grown throughout much of the season in the Northeast, but because of pressure from flea beetles and the plants' tendency to go to seed in hot weather, we prefer to grow most of them in fall. We seed kale and collards in our greenhouse in the first few weeks of June and transplant them in late July and early August. We seed Asian greens — such as mizuna, mispoona, tatsoi, and the mustards — a couple of weeks later and transplant them in mid-August.

How. With the exception of our mesclun greens, we seed our open-leaf brassicas in 128-cell flats. We usually place one or two seeds in each cell, then thin these down to one. These crops are transplanted using our water-wheel transplanter — three rows to a bed, 18 inches between rows, and 12 inches between plants. Occasionally, if we're pressed for space, we'll go down to 12 inches between rows. Because they germinate quickly in warm soil, kales can also be direct-seeded in the field with some success.

All the open-leaf brassicas have moderate water and fertility needs. Weeds are relatively easy to control with wheel hoes and hand hoes, but we've often thought about getting a small cultivating tractor to deal with these fairly large plantings.

Pests. Unless controlled with Bt, cabbage worms and loopers will damage most brassicas. Also, during warm weather, the more delicate brassicas are highly attractive to flea beetles (small, shiny black beetles that jump like fleas). If left unchecked, the beetles will riddle the leaves with tiny holes. These pests are especially attracted to mustards and arugula and to Asian greens such as mizuna and tatsoi. Depending on when the plants are set out in the field, we may use hoops and lightweight row cover to exclude flea beetles.

If using row cover, apply it immediately after planting. If you wait too long, you'll increase the chance that beetles will get trapped underneath, which will only result in worse damage. One could also leave the plants uncovered and spray with Entrust or Monterey Garden Spray.

Lettuces

Why. Lettuces are fast growing and can be planted throughout the season. They are relatively disease and pest free — that is, if you can keep rabbits, woodchucks, and deer away. Almost everyone eats lettuce. A large display of lettuces at our stand always brings in customers. On the downside, lettuces are flimsy — at an outdoor market, when exposed to direct sun or wind, they quickly wilt and become unattractive. Also, lettuces in the field tend to go to seed in hot weather.

Varieties. There are literally hundreds of varieties of lettuce on the market. Some grow faster than others, some grow bigger; some are green, some are red, and some are multicolored; some are crisper, some sweeter; some are easier to pack in crates, some are more flimsy and delicate and don't hold up well at market. We've tried at least 40 or 50 of them and settled on a handful of favorites, but still, we're always on the lookout for new varieties that are more tolerant of summer heat and slower to bolt. Two of our standbys in this respect are 'Red Salad Bowl' (a red oak leaf lettuce) and any of several types of French Crisp lettuce ('Magenta', 'Sierra', and 'Nevada'). 'Jericho' is a moderately heat-tolerant romaine. 'Tin Tin' is another lettuce we like. It's a small, dark-green romaine type with good flavor.

When. We seed lettuces almost every week from early March until early August. In midsummer, the turnaround time (from transplanting to harvest) is short — usually 5 weeks or less. If left growing too long in the summer heat, lettuces will become bitter and go to seed. Lettuces transplanted in August or September take longer to reach a harvestable size, but they can happily remain in the field for a month or more without losing quality.

We seed a lot of lettuces — about 15,000 every year. Not all of them make it to market, though. Some flats don't germinate well in the greenhouse. In the field, some lettuces get eaten by woodchucks or deer; some bolt in the summer heat; some are damaged by excessive rain and develop bottom rot. I estimate that about 60 percent of the lettuces we seed make it to market and are sold.

How. We start lettuces in our greenhouse in 128-cell flats (two seeds per cell, thinned down to one after germination), where they remain for about 5 weeks before being moved to the field. Once they have developed good root systems, lettuces should not be left in their flats for more than another week or two. The cramped quarters of the flat can induce bolting soon after they are moved to the field.

Our lettuces are transplanted with a water-wheel transplanter — four rows to a bed, 12 inches between rows, and 12 inches between plants. They are weeded with wheel and hand hoes and irrigated during dry spells, though their water needs are not as great as one might expect.

Lettuces can be finicky germinators. They need only a light covering of moist soil. Most varieties germinate well enough when the soil temperature is between 70 and 80°F (21 and 27°C). Above 85°F (29°C), germination drops off dramatically. This can be a real problem in mid-summer. To meet these challenges, we seed lettuces when we have a few cooler days in the forecast and place freshly seeded flats in cooler, shaded spots in the greenhouse — such as under the benches or outside under the shade of a tree. Putting a moist, white paper towel on top of a seeded flat also seems to help. When we do this, the lettuce seeds need just the barest covering of soil. The towel must be kept moist at all times and should be removed as soon as germination occurs. This technique works especially well with various varieties of romaine lettuce, which can be particularly difficult to germinate in hot weather.

Mesclun

Why. Mesclun is a washed, ready-to-eat salad mix comprising baby lettuces and a variety of other greens. When we started making our own mix in 1989, mesclun was an uncommon item seldom seen outside a farmers' market, and even then it wasn't easy to find. Today, there's hardly a supermarket in the land that doesn't offer a mass-produced, industrial-grade version of mesclun. Prices have come down, and so has quality.

Growing and preparing high-quality mesclun is demanding and time consuming, and I've periodically thought of dropping it from our crop plan altogether and leaving it in the hands of agribusiness. But there are some good reasons it still has a place on our farm. Mesclun is a value-added product, and it fetches a good price, as long as it is fresher, lasts longer, and is better tasting than the mass-produced versions that sell for considerably less. Relative to its high value, mesclun takes up a small amount of space in the field, on our truck, and at our stand. These last two points are especially relevant when we have a full load and a crowded display.

We take mesclun to market in coolers with cold packs, which keep it fresh and good looking all day long. It continues to sell well when lettuces and other leafy greens in open lugs are sometimes stressed out by sun and wind and therefore less desirable. Mesclun is especially appealing to people on their way home from work who are looking for a diverse salad but don't want to go to the trouble of buying different greens, then washing and mixing them.

A final reason we like mesclun is that we get to wash and prepare it in a cool room in our lower barn in the middle part of the day. It's one job that gives us a welcome break from field work in hot summer weather.

Mesclun ingredients. We try to include seven different categories of greens in our mesclun mix, with each category having a proportional representation and often containing more than one component. Here's what our typical mesclun mix looks like:

   Lettuces (35 percent of mix): 'Red Oak', romaine, lettuce mix*

   Asian greens (20 percent of mix): mizuna, mibuna, tatsoi

   Spicy greens (20 percent of mix): arugula, red mustard, green mustard

   Other brassicas (10 percent of mix): 'Red Russian' kale, 'Lacinato' kale, collards

   Bitter greens (5 percent of mix): Italian dandelion, chicory frisée

   Beet family greens (5 percent of mix): red chard, golden chard, beet tops

   Sour greens (5 percent of mix): French sorrel

*Many seed catalogs offer a mesclun lettuce mix, which contains several different varieties of lettuce.

Depending on availability, a few other wild-crafted greens, which are essentially edible weeds, can find their way into our mix. These include lamb's quarters, purslane, and wood sorrel. Whether wild-crafted or intentionally grown, all mesclun greens should be small, complete leaves. Large leaves or chopped-up leaves do not qualify.

Arugula is an excellent ingredient in mesclun. It is also a stand-alone green that sells very well by itself at our markets. Accordingly, we seed extra arugula for fall harvest and treat it as we would any mesclun ingredient — we cut whole leaves, wash them, dry them in a salad spinner, and pack them in a cooler. We usually sell our arugula in ¼-pound bags.

When. We try to have mesclun at our stand throughout the marketing season, which, for us, is late May to late December. Starting at the end of March, we seed a batch in our greenhouse every couple of weeks until mid-July. We usually transplant each batch into the field 3 or 4 weeks after seeding. Through August and the first half of September, we seed mesclun directly in the field.

In late September, we do one more batch in our greenhouse. This final seeding is transplanted into a high tunnel after tomatoes or basil or some other cold-sensitive crop has been removed. It provides us with something to harvest and sell when most of the field greens are finished.

How. Our early mesclun is seeded in 144- or 162-cell flats — three or four seeds per cell — and left unthinned. We transplant the little clusters of seedlings to the field with the water-wheel transplanter. We use four wheels with 12 inches between rows and 6 inches between holes. We then press additional seedlings between the rows, even though we have made no holes there with the transplanter (our transplanter can handle only four rows). These additional seedlings are hand-watered after the transplanter leaves the field. This way, we end up with seven rows and 6 inches between seedlings on all sides. If everything goes according to plan, as the plants grow in both height and width, we arrive at a solid carpet of mesclun with no room for weeds.

At the beginning of August, our approach to growing mesclun changes. Instead of starting plants in the greenhouse, we sprinkle seeds by hand in freshly tilled, 5-foot-wide beds in the field. We lightly rake soil over them, using either hand rakes or a blind cultivator drawn behind a tractor. Whenever possible, we seed just before rain. If rain doesn't come, we sprinkler-irrigate to hasten germination. It takes a little practice to get the right concentration of seeds per square foot when hand-seeding. Sowing too thickly gives you seedlings that are spindly and deprived of light; too few seeds causes patches of bare soil and openings for weeds.

In general, the drawback to direct-seeding mesclun, or any other crop, is that there are usually more weeds to contend with. This is a bigger problem for us in May, June, and July. In August and September, with shorter days and less weed pressure, we've found that direct-seeding can work very well. And it's definitely a lot faster.

Pests. As noted above in the section on growing kale and other brassicas, the Asian greens, mustard greens, and arugula are popular with flea beetles. These insects are a problem in summer but not in fall. The best, nonlethal defense against them is row cover over hoops.

Onions and Shallots

Why. Organic onions and shallots are a challenge, and you might want to think twice before planting them, or planting a lot of them, at least in your early years. They grow slowly, have thin leaves that compete poorly with weeds, and are susceptible to several diseases.

But organically grown onions and shallots have their pluses, too: They are not easy to find; they taste good; they are in high demand — at least, at our markets; and, perhaps most importantly, they fetch a good price. Most cooking onions store well, so once you've got them, they're a solid saleable asset. Salad onions don't store as well, but even these can last for 2 or 3 months. Shallots have a very long storage life.

Another big plus for us is that the nonhuman inhabitants of our farm (deer, woodchucks, and rabbits) do not include onions in their diet. There is one exception here — a tiny insect called thrips — but these seldom present a significant problem. Because they aren't palatable to our mammalian relatives, we can safely plant onions in the farther reaches of the farm where many of these critters hang out. If we put beans, peas, lettuces, kale, collards, and many other leafy greens in these distant areas, they would get eaten in no time — unless we erect serious fencing.

Varieties. There are plenty of onion varieties to choose from, and you should try a good number of them so you can determine which perform best at your latitude and in your climate zone. Our preferred cooking and storage onions are 'Cortland', 'Copra', and 'Prince'. 'Redwing' is an excellent red onion with a long storage life. 'Ailsa Craig' is a mild salad onion that can get quite large, but it has a short storage life. We also like 'Borrettana Cipollini', an unusual Italian braising and boiling onion that grows into a flattened sphere. For our seed-grown shallots, we've had good results with the varieties 'Ambition' and 'Saffron'.

When. Onions and shallots are the first crops we start in our greenhouse. We seed them in the last half of February and transplant them to the field in late April and early May. Onions are sensitive to day length, or photoperiod. There are two main types grown in North America: long-day onions and short-day onions. Long-day onions are suited to northern regions. Regardless of when you plant them, they will begin bulbing when they receive 15 or more hours of daylight. For us, this is around the time of the summer solstice, after which the days begin to shorten. If you live in southern regions, where the days are shorter, you need to grow short-day onions, which begin bulbing when they receive 11 or 12 hours of daylight. There are some intermediate-day onions.

The more aboveground photosynthetic growth you can get before bulbing occurs, the bigger the final product. This is why we start our onions so early. Depending on the variety, bulbs are ready for harvest 5 or 6 weeks after the solstice but can stay in the field much longer than that.

In addition to starting onions from seed, we also plant onion sets. These tiny onions (about the width of a nickel) are pressed into the soil,

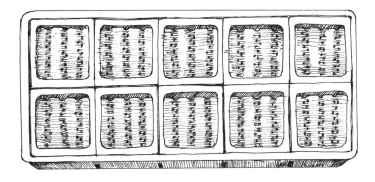

We use a 10-cell flat for starting onions. Seeds are sprinkled in three shallow furrows per cell, about 18 seeds per furrow.

sprout end facing up, as soon as it can be worked, which for us is usually mid- to late April. They grow more rapidly than onions started from seed and are ready for market in June and July. We sell them in bunches of four or five and label them "green or spring" onions. The bulbs are usually small, but both bulbs and tops are edible and quite tasty. The problem with set onions is that they seldom get very large before going to seed, but they are a good early-season item to bring to market, before the larger, more traditional onions come in.

Shallots can also be grown from seed or sets. Set-grown shallots are less likely to go to seed, which makes them more desirable than their onion counterparts. Set-grown shallots usually produce clusters of five or more shallots, whereas those grown from seed usually yield just one large shallot.

How. We seed onions and shallots in 10-cell flats. Each cell is 3½ inches by 4¾ inches by 2¼ inches deep. We place about 55 seeds in each cell, spread across three parallel furrows, ¼ inch deep and 1 inch apart — that's about 18 seeds to a furrow. Once the seedlings reach a height of 3 or 4 inches, we use large scissors or very sharp hedge clippers to trim off the top couple of inches. This is usually done two or three times, at 2-week intervals, to stimulate root development and create a stockier seedling that is easier to transplant.

Using the water-wheel transplanter, we transplant our onions four rows to a bed, 12 inches between rows and 6 inches between plants. We plant about half of them as individuals and the other half in clusters of two or three. The single onions form larger bulbs, but the clusters produce more weight of onion per given area. This wouldn't work if we were growing for a wholesale market, which would want uniformity and large size; however, at a retail farmers' market, the different onion sizes are just fine. They offer our customers more choice.

The set onions are planted at the same spacing as our seed onions and are pressed about ¾ inch into the soil. They grow more rapidly than the seed onions.

As already noted, weed control is a big issue with onions — more so with transplanted onions than with those planted as sets. But even with set onions, it is critical that we eliminate weeds during the first several weeks. If weeds gain the upper hand, we can expect very small bulbs. We use wheel hoes and hand hoes and often resort to hand weeding, usually on hands and knees. Onions are a closely spaced crop, and one has to be careful not to nick them with steel tools. Any damage to the stems or bulbs gives disease organisms an entry point and will often result in mature bulbs with soft or rotten spots.

To beat the weeds, some growers transplant onions into plastic mulch with drip lines underneath. We've done this before and might yet do it again. It is more difficult to transplant the onions this way, and one has to be careful not to let the flimsy seedlings get caught under the plastic in their first week or two in the field. If this happens on a sunny day, they will quickly become overheated and die. Even with plastic mulch, you must do some hand weeding around the developing plants.

Harvesting and curing. Once the bulbs reach a good size, we harvest onions for market, even though the tops may still be green. We also begin harvesting for long-term storage, but before doing this, we break the onion stems. This can be done with the back of a rake or by rolling a round object, such as a 4-foot-wide spool, over the plants. We've found that the core spool from rolls of black plastic works quite well. We simply thread a piece of rope through the spool, and two people walk down the length of a bed of onions, one on each side, dragging the spool behind them. If the spool is too light, we place a piece of pipe or a crowbar inside it to add heft.

To cure onions for storage, the tops must be dried before they are cut off. If the weather is dry, we stack half-full lugs of onions in the field and let the wind blow through them. If rain threatens, we move them into a barn and blow fans on them. Depending on relative humidity, the curing process can take a month or more.

Peas — Edible Pod

Why. We grow sugar snap and snow peas because they come in early when there's not much else around. They taste good, fetch a good price, and fix atmospheric nitrogen, but their season is relatively short, and they require a fair bit of work in terms of trellising and picking. Peas like moist, cool conditions. They are a spring and early-summer crop; a string of hot days will send them into decline.

When. We seed peas once a week for a few weeks, starting at the beginning of April, or as soon as the ground can be worked. When the weather cooperates, we sell them from early June until mid-July. Our total pea planting for the season amounts to just 800 to 1,000 row feet. Our preferred snap varieties are 'Sugarsnap' and 'Sugar Ann'. Our preferred snow peas are 'Oregon Giant' and 'Mammoth Melting Sugar'.

How. We inoculate our peas with rhizobia bacteria and seed them with a walk-behind seeder—two rows per bed, about 40 inches between rows, and 1 inch between seeds within the row. We make sure they have at least ¾ inch of soil on top of them. After the plants emerge (usually a couple of weeks after seeding), we easily control weeds with a wheel hoe. Next comes the hard part: trellising.

Trellising or staking peas makes them more productive and easier to pick. If you're going to grow varieties that reach a height of at least 3 feet, it's definitely worth doing. We've used a few of the different trellis products on the market. All of them require that stakes be set firmly in the ground every 6 feet or so. Some peas can grow to a height of 6 feet or more and become quite heavy; they need substantial support. The trellises and stakes are costly, and this is one of the reasons we seldom grow more than 1,000 row feet of peas a year. Once the plants stop producing, we must dismantle the trellis system and remove the vines.

Peppers

Why. When planted into a plastic mulch, with drip irrigation lines underneath, peppers are a relatively low-maintenance and problem-free crop. They thrive in hot weather. They are attractive and add color and variety to our stand, but they are not a big money maker for us. In New York City, and I suspect in most other parts of North America, people don't buy peppers the way they buy tomatoes.

Varieties. Peppers come in a seductive range of shapes, sizes, and colors, as well as degrees of sweetness and hotness. If you're looking to earn a reputation for peppers, you'll have plenty of varieties to work with. In our modest planting, we've settled for just a few.

Our preferred hot pepper is 'Hot Portugal'. It is open-pollinated, easy to grow, and highly productive. It produces beautiful, elongated fruits that are not superhot but pack enough of a punch for most tastes. We also like to grow at least one type of Italian frying pepper. Lately, we've settled on a hybrid called 'Carmen'. It is a generous producer of substantial, tapered fruits that are great for roasting. 'Carmen' does particularly well in the high tunnel. Other specialty sweet peppers that produce well for us and have good flavor are 'Lipstick' and 'Feherozon'.

When it comes to bell peppers, we've tried quite a few over the years, hoping to find the perfect one. We're always on the lookout for early ripeners. We've had reasonable success with 'King of the North', an open-pollinated variety that performs well in lower temperatures. And lately, we've gone back to 'Sweet Chocolate', which is a good early producer with mild flavor. If allowed to ripen, it turns a stunning chocolate brown.

When. We start peppers in our greenhouse in the second week of March. They are potted-on a few weeks later or as soon as they have adequate root systems. Some are transplanted into a high tunnel in the first or second week of May. The rest are planted in the field at the end of May, after all danger of frost. We are usually harvesting hot peppers by mid-July, then bells and Italian frying peppers by the end of the month.

How. We seed our peppers in 144- or 162-cell flats. We put one or two seeds in each cell, depending on the price of the seed and the germination rate given on the seed packet. The flats are placed on heat mats that are kept to a temperature of about 85°F (29°C). Germination usually occurs within 10 days. Once their roots have occupied the small starting cells, the seedlings are potted-on to 72-cell flats, where they remain until transplant time.

Our first one hundred or so peppers (out of a total of around four hundred) are transplanted into 6-foot-wide weed barriers in one of our high tunnels — three rows, 18 inches between rows, and 18 inches between plants. Peppers like a well-drained soil with ample phosphorus and calcium. We make sure the soil is adequately limed and at planting time give each pepper a tablespoon of rock phosphate and a tablespoon of organic 4-2-4 fertilizer. On cold nights and overcast days, the tunnel peppers are protected with hoops and row cover; however, it's important that peppers not be allowed to get too hot under the row cover. At temperatures above 90°F (32°C), the flowers can abort.

Our field peppers are planted in 4-foot-wide black plastic with one line of drip tape underneath. We plant two rows in a zigzag fashion in each strip of plastic, 18 inches between rows and 18 inches between plants within each row. Every pepper receives a tablespoon each of rock phosphate and organic 4-2-4 fertilizer, just as the tunnel peppers do.

Too much nitrogen in the soil can result in large pepper plants with plenty of foliage but not many fruits. Good fertility all around will give the best yields. Peppers like warmth and can be set back by cold nights, so it's important not to rush them into the field until the weather has settled into a summer pattern. The ideal pepper at transplant time will have developed some buds but not open flowers or fruits.

Peppers like water; it's hard to give them too much. We hand-water our tunnel peppers immediately after planting, then provide drip irrigation as follows: 1½ hours, three times a week, for the first 4 weeks; thereafter, 3 hours three times a week.

Our field peppers are also hand-watered at planting time. They then receive water 2 hours twice a week for 4 weeks, then 3 hours twice a week.

Peppers are shallow-rooted and somewhat fragile plants. They can topple over, especially when laden with fruit. If we see this happening, we drive narrow 4-foot stakes into the ground a few inches away from each pepper and tie each plant to a stake with string. A basket weave system can also be used to support peppers.

Harvesting. The fruits of most peppers start out green (exceptions include white and pale-yellow varieties), then ripen to all manner of colors, depending on the variety in hand. Green peppers are quite edible, especially when cooked, but they are usually not as sweet and full flavored as their riper versions.

Unfortunately, as peppers ripen, they are more prone to developing blemishes and bad spots. These may be caused by the European corn borer, a pest that can be controlled with *Bacillus thuringiensis*. Whatever the cause, it's a good idea to remove blemished fruits early on rather than let them continue to take energy from the plant. We sell a fair number of our bell and frying peppers while they are still green. It's always a good idea to pick some early peppers in the green stage to stimulate more fruit set. We get fewer peppers when we let all of them ripen to full color.

Eagle Street Rooftop Farm

Proprietor: Annie Novak

Location: Brooklyn, New York

Year Started: 2009

Farm Size: 6,000 square feet

Crops: Mixed vegetables and herbs

ANNIE NOVAK WAS BORN AND RAISED in Evanston, Illinois. She came to work at our farm in September 2005 with a degree from Sarah Lawrence College and a strong interest in learning how to grow food and passing that knowledge to others. For a couple of months, Annie helped us harvest and market our fall greens. When the days got cooler and shorter, she planted garlic with the rest of the crew. When the season ended, she moved to New York City and landed a job with the New York Botanical Garden, at the 2-acre vegetable-growing Ruth Rea Howell Family Garden, a position she still holds.

Not one to sit still for long, Annie soon founded Growing Chefs: Food Education from Field to Fork, a group dedicated to teaching eaters of all ages how to grow and prepare food. Today, Growing Chefs boasts among its ranks an assortment of green thumbs, educators, chefs, and nutritionists. The organization writes and teaches curricula throughout the five boroughs of New York City and offers workshops on such topics as composting, bee-keeping, and the culinary arts.

In 2009, Annie joined with Ben Flanner (he has since moved to another urban farming project) to cofound the Eagle Street Rooftop Farm in Greenpoint, Brooklyn. Three stories up, atop a flat-roofed warehouse overlooking the East River and Manhattan skyline, the Eagle Street Rooftop Farm has become an oasis of vegetables and herbs in an aged urban landscape.

The idea of a rooftop farm was first proposed by the building owners, Broadway Stages (a sound stage company) and Goode Green (a New York City–based green roofing company). Broadway Stages funded the installation at a cost of $10 per square foot (the rooftop farm encompasses 6,000 square feet) and remains a partner in the project. Goode Green designed and executed the project. Annie and Ben advised on the growing medium and planted the crops.

Underlying the Eagle Street Rooftop Farm are sturdy steel girders and a "green" roof — a 2-inch polyethelene drainage mat, along with retention and separation fabrics that can absorb and hold 1½ inches of rain water. On top of this base are 200,000 pounds of growing medium/soil comprising largely compost, rock particles, and shale trucked in from Pennsylvania. A crane lifted the growing medium to the building's roof.

The growing area of the rooftop farm is divided into 16 raised beds that range from 30 to 48 inches wide. Each bed has 4 to 7 inches of soil. Access aisles between the beds are lined with a thick layer of bark mulch.

In 2009, the rooftop farm's first year of operation, Annie and Ben planted more than 30 distinct crops. Some did a lot better than others. The next year, Annie trimmed the number down, eliminating the poor performers such as summer and winter squash. She concentrated her efforts on tomatoes, peppers, eggplant, cucumbers, carrots, peas, beans, radishes, kale, chard, lettuces and other salad greens, and a variety of herbs. In 2012, the rooftop focused on hot peppers, which thrive botanically and are particularly appealing to Brooklynites. In addition to the vegetables and herbs, the farm is home to three bee hives and a small flock of chickens.

Work at the Eagle Street Rooftop Farm is done by Annie and an assortment of part-time employees, volunteers, and apprentices. Instead of tractors and rototillers, they use hand tools such as shovels, forks, rakes, trowels, hoes, and wheelbarrows. Seedlings and transplants are watered with New York City tap water and a garden hose. Established plants depend largely on rainfall for their water needs.

The farm makes its own compost using vegetable waste from area restaurants and leftover food scraps delivered by neighbors who often volunteer their labor. The Eagle Street Rooftop Farm sells its produce at an on-site Sunday farmers' market and to several local restaurants. It also offers CSA shares in collaboration with an upstate New York vegetable farm.

The farm is a for-profit venture and it does generate a modest amount of income — Annie estimates a return of between $1 and $2 per square foot of growing area. But there's a lot more to it than that. Every year, thousands of New Yorkers and others, including multiple groups of school children, visit the rooftop farm and get a glimpse, and often some first-hand experience, of what urban farming and food production can be. The Eagle Street Rooftop Farm is a true island of green that furnishes both food for the table and food for thought as it reaches for a healthier planet.

Potatoes

Why. In our early years, we didn't give much thought to the common and lowly spud, preferring to focus our energies on more exotic and expensive fare. Then I noticed other farmers at my market doing a brisk business with potatoes and getting good prices to boot. I decided to give them a try and have never looked back. Now we plant about ¾ acre of potatoes every year. They taste much better than their distant supermarket cousins; they store well; and they are not too difficult to grow, as long as you have a tractor and a set of hilling discs. We nearly always make money on potatoes.

Varieties. There are a lot of different potatoes to choose from. Some take longer to grow and produce larger tubers than others. Catalogs usually divide their offerings into early-season (65–80 days), mid-season (80–90 days), and late-season (over 90 days) varieties. We usually select a few different early- and mid-season types and go for a range of colors — both of skin and flesh. Our favorite staple, white-fleshed potato is 'Kennebec'. It's our best producer.

Yellow-fleshed potatoes, such as 'Yukon Gold' and 'Carola', don't produce as well but excel in flavor, giving them the right to command higher prices. 'Keuka Gold' is a yellow-fleshed variety that yields better than 'Yukon' or 'Carola', though perhaps it does not taste quite as good. We've been happy with it in recent years.

Red-skinned potatoes, such as 'Red Gold' and 'Rose Gold' (these two also have yellow flesh), are popular with our customers. They are early producers but do not store well. Blue-skinned and blue-fleshed potatoes are eye-catching and have an interesting flavor. Our favorite among these is 'Adirondack Blue'.

When. Potatoes grow nicely in cooler weather but need warm soil to get going. We wait until the soil temperature is 60°F (16°C) day and night, which in our part of the world is usually the end of April, or when the dandelions begin to bloom. If we plant while the soil is still cold and damp, the seed pieces will lie dormant in the ground and sometimes rot before they get a chance to emerge.

As soon as our potatoes reach harvestable size (the early-season varieties are ready by mid-July), we use garden forks to dig just enough spuds to meet demand for market the next day. Later in the season, usually by the end of September, we use a tractor and middle buster, sometimes referred to as a potato plow, to bring the tubers to the surface, then we do a fair amount of sifting through the soil on hands and knees. The middle buster is a simple three-point-hitch implement with a single, roughly triangular blade that cuts a wide furrow several inches deep. A tractor-powered potato digger would make the work a lot easier, and we might yet get one. But, because there are so many rocks in our soil, I'm sure we'd need a sturdy model.

How. Our annual ¾-acre planting of potatoes requires about 1,500 pounds of seed stock. To plant potatoes, we cut two furrows about 6 inches deep and 36 inches apart with a tractor-drawn cultivator. We drop the tubers into the furrows by hand, 10 inches apart, and press them into the ground with our feet, then rake soil over them. We want 2 to 3 inches of soil on top of each tuber or seed piece.

Small potatoes, about the size of a kiwi fruit, are planted whole. We prefer these because they are less likely to rot before emerging. Larger tubers are cut into two, or sometimes three, pieces. Each seed piece should have at least two eyes and weigh a minimum of 2 ounces.

Before planting we treat our potato seed pieces with an inoculant called T-22 Planter Box. This liquid drench contains the beneficial fungus *Trichoderma harzianum*. It protects against harmful fungus-borne root diseases, including rhizoctonia, pythium, and fusarium, and reduces the incidence of seed-piece decay. It also helps the growing potato plants access soil nutrients more efficiently. It's an expensive brew, but it's worth it.

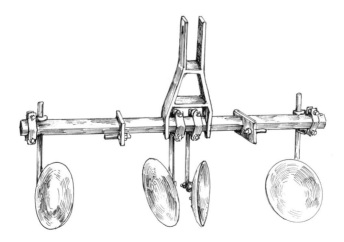

Potato hilling discs

As soon as potatoes reach a height of 6 or 8 inches, they should be hilled, meaning that soil should be placed around the stem of the plant to within a few inches of the top of the leaves. We use a tractor and hilling discs to do this. If you don't have a tractor at your disposal, you'll have to do the job by hand. Hilling keeps the tubers at a safe distance from sunlight, which can cause them to turn green. It also controls weeds and warms the plant roots, promoting more rapid growth. We do a second hilling when the plants reach a height of 12 inches, and sometimes we do a third.

Potatoes are shallow rooted and therefore sensitive to changes in soil moisture. To produce well, they

Green Sprouting

Warming the tubers before planting, then exposing them to light, is called green sprouting. It encourages the development of sturdy, stubby sprouts and promotes more vigorous early growth. For details on green sprouting, see page 420.

need ample water, especially after their flowers bloom and during tuber set and development. Unless we are getting at least 1 inch of rain a week, we irrigate our spuds with drip lines draped over the top of the plants.

Scallions

Why. Scallions are a popular, versatile, and widely used allium. They can be grown for spring, summer, and fall sales. We like scallions because they don't require a lot of work and they sell well at our market.

When. Traditionally, scallions have been a direct-seeded crop, and we used to follow that tradition; now we seed most of our scallions in the greenhouse from early April until mid July, at 3-week intervals, and transplant them to the field 5 or 6 weeks later. If direct-seeding this crop, our first planting occurs at the end of April, or as soon as we are able to create a satisfactory seedbed, and our last planting occurs in late July.

How. Now that we have a water-wheel transplanter, we prefer to start scallions in the greenhouse and use the transplanter to move them to the field as seedlings. We do this because germinating direct-seeded scallions in the field can be slow and spotty, and rescuing the tiny plants from weeds is often time consuming.

We seed our scallions in 162-cell flats, six to eight seeds per flat, and leave them unthinned. Germination occurs most rapidly (within a week) when the soil temperature is between 60 and 85°F (16 and 29°C). The contents of each cell are planted in the field as clusters, four rows to a bed, with 12 inches between rows and 6 inches between clusters within the rows. After planting, a few quick passes with a wheel hoe between the rows and a hand hoe between the clusters are usually enough to control weeds. We plant 300 to 400 row feet of scallions at a time. Our standard variety is 'Evergreen Hardy White'.

At harvest time, we pull out each cluster in its entirety. Then, we usually separate the individual scallions as we dunk the bottom 2 or 3 inches of the roots in water to make them look a little cleaner. We then bunch them with rubber bands.

Summer Squash, Zucchini, and Cucumbers

Why. These heat-loving cucurbits are prolific growers in midsummer. They add color and substance to our stand, but because of cucumber beetles and other cucurbit pests, they are not as easy to grow as they used to be.

Varieties. 'Dark Green' zucchini is our standard green zucchini. 'Soleil' is a good yellow variety. 'Yellow Crookneck' is an open-pollinated summer squash we've been growing for many years. We grow two types of pattypan squash — 'Sunburst' and 'Benning's Green Tint'. 'Marketmore 76' is our preferred cucumber, but we also like 'Poona Kheera'.

When. The summer cucurbits can be direct-seeded in late spring or early summer, once the soil is warm (over 60°F [16°C]) and all danger of frost has passed, but we prefer to start them in our greenhouse between mid-April and mid-May. We move the seedlings to the field in June as soon as they have good root systems, but before they get too tall and leggy (about five inches is a good height). We usually do just two seedings a year, for a total of about five hundred plants.

How. We seed cucurbits in 72-cell flats, placing the seeds ½ inch deep, and keep the potting soil at around 75°F (24°C) until germination. It's always fun to watch the chubby little cotyledons emerge. In the field, we plant squash, cukes, and zukes by hand into 4-foot-wide black plastic — one row down the middle and one line of drip tape underneath. The plants are spaced 15 inches apart, with 6 feet between rows.

Immediately after transplanting, we put 40-inch-high hoops over the plants and cover them with row cover. The row cover creates a warm microenvironment that promotes rapid growth and, more importantly, keeps cucumber beetles, squash bugs, and squash vine borers out. Each of these pests can do serious damage to the young plants, but for us, the striped and spotted cucumber beetles are the worst because they are vectors for a disease called bacterial wilt. As the name suggests, this disease causes the leaves to wilt and the plants to die soon after.

As soon as the first blossoms appear, we remove the row cover so that bees and other insects can pollinate the plants.

To produce well, the cucurbits require good fertility and plenty of water, especially during fruit set. As we set them in the ground, we give each plant a tablespoon of organic 4-2-4 fertilizer and a good shot of water. Starting in their second week, we provide drip irrigation for two hours, twice a week. In the fourth week after transplanting, we increase the quota of water to three hours, twice a week. During warm weather, these crops need to be harvested frequently — every other day is not too often — otherwise, the plants will put most of their energy into producing a few very large fruits. What we're looking for is lots of small fruit; they are more tender and taste better. No one, or almost no one, wants a baseball-bat-size zucchini.

Swiss Chard

Why. We can grow Swiss chard through most of the season, though it prefers the cooler weather of spring and fall. Neither flea beetles nor imported cabbage worms attack it, and it doesn't seem to be popular with woodchucks, rabbits, and deer, though this could change. If harvested

Perpetual Spinach

This member of the chard family has narrower stems and smoother, more tender leaves than regular Swiss chard. It sells well at our stand, but we need to remind our customers that it is a chard and not a spinach. We use the same seeding, spacing, and cultivation methods as for Swiss chard.

2 or 3 inches above soil level and then supplied with water, through rainfall or irrigation, it will grow back for a second and sometimes a third cutting. Though it is not an outstanding seller at our markets, Swiss chard performs well enough to retain a place on our crop list.

Varieties. 'Fordhook Giant' is our standby green chard. We also like 'Golden Chard', though it doesn't grow quite as big.

When. We start Swiss chard in the greenhouse every few weeks from early March until mid-July. We do larger and more frequent seedings in June because as the days shorten and become cooler, this crop really comes into its own. After 5 or 6 weeks in the greenhouse, we move plants to the field.

How. The chards are seeded in 98-cell flats, one or two seeds per cell, and thinned down to one after germination. Chard seeds are large and should be covered with nearly ½ inch of soil. They germinate best between 75 and 85°F (24 and 29°C).

We transplant chard with the water-wheel transplanter, usually with a spacing of three rows per bed, 18 inches between rows, and 12 inches between plants. Sometimes, when going for a quick turnaround time in early summer, we do four rows per bed, with 12 inches between rows, but plants destined to be harvested in fall definitely need the wider row spacing — they are capable of impressive growth, both in height and girth. Timely visits with wheel hoes and hand hoes take care of the weedy competition. Swiss chard can handle light to moderate frosts, and in our region it remains marketable until early November.

Tomatoes

Why. Almost everybody loves a good, local tomato, which is why this fruit is one of the best money makers for small growers — especially those who direct-market their wares. Well-grown, vine-ripened tomatoes are invariably superior to the industrial models and everyone knows this. If only we small growers had a longer season for this crop, but unfortunately, we don't — in most parts of North America the time for local tomatoes is quite short, often no more than 8 or 10 weeks. This is probably why customers are willing to splurge because, let's face it, these days in many places, local tomatoes are not cheap.

Though vulnerable to several diseases, tomatoes are relatively easy to grow, once you know what you're doing. They are highly productive.

When ripened in hot sun, they have a winning flavor, and they sell extremely well. From midsummer into early fall, our stand would feel naked without them.

Varieties for tunnel culture. Seed catalogs will tell you which these are. There are plenty to choose from, and almost all are recently developed hybrids. Traits to look for in a tunnel tomato include earliness (if you're trying to capture the early market); good flavor (always a must); disease resistance (very important in the close environment of a high tunnel); crack resistance; and good, uniform size. We've had reasonable success in our tunnels with 'Jetstar', 'Cobra', and 'Arbason', which are standard red tomatoes in the 6- to 9-ounce range, and 'Apero', which is an elongated cherry. We've heard good things about 'Mountain Spring' (known for its disease resistance) and 'Mountain Fresh', but we haven't tried either of these yet.

Varieties for field tomatoes. Of course, all of the above qualities listed for tunnel tomatoes are good to have in field tomatoes as well, but the varieties are different.

We're always on the lookout for those varieties that will perform best on our farm. Here are some we've had good success with:

- Hybrid Reds: 'New Girl'

- Open-Pollinated Reds: 'Cosmonaut', 'Glacier'

- Hybrid Cherries: 'Sungold', 'Super Sweet 100'

- Heirlooms: 'Brandywine', 'Cherokee Purple', 'Paul Robeson'

- Plums: 'Roma', 'Amish Paste'

When. Because we want to sell tomatoes as long as possible, we seed this crop in our greenhouse on four separate occasions — around March 10, March 25, April 10, and the end of April.

The first group is just three hundred seedlings in number, all of which are transplanted into one of our high tunnels at the end of April or beginning of May. The three subsequent seedings (Groups 2, 3, and 4, which total about eight hundred, eight hundred, and five hundred plants, respectively) are transplanted in widely different locations in the field in mid-May, late May, and mid-June.

Tomato Planting Schedule

Group, Number of Plants	Approx. Seeding Date	Approx. Pot-On Date	Approx. Transplant Date	Location
Group 1, 300	March 10	April 5	May 1	High Tunnel
Group 2, 800	March 25	April 20	May 15	Field
Group 3, 800	April 10	May 5	May 30	Field
Group 4, 500	April 30	May 20	June 15	Field

How. Seeds are sown in 144- or 162-cell flats, one or two seeds per cell. The flats are placed on heat mats with thermostats set to maintain a soil temperature of 75°F (24°C). Germination usually occurs within 7 days. Approximately 3 weeks later, when the seedlings have reached a height of about 3½ inches and have their first true leaves, they are potted-on to larger 72-cell flats, where they remain for another few weeks. The larger cells, fresh potting mix, and additional light help build sturdy and stocky plants with well-developed root systems, which is what we want. Seedlings that are left in tight, crowded quarters grow tall and leggy in their search for light and do not transplant well.

Tunnel Tomatoes

The tunnel tomatoes are transplanted into 6-foot-wide strips of landscape fabric, which acts as a very effective weed barrier. In the tunnels, we prefer this product to black plastic because it can be reused for many years and is a good width for our purposes. A 90-foot strip of fabric is wide enough to take two rows of tomatoes, 4 feet apart, with 20 inches between plants. That's 108 plants per strip. We can fit four strips in our 30-foot-wide tunnel, leaving a 2-foot access path between strips. Usually, we do just six rows of tomatoes. This leaves us with 25 percent of the space for other crops.

Each row of tomatoes has its own dedicated drip line underneath the landscape fabric. For their first few weeks of growing, the tunnel tomatoes are protected with hoops and row cover at night and on overcast days. The goal is to keep them warm without using artificial heat. On sunny days, we promptly remove the row cover so that the plants can be well aerated.

(331)

Trellising. As the tunnel tomatoes grow, they are trellised and trained upward to a single stem. Each plant has its own dedicated line suspended from horizontal lengths of conduit resting on cross braces 8 feet above soil level. We use tomato trellis clips to attach the plants' main stem to the line. Every week we remove all sucker shoots and affix additional trellis clips, as needed, to support the growing plants.

These trellised tomatoes growing in a high tunnel are trained to a single stem.

Tomato trellis clip

Irrigation. Tunnel tomatoes grow rapidly, and some varieties can easily reach a height of 8 feet or more within 3 months. They need a lot of water — more than field tomatoes — because they are in a warmer environment that receives no rainfall. We drip-irrigate our tunnel tomatoes using 8-mil drip tape with 12 inches between emitters and a flow rate of 0.45 gallon per minute.

We irrigate three times a week. In the first 4 weeks after transplanting, the plants receive 1.5 hours of irrigation per session; in the next 3 weeks, they receive 2.5 hours per session; thereafter, they receive 4 hours per session, bringing the total to 12 hours per week. Less water would result in less growth, flowering, and fruit set and might lead to blossom-end rot — an environmental condition caused by uneven or inadequate uptake of water and calcium.

Ventilation. We ventilate our high tunnels by opening the doors at each end and rolling up the sides. In the tunnel's protected and congested environment, humidity levels can quickly get too high, especially when the doors are closed and the sides rolled down at night — measures that many growers take to hold on to warmth. The humidity and lack of air movement during these times create an environment in which leaf mold and other foliar diseases can gain a footing. The risk is greatest in the few hours before dawn. This problem can be ameliorated by installing overhead fans in the tunnel and setting them (with a timer) to run for a few hours before the sun comes up. We haven't yet invested in fans, but we probably will soon enough, given that we've already had a scary run-in with leaf mold in one of our tunnels.

As you would expect, the high-tunnel tomatoes are the first to ripen and go to market. In our southern New York climate, they usually bear harvestable fruit by early July — a few weeks before anything significant comes out of the field. These first fruits capture the early market and fetch a premium price.

Protection from the elements. Unlike their field-grown counterparts, tunnel tomatoes do not have to withstand the ravages of rain, hail, and wind. Each of these, but most often rain, can damage or compromise plants in the field. During periods of heavy or extended rainfall, field tomatoes take in too much water, diluting their flavor and often causing them to split (some varieties, especially heirlooms and thin-skinned cherries, are more vulnerable to splitting than others). It's heartbreaking to see a field of ripe tomatoes rendered largely unsalable by a torrential summer

downpour. This is never a problem in high tunnels, where plants are protected from the unpredictable forces of nature.

Shelter from rain has other benefits, too. Diseases can spread easily when the tomato foliage remains wet for prolonged periods. Tunnel
tomatoes are spared this danger.

Field Tomatoes

As already noted, tomatoes are a lucrative crop, and for this reason we
like to place multiple bets on them. These bets take into account four
different variables: Determinate versus indeterminate varieties, time of
planting, location of planting, and method of planting.

Determinates vs. indeterminates. Determinates are more compact, are
less sprawling, and bear heavily for a shorter period of time. They need
less support from staking or trellising, but they will usually benefit from
it, nonetheless. Most hybrids are determinate.

Indeterminates have a sprawling growth habit. Unless you're looking for an impenetrable jungle of tomatoes, this type definitely needs
staking, trellising, or caging. Indeterminates also benefit from pruning —
that is, removing suckers. They bear and ripen fruit more gradually and
over an extended period of time. Heirlooms, which are always open pol-
linated, and cherries are generally indeterminate.

Seed catalogs always identify their tomatoes as determinate or indeterminate. We plant both types to take advantage of the benefits that
each has to offer.

Time of planting. By allowing at least a couple of weeks to elapse
between tomato plantings, we increase the length of our harvest period.
This is especially relevant with the determinate varieties that concentrate their fruit production into a shorter period of time.

There is also the unfortunate fact that tomatoes are vulnerable to an
assortment of diseases, some of the most common being: early blight;
anthracnose; botrytis gray mold; *Septoria* leaf spot; verticillium wilt;
and the big daddy of them all, late blight. Many of these diseases are
somewhat endemic and, sooner or later, are likely to pay you a visit.
Most conventional tomato growers spray preventively with synthetic
pesticides to keep the nasties at bay. Organic growers can do the same
with approved natural pesticides, but these are much less effective and
often have some negative side effects. A lot of organic growers prefer not
to use the pesticides available to them, except in extreme circumstances.

On the bright side: Many tomato diseases are not the end of the world. They might slow growth, reduce production, and eventually cause plants to wither and die, but along the way, there can still be plenty of good fruit to harvest. Add multiple, consecutive plantings to this strategy and you have a defense, of sorts, against disease.

We've found that as our early plantings go into decline, either because of their determinate nature or the onset of disease, our later plantings usually come in to save the day. The notable exception here is the dreaded late blight — the disease that caused the Irish Potato Famine. If late blight is in the air and you are not willing to spray heavily and preventively, you can pretty much close the book on tomatoes for the season. In 25 years of tomato growing, as already noted, we've had late blight just once. It was a painful experience.

Rotation and location. Most organic growers rotate their tomato plantings, along with other members of the nightshade family (peppers, potatoes, eggplant), on a 4-year schedule to get away from pests and diseases that overwinter in the soil and are troublesome to all of these crops. We do this. We also try to space our four different plantings as far apart as possible. If planting number 2 picks up a mild case of early blight or *Septoria* leaf spot, it might be a few weeks before the disease makes it over to planting number 3 or 4.

As already noted, we plant our first tomatoes in 6-foot-wide strips of weed barrier stapled into the ground in a high tunnel. Our subsequent plantings go into the field in 4-foot-wide strips of black plastic — one row of tomatoes per strip of plastic. Plants are spaced 22 inches apart within the rows.

We like to leave at least 6 feet between rows of tomatoes, to allow for easy access when harvesting. Soon after planting, the bare-soil pathways between the strips of plastic are seeded to white clover. This low-growing clover provides a good, nitrogen-fixing ground cover that tolerates foot traffic and mowing. The more we mow, the more the weeds succumb and the clover takes over.

Trellising. We support all our field tomatoes, be they determinate or indeterminate, using a trellising system known as basket weave or Florida weave. We drive 6- to 7-feet-high stakes at least 1 foot into the ground at the beginning and end of each row and again after every third plant. We loop (or weave) lines of twine and tie them around the stakes to hold the plants in an upright position. This weaving operation is performed at least three or four times — first when the plants are about 15 inches high and again with each additional foot of growth. Some varieties of cherry tomatoes can easily reach a height of 6 or 7 feet. These will benefit from an 8-foot stake, allowing for 1½ feet in the ground.

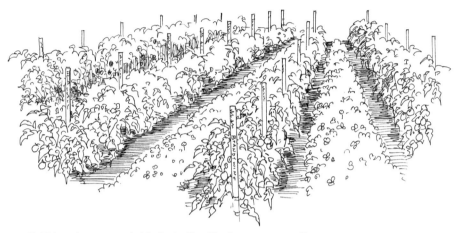

Field tomatoes grown in black plastic with clover access paths

Florida weave (or basket weave) system for supporting tomatoes

Some tomato diseases can survive winter on trellising materials. We throw away the Florida weave twine, but we do reuse the stakes — only after letting them sit for half an hour in a 10-percent bleach solution. We do this between-planting sanitation to avoid infecting the next year's crop.

Irrigation. Each row of tomatoes receives its own drip irrigation line, which is laid under the black plastic. Our pump and well capacity allow us to drip-irrigate about 1,000 row feet at a time. This means we can irrigate ten 100-foot rows or five 200-foot rows, or any combination that does not exceed 1,000 feet. In the first few weeks after they are planted, our field tomatoes receive 1 hour of drip irrigation twice a week. In weeks 3 to 6, they receive 1.5 hours three times a week, and thereafter 3 hours thrice weekly. After periods of heavy rain, we can reduce the amount of drip irrigation we give, but before doing so, we use a soil probe to ascertain how much moisture is present underneath the black plastic — often it's less than we expect.

Harvesting. When you're harvesting a tomato that will be eaten within a few hours, it'll taste best if it's fully ripe, but when you're harvesting for sale the next day, fully ripe can be too ripe. We like to pick tomatoes for market when they have good color but are still firm. We often find that our customers like to buy several tomatoes at different stages of ripeness — one or two ripe ones to eat for lunch or that night, and a few others that are a little less ripe for eating in the days ahead.

Heirloom tomatoes are thin-skinned and do not travel well. If you want them to look okay at market, or in your CSA box, they should be picked well before they are fully ripe. Picking them a little early does not seem to compromise their flavor at all. Thin-skinned cherries, such as 'Sungold' and 'Super Sweet 100', should also be picked before full ripeness; otherwise, some of them will split before your customers get them home. We always pick our cherries directly into pint containers in the field and try to handle them as little as possible.

As already noted, too much water, usually from rain, can cause perfectly beautiful, ripe tomatoes to crack or split. Knowing this, we always try to harvest tomatoes before a significant rain event — sometimes as many as a day or two before. The fruits seem to be especially vulnerable to splitting when rain follows a really hot period.

Turnips and Radishes

Why. Turnips are an easy-to-grow fall crop. They don't require much work, and they make a colorful display. Radishes are a quick-turnaround crop that produces a harvestable yield in less than a month. Both are moderate sellers at our stand.

Varieties. We do well with a fresh-eating, fast-growing, short-term-storage Japanese turnip called 'Hakurei'. More traditional varieties, such as 'Purple Top White Globe' and 'Golden Ball', which have a longer storage life, are also on our crop list. These cooking turnips fill out our stand in late fall, when many other crops are getting scarce. They are modest but not great sellers. We like the radish varieties 'Purple Plum', 'White Icicle', and 'Watermellon'.

When. Turnips and radishes germinate rapidly in warm soil — anywhere from 60 to 85°F (16–29°C) will suit them just fine — and they grow quickly. Both crops can be grown throughout the season, starting in early spring, but we've found they are happiest, taste best, and sell best in the cool days of fall. We harvest weekly, for each of our markets, from late September until late November.

How. We direct-seed turnips and radishes with a Planet Junior walk-behind seeder — four rows to a bed, 12 inches between rows. Often, we do a couple of passes with the seeder, making the rows 2 to 3 inches wide. After germination, if necessary, we thin the 'Hakurei' turnips to about 1½ inches between plants. They can grow quite large but are at their peak of crispness and sweetness when no more than 2½ inches in diameter. The 'Purple Top White Globe' and 'Golden Ball' turnips require more space — they are thinned to 2½ or 3 inches between plants and harvested when they reach a diameter of 3 to 4 inches.

Radishes don't need much more than 1 square inch of space to attain optimum size. They are best when grown rapidly and harvested promptly. Left in the ground too long, they will crack, lose their crispness, and become corky.

One of the nice things about growing turnips and radishes in late summer and fall is that weeds don't usually pose much of a problem. A couple of quick passes with a wheel hoe and one or two light hand weedings while the plants are still small, should be enough. As the tops grow larger, any remaining weeds are shaded out. Turnip greens are edible and nutritious — and most appealing — while in their tender green stage.

In warm summer weather, flea beetles can damage the young and tender tops of both turnips and radishes. Floating row cover will keep them out.

Winter Squash and Pumpkins

Why. Like summer squash and zucchinis, these more durable members of the cucurbit family can be highly productive and relatively low maintenance once they are established in the field. When harvested at the right time and stored in a cool and dry environment, they can last up to 3 months. In fall, they provide theme, color, and substance to any display.

Pumpkins sell well around Halloween and Thanksgiving but not at other times. Winter squash sell moderately throughout fall and peak just before Thanksgiving. One word of caution: These crops are heavy, and they don't fetch a very high per-pound price. If you'd rather not be hauling around 50-pound crates, you might want to pass on growing them.

Varieties. 'Waltham Butternut' is our principal winter squash. It's a good producer and a good seller. We grow lesser quantities of 'Ebony Acorn', 'Delicata', and 'Sweet Dumpling'. 'Delicata' and 'Sweet Dumpling' are very good-tasting squash, but they are more temperamental and don't yield as well. 'Ebony Acorn' is a fair producer, but it's not at the top of everybody's squash list.

For pumpkins, we like 'New England Pie' and 'Baby Pam'. Both produce smallish, uniform fruits that are good for pies. 'Jack-Be-Little' is a miniature ribbed pumpkin that is very cute and also good for eating. We don't bother with large carving pumpkins.

When. We seed winter squash and pumpkins in our greenhouse in mid-May and move them to the field 3 or 4 weeks later. The plants grow rapidly and should not be allowed to become too leggy before transplanting.

How. We seed our squash and pumpkins in 98-cell flats (one or two seeds per cell) placed on heat mats set to maintain the potting soil at 75°F (24°C). Once the seedlings are 4 or 5 inches high, they are set in the field with our water-wheel transplanter, in rows 6 feet apart. Compact bush varieties and small-fruited squash, such as acorns, are spaced 18 inches apart within the row; sprawling, large-fruited varieties, such as butternuts and pumpkins, are spaced 36 inches apart. We give each seedling plenty of water at transplant time.

These crops can also be directly seeded in the field using the same spacing between rows but much closer spacing (6 to 8 inches) between seeds within the rows. After germination, compact bush and small-fruit varieties should be thinned to 18 inches apart within the row and sprawling, large-fruited varieties should be thinned to 36 inches apart.

A couple of weeks after transplanting (or germination of direct-seeded squash), we use hand hoes and wheel hoes to control weeds — both between plants and in a 12-inch-wide strip on either side of them. A week or two later, a tractor and cultivator are used to knock out weeds in the larger space between the rows. We repeat this tractor cultivation at least once.

Immediately after the last cultivation, and before a significant rain event, we broadcast New Zealand White clover over the top of the squash plants. With help from the rain, the tiny clover seeds find their way into openings in the soil. If all goes well, by the time the squash are harvested, there's a nice cover of clover surrounding them. For more detail on over-seeding clover into existing cash crops, see page 251.

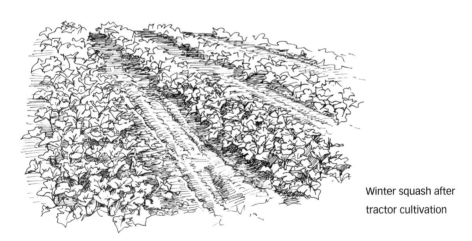

Winter squash after
tractor cultivation

A final point. After 25 years of farming, we've developed systems and methods that work for us and that we are more or less comfortable with, but this certainly doesn't mean we've smoothed out all the kinks or that our days of experimentation are over. Each year, we try new varieties and new methods. There's always more to learn and always room for improvement. It might be hard to teach an old dog a new trick, but I'd like to think that an old farmer is different.

Chapter 14

Garlic — Our Signature Crop

A SIGNATURE CROP IS A SPECIAL CROP — one that people come to associate with your farm over and above the other vegetables and herbs you offer. It should be a crop you are able to grow successfully and in sufficient quantity to meet ongoing demand. It should be a crop you enjoy growing and are willing to give priority to, whenever necessary. And preferably, it should be a crop with a decent profit margin. A signature crop might be a staple vegetable that does particularly well on your farm and that you have chosen to specialize in. It could also be a family of vegetables, such as brassicas, alliums, or nightshades; or an unusual, hard-to-find heirloom variety. There are many possibilities.

A signature crop is one that you will want to promote aggressively using any means at your disposal, such as creative signs, special offerings, free samples, recipes, and prime display space. If a newspaper or TV reporter comes along and asks what you grow, the signature crop should be the first one you mention, and the last one, too. A signature crop will help create an identity for your farm.

Farm identity is what differentiates you from the competition. It doesn't necessarily mean that you are better but, rather, that you are in some way different, in some way memorable; that your farm has its own personality. For many people, buying directly from small farmers or producers is a meaningful dimension of the eating experience. It involves ongoing, personal interaction between growers and eaters. It involves trust and accountability. And it involves wholesome, fresh food. These and other factors come together to shape your farm's identity. They are also areas in which you should be able to outcompete corporate agribusiness every day of the week.

Agribusiness' identity is expressed through TV commercials and glossy packaging, which bear little semblance to reality and which many people have grown suspicious and weary of. Your task as a small farmer is to provide a real alternative — namely, fresh, healthy food with a human face on it. Give your customers a positive, personal experience, as well as a good product, and they will stand by you through thick and thin.

You can, and should, express identity in multiple ways: through the name you give your farm; the ambience you create at your stand or CSA pickup point; the quality and range of the produce you provide; your signs; your wit; the way you treat your customers. All of these create an impression, hopefully favorable, of who you are and what you do. They make up your identity. Another good way to make your farm stand out is to cultivate (in all senses of the word) a signature crop.

Our signature crop is *Allium sativum*, a.k.a. garlic. Garlic didn't always occupy this lofty position. In the beginning, it might have been tomatoes, or lettuces, or potatoes, or mesclun salad mix. All of these were early contenders, but then the winds of fortune carried us down the garlic path, and I have to say, it's been a good path to be on.

Our Garlic Story

I planted some garlic of unknown origin in my second year of farming, with disappointing results — the bulbs were diminutive and had many small, overlapping cloves that were hard to peel. It didn't help that the flavor was excessively sharp and somewhat bitter. My customers bought the small amount I had to offer but didn't ask for more.

At that time — it was the late 1980s — most of the garlic consumed in the United States came from California. In the Northeast, the plant was seldom grown commercially. Today, the picture is very different. China is now the principal supplier of supermarket garlic in America, and the California garlic industry has gone into severe decline. It boiled down to one thing — price. The imported Chinese garlic was a lot cheaper, but that's not the end of the story. While this was happening, and even before the Chinese takeover, a countertrend was already under way: Lovers of *Allium sativum* had been noticing that local and organic garlic, when they could get it, tasted a lot better than the mass-produced stuff from both California and China. Before long, the secret was out, and demand for high-quality, local garlic began to climb. It has climbed ever since. Today, garlic is a hot crop for many small, organic growers in the Northeast and other parts of the country, but back in the late 1980s, garlic seemed to be a poor bet. I was ready to drop it from my crop plan altogether, but then something happened that changed the picture entirely.

On one of her landscape-painting excursions, my wife, who is of Italian stock, was befriended by an older Italian-American man by the name of Andy Burigo. Andy, who lived about a mile from us, had been an accomplished gardener for many years. His most prized crop was an heirloom hardneck garlic that a friend had brought over from Calabria, Italy, and given to Andy some decades earlier. During one of my wife's several visits to paint a picturesque old barn on his property, Andy learned that I was a novice organic farmer and presented her with a brown paper bag containing some 25 or 30 bulbs of his Calabrian garlic. "Tell your husband to plant these," he said. And she did. Moved by our neighbor's generosity and at the same time intrigued, I decided to give garlic another try.

In late October, I divided the bulbs into their constituent cloves (there were about seven in each bulb) and pressed them into freshly tilled ground. I covered them with a couple of inches of soil and a similar amount of straw mulch. Soon winter arrived, and I thought no more about garlic.

In late March the following year, I was pleasantly surprised to see a couple hundred perky little spear-like shoots poking through the mulch. They grew rapidly and within a few weeks had transformed themselves into vigorous, dark-green leafy plants several inches tall. I pulled the few weeds that came up through the mulch and provided water when I thought it necessary. By mid-June, when the plants had reached a height of about 20 inches, they developed a flower stalk that I soon learned is a standard feature of hardneck garlic and is known as a scape. Following instructions from Andy, I removed all the scapes.

About a month later, I dug up my first bulb. It was a stunning sight: large, vibrant, well formed, and streaked with a beautiful reddish-purple hue. It made a strong impression. I brushed off the dirt and stuffed the entire plant, leaves and all, into my pocket, feeling quite pleased with myself. At the end of the day, I presented it to my wife, who was preparing dinner. She pried loose a fat clove, peeled it, sliced it into several pieces, and threw them into the frying pan with some zucchini and Swiss chard, also fresh from the field. A few minutes later, after a quick taste test, she turned to me and said, "This is good garlic." And I seem to remember she then added, "and you've got me to thank for it." She was right.

At my next farmers' market, I put out some 30 bulbs for sale at $1.00 each, which, at the time, seemed to me a fairly high price. It took most of the day, but every one of them sold. The following week, I put out another 30. They were gone within 2 hours, and for the rest of the day, customers kept showing up, asking for the garlic they had bought the week before. That's when I knew I was on to a good thing. I determined right then not to sell any more of Andy Burigo's garlic but instead, to save all I had left to plant the next year.

In fall, the hundred-odd remaining bulbs were divided into about seven hundred cloves and carefully planted in well-rested ground. The following summer's harvest did not disappoint. We sold a few hundred bulbs, again to rave reviews, and set aside the rest for planting. Now, more than two decades later, garlic has become our major crop, and its compelling growth cycle is repeated every year. The only difference is that the numbers have increased exponentially. Today, instead of setting aside 100 bulbs from the summer harvest for our fall planting, we set aside 10,000 — all of them descendents of the original 25 or 30 bulbs given to us by Andy Burigo. I know this sounds like a tremendous increase, but if you do the arithmetic, you'll see it can be done surprisingly quickly (see box on page 345).

Our farm's reputation for garlic was greatly enhanced by some good press that started a few years after our first bulbs were sold. At that point in time, there was almost no other high-quality garlic in the markets I attended. This worked to our advantage. The first positive review appeared in *New York* magazine and included a stunning photograph of my wife and me sitting in front of some very large bulbs of curing garlic. We were delighted. A couple of years later, we ended up on a Food Network TV show called *Follow That Food,* which was aired repeatedly across the nation. Then came the big score: *The New York Times* ran a half-page photo and description of our garlic on the front page of its Food Section. This really ramped up the stakes, especially in New York City, which is the heart of our market. Droves of people I had never seen before started showing up at our stand asking for garlic, and many of them bought other vegetables as well. Our reputation was sealed. All we had to do then was maintain the high quality of our garlic and grow a lot more of it. Of course, we made good use of the positive publicity — we copied, enlarged, and laminated the best of it and conspicuously posted it in front of our stand.

The 10,000 bulbs we now set aside for planting are divided into about 65,000 cloves, each of which gives us a plant the next year (the very tiniest cloves are usually eaten or discarded). We've kept this number steady for several years because to grow more might severely tax our resources.

Remember, for every one thousand cloves planted, there will follow a lot of mulching, weeding, irrigating, harvesting, and curing. Some of these tasks, most notably harvesting, must be performed in a short time-frame — usually 2 to 3 weeks. In midsummer, we must plant, tend, and bring to market numerous other crops each week. We can't just drop everything and put all of our energy into harvesting garlic. Another few thousand bulbs might jeopardize other important work and could contain within them the clove that breaks the camel's back, so to speak.

Our signature crop deserves special attention, but unless we decide to move in the direction of monoculture, we can't allow it to consume us entirely. As with so many aspects of organic farming, the important thing is to find the right balance, taking into account the entire farm system, not just one part of it — however important that part might be.

Save Your Own Garlic for Planting

Once you've found a variety of garlic that performs well for you and that your customers like, resist the temptation to sell all of it. Always save enough for the following year's crop. It can be expensive and risky to import new garlic onto your farm, especially from unknown sources. The most recent and glaring example of such a risk occurred in 2010, when many U.S. growers were grievously harmed by garlic from Canada, which was infected with a microscopic worm called the stem and bloat nematode. Another good reason to keep a closed loop when it comes to planting stock is that garlic tends to adapt, over time, to the locale in which it is grown – specifically, the soil and the microclimate.

It doesn't take long to build up a good supply of your own planting stock. Starting with just one bulb with seven cloves, and assuming you have enough resolve to neither eat nor sell any of it but only replant what you grow, you can end up with over 100,000 bulbs in 5 years ($7 \times 7 \times 7 \times 7 \times 7 \times 7 = 117,649$).

Depending on the size of the cloves you are planting, 1,000 to 1,500 pounds of seed stock should be enough to plant a full acre.

A Step-by-Step Guide to
Growing Hardneck Garlic

In case you, too, are bitten by the garlic bug, here's a step-by-step account of how we grow this most admirable plant on our farm. I believe that the methods I'm about to describe will work reasonably well for garlic growers throughout much of North America, especially those who are planting hardneck garlic. The timing of the different steps, however, might vary significantly depending on the region in which the garlic is grown and, to a lesser extent, the prevailing weather patterns. So it would be more correct to say that what follows is a step-by-step guide to growing hardneck garlic in the Northeast and Central United States.

Step 1. Decide When to Plant

Determine the optimum time to plant for your hardiness zone. In our southern New York climate (Zone 6), the window for planting garlic is roughly October 20 to November 10. We want the cloves to have enough time to develop roots before the ground freezes, but not enough time to send up top growth. Roots will help hold the cloves in place and reduce the chance of frost heave, which happens when the ground repeatedly freezes and thaws. Existing roots will also give the clove a head start in spring as soon as conditions are right for aboveground photosynthetic growth. Any shoots that come up in fall will most likely die back when nighttime temperatures drop into the single digits, rendering such growth a wasteful expenditure of the clove's stored energy; therefore, it's better not to plant too early. Depending on where you live, the optimum window for planting may differ from ours by up to a few weeks on either side.

If for some reason you are unable to plant your garlic in fall, spring is the next best option. Plant as soon as the ground is thawed and you can get onto it. Assuming your planting stock is in good condition, you can still grow garlic that is respectable, though probably not quite as big.

Step 2. Prepare the Ground

At least several months before pushing the first clove into the ground, field preparation should be on your mind. Garlic always does best in a well-rested, fertile soil that has plenty of organic matter. Because you won't be planting until fall, you have spring and summer to beef up organic matter content. You can do this by growing a high-biomass-producing cover crop such as sorghum sudangrass; Japanese millet; or, for shorter periods of time, buckwheat. We've used all of these. We've also preceded our garlic with hairy vetch and a few different clovers. If you can leave the land fallow for more than one season, a leguminous cover crop, such as clover or vetch, is a good choice because it will give you both organic matter and nitrogen. For a lot more information on cover crops, see chapter 10.

Step 3. Divide Bulbs into Cloves

Divide your planting bulbs into cloves a week or two before you intend to plant them. This is called "clove popping." Depending on the amount of garlic you are going to plant, it can be a time-consuming job.

Line up several baskets and roughly separate the cloves into three or four categories, depending on their size. We do large, medium, and tiny. We also have a basket for damaged or diseased cloves, which are neither planted nor composted — we usually burn them. Be careful not to tear into the flesh of the cloves while separating them, and try to leave the protective skins intact. Any wound is a possible entry point for bacterial or fungal organisms.

Step 4. Plant

Plant cloves by hand in shallow furrows. The pointed end should always be facing up and the flatter end (from which the roots will emerge) should face down. Push the cloves into the ground so that when they are covered, there will be about 2 inches of soil on top of them.

There is no hard-and-fast rule regarding spacing. We plant our garlic three rows per bed, with 18 inches between each row. A tractor and a cultivator with three tines are used to define the rows and create furrows. Spacing within the rows depends on clove size: The largest cloves, which usually give us the biggest bulbs, are spaced 6 inches apart. Our medium- and smaller-size cloves, which constitute about 75 percent of the total, are spaced 5 inches apart.

The very small, or tiny, cloves are given just 3 inches. These smallest cloves are usually not allowed to grow into mature bulbs; instead, they are harvested throughout June and early July, bunched in threes, and sold as green garlic. The entire plant, minus the roots and outer skin, may be chopped up and stir-fried or steamed. The flavor is mild, fresh, and a

Tractor and three-tine cultivator for cutting parallel furrows for planting garlic

Planting garlic using 5-inch spacer sticks

nice foretaste of the mature garlic that is to come. Selling this immature green garlic at our early markets establishes us as serious purveyors of the "stinking rose." It also boosts early season cash flow. As befits a signature crop, we want to have garlic at our stand, in one form or another, from the first market we attend until the last. Using the small cloves for green garlic also eliminates them from the planting stock, leaving us with larger cloves that will produce larger bulbs in the next generation.

While planting, we use spacer sticks to make sure our spacing is consistent and accurate. These simple sticks are usually 5 feet long and have bold marks every 3, 5, or 6 inches, depending on the size of the cloves we are pushing into the ground.

Step 5. Mulch

After you rake soil over the tops of the cloves, it's a good idea to put down 2 or 3 inches of mulch. We use aged horse bedding from a neighboring farm that boards a couple dozen horses. Straw, wood shavings, or some other light organic material would also do the job.

Mulching provides four benefits:

1. It greatly reduces the possibility of frost heave, which can literally push cloves out of the soil. This is most likely to occur in winters when there is little or no snow cover and the upper few inches of the soil repeatedly freeze and thaw.

2. A layer of mulch reduces early season weed pressure. Any weeds that do make it through the mulch should be pulled by hand. If you go to work with a hoe, you'll disturb the mulch, and this will make it easier for weed seeds near the surface to germinate, but by June and July, some of the mulch will have decomposed, and there might be enough weeds coming through that it's necessary to do a good hoeing, which means disturbing the mulch.

3. Mulch helps retain moisture. This is especially important if you don't have adequate irrigation, or if you do have access to water for irrigation but don't get around to using it. After poor fertility, lack of water during the crucial growing period is usually the cause of undersized bulbs.

4. It breaks down and contributes organic matter to the soil after it has served its purpose as a weed barrier.

Step 6. Weed

As I have already mentioned, none of the alliums (onions, garlic, shallots, scallions, leeks) compete well with weeds. For one thing, they are slow growers; for another, they have slender leaves that allow plenty of sunlight to reach the soil — many weeds germinate when light reaches them. Bottom line: If you're going to grow garlic, be prepared for plenty of weeding. If you let weeds gain the upper hand, they will rob your garlic of water, nutrients, and sunlight, and the end product will be disappointing.

Once the mulch around our garlic has outlived its usefulness as a weed barrier, we use stirrup wheel hoes or hand hoes to control the weeds between the rows. It's more challenging to remove weeds between the plants because you risk nicking or otherwise damaging the developing bulbs. You can use a very small hand hoe with caution. A safer approach is to simply hand-pull any in-row weeds.

If you are growing a fairly large amount of garlic (let's say, more than 0.5 acre), you might consider investing in a spring-tine weeder, also known as a blind cultivator. These implements, which must be pulled behind a tractor, have numerous, closely positioned tines that are a bit thicker than a wire coat hanger. They can be set to different tensions, then dragged right over your crop. They do a good job of uprooting very small weeds but will do virtually no harm to garlic, which is a sturdy crop, even though they will drag vigorously over the tops of the plants. We've had one of these devices (a Lely Weeder) for at least 15 years and find it very effective on garlic. The trick is to use it early, when the weeds are barely visible — not more than 1 or 2 inches high. At this stage they can be easily dislodged and left to dry out in the sun. Lely and Steinbock are two well-known manufacturers of spring-tine weeders (see pages 107–109 for more on these weeders).

Step 7. Irrigate

Garlic needs water most during its active growing phase. In southern New York, this takes place from early spring, when the buried cloves send their first shoots aboveground, until the longest day of the year — the summer solstice — which occurs in the Northern Hemisphere on either June 20 or June 21, depending on shifts in the calendar.

During this period, we make sure our garlic receives at least 1 inch of water per week. In early spring, there's usually enough moisture in the soil (from snowmelt and spring rain) to meet the need, but when May and June roll around, garlic growers should closely monitor rainfall. If the crop is getting less than 1 inch of rain a week (averaged over a 2-week period), you must irrigate if you want bulbs of a respectable size. On our farm, when rainfall is inadequate, we use drip irrigation to make up the deficit.

After the summer solstice, our garlic seems to notice the gradual decrease in day length. This is a signal to the plant to slow down its aboveground growth and begin the process of replication. The energy the leaves have captured through photosynthesis is directed underground to form the new bulb. This process takes 3 to 4 weeks, and during this period, the plant's water needs diminish. The equivalent of ½ inch of rain a week in the last week of June and the first week of July is sufficient; thereafter, we withhold irrigation and hope we don't get a lot of rain. Too much water during the bulb-formation stage can cause cloves to swell and burst out of their skins, which reduces storage life.

Step 8. Remove Scapes

On its way to maturity, hardneck garlic produces a type of seed stalk known as a scape. If left on the plant and allowed to develop, the scape will gain substantial height and undergo some impressive spiraling contortions — at first, curling up on itself and later straightening out and forming a cluster of seeds at its top, known as bulbils. These bulbils (the number and size will vary depending on the variety of hardneck garlic) are not true seeds, because garlic does not engage in sexual reproduction via cross-pollination, as most other plants do. Instead, the bulbils are an early clonal representation of the bulb that is forming below ground.

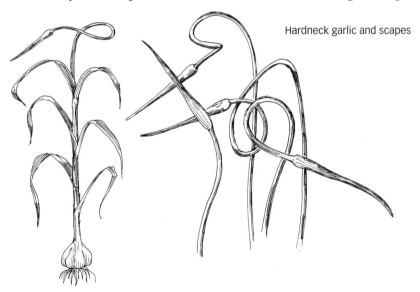

Hardneck garlic and scapes

Most garlic growers cut off the scapes early in their development, believing that this will result in a larger bulb, but there is some difference of opinion on this matter. Some growers prefer to leave the scapes intact until harvest, thinking that the bulbs will store better if allowed to complete their natural cycle.

We remove most of the scapes on our garlic over a 2- to 3-week period, usually starting in early or mid-June. We sell them in bunches of five at our farmers' market. When small, the scapes have a crunchy, mild garlic taste. As they grow larger, the stems of the scapes toughen and become fibrous, but the bulbils at the top grow bigger and develop a strong garlic flavor and texture. If allowed to reach full maturity, the bulbils can be saved and planted in fall or the following spring — they store remarkably well. The only problem with planting bulbils is that they produce very small bulbs — up to about the size of a cocktail onion. When you consider it takes about the same amount of time to plant, irrigate, weed, and harvest bulbils as it does cloves, it makes better sense to plant cloves, if you have them. You'll arrive at a much larger final product.

Step 9. Harvest

After removing scapes, exactly when to dig or pull the crop is a question on every garlic grower's mind. The correct time will vary depending on the strain of garlic you are growing and the region and latitude in which you are growing it. Harvest too soon, and your bulbs will not have reached their optimum size. Harvest too late, and they will get too big for their own skins and start to break apart, resulting in a shorter shelf life. One approach is to count the hours of sunlight in a day. When day length exceeds 13 hours, garlic stops growing aboveground and directs its energy toward bulb development. The bulbs should develop over the subsequent 3 or 4 weeks.

Our guideline for harvesting is quite simple: When about half of the plants' leaves have turned brown, or a brownish yellow, and the other half are still green, we set out to pull our garlic. In southern New York, this is usually mid-July; however, I've noticed that over the last 10 years our garlic seems to be ready a little sooner each season — perhaps courtesy of global warming. But, again, the prevailing climatic conditions are highly relevant. I know someone who is growing our strain of garlic in Maryland and harvesting it almost a month earlier than we do.

Around garlic harvesting time, I closely watch the weather. For one thing, we don't want to spend a lot of time harvesting in the rain. But more importantly, if we pull garlic from wet soil, it takes longer to dry and cure and is more likely to be afflicted with fungal or bacterial diseases. For this reason, when there's a high chance of rain in the long-term forecast, we often start pulling garlic a little earlier than we would otherwise — preferring to bring in bulbs that are a little smaller but dry rather than wet.

Another factor is the large amount of garlic we grow (large, at least, for a small, diversified farm). We need to take into account the amount of time it will take us to complete the harvest — usually about 2 weeks. This means that we are not able to bring in all of our garlic at the optimum time, which might be a window of only a few days. Accordingly, we usually start harvesting about 5 days earlier than is optimal, even when there's no rain in the forecast.

Harvesting with a fork. If your cloves were not planted too deeply, and your soil is friable and not too dry, garlic can usually be pulled by hand, though at day's end your back will feel like it has done a lot of hard work. It will be easier to pull if you use a garden fork to loosen the soil around the bulbs, but forking is an extra step that also requires a fair expenditure of physical energy and carries with it some risk of piercing bulbs.

Harvesting with a cultivator. When our soil is too dry for easy pulling, instead of using a fork, we employ a tractor and a simple cultivator with six narrow tines — two for each of our three rows of garlic. Each set of tines straddles a row, staying 4 or 5 inches away from the bulbs and cutting at least 5 inches deep.

The cultivator makes quick work of loosening the soil on either side of the plants, which can then be lifted without much effort. But a word of warning: If your tractor strays off course, even a little, the cultivator tines can easily damage a lot of bulbs. The best approach is to drive slowly and cautiously and to stabilize the cultivator so that it cannot move abruptly to one side or the other when it encounters an obstacle such as a rock.

After it is uprooted, garlic should be brought into a barn, an open shed, or some other protected space for curing. It's important to get the bulbs out of hot, direct sun within minutes of removing them from the ground. They've experienced a fairly stable soil temperature for the previous few months and can be seriously damaged by being suddenly exposed to searing heat. You can avoid the danger of sunburn by standing the plants in crates immediately after pulling, so that the leaves provide shade for the bulbs. Or, for a short period of time, bulbs can be laid on the ground in windrows, using the leaves of one clump of plants to cover the bulbs of another.

Step 10. Cure and Remove Tops

Garlic can be taken straight from the field to the kitchen and enjoyed, with no intermediate steps. It can also be sold directly as freshly dug garlic. It will be crisp and juicy and will have a delightful fresh flavor. Your customers will love it. But if you plan on selling (and eating) garlic 3, 6, or even 9 months into the future, you must cure it. Through curing, the bulb's outer layers of skin and its stems are dried so the flesh inside is protected and less vulnerable to being attacked by fungal and bacterial organisms.

The easiest way to cure fresh garlic is to hang the plants, in their entirety, in a dry, airy place, out of direct sun. We wrap lengths of bailing twine around sheaves of 10 to 12 bulbs (even 15, if they are small)

Garlic curing in an open pole barn

and hang them from the rafters of every available shed and pole barn on the farm. If the garlic has come out of dry ground, a month of curing might be enough. If it was harvested wet, the bulbs might need to hang for 2 months. Using fans, especially large, commercial barn fans, will speed up the process.

We judge our garlic to be well cured when the leaves are completely dry and the stems also are dry and almost woody. At this point, to conserve space, we un-hang the sheaves, separate the plants, and clip off the tops, leaving 2 to 3 inches of stem on each bulb. The bulbs are then placed in mesh bags or crates.

Step 11. Grade and Sell

We sell all of our garlic by the unit — mostly by the piece, but sometimes by the bag, bunch, or braid. Never by the pound. The main reason for this is that a *per-pound* price might seem too high. Depending on size, most of our garlic bulbs sell for anywhere from $1.00 to $1.75 apiece. Occasional monsters go for $2.00. As far as I'm concerned, this is not unreasonable. In the kitchen, a large bulb of garlic can go a long way. It is a plant packed with flavor and health benefits. And organically grown garlic is a labor-intensive crop. The price we charge is what we need to make a decent profit and stay in business.

The problem arises when a person walks into a supermarket and sees imported Chinese garlic selling by the pound for roughly half the price. Suddenly, our stuff looks like a bit of a rip-off. That's the global marketplace for you: cheap labor, industrial-scale production, and questionable soil health and growing conditions. We can't compete with that model. All we can do is offer a superior product and use a different pricing system. Fortunately for us, a single bulb of garlic selling for anywhere from $1.00 to $2.00 seems like a reasonable buy. It's a matter of perception. And the principle applies to many other crops as well.

The garlic sizer. In his excellent book *Growing Great Garlic*, Ron Engeland describes a very simple method of grading garlic according to size. He uses a piece of wood (a 15-inch piece of 2×6 will work fine) with several pairs of nails along the length of it. Each set of nails is spaced a bit closer than the pair above. The first two nails might be 2 inches apart, the next two nails 1¾ inches apart, the next two 1½ inches, and so on. When a bulb of garlic is passed between the nails, its size is determined by the point at which it comes to rest and can pass no farther. An experienced worker should not need to pass every bulb between the nails. To the trained eye, many bulbs will clearly fit into a particular size category; it's those that are on the edge, close to the upper or lower threshold for a particular category, that definitely need to be run through the sizer. But even with the easy choices, I'll frequently go back to the sizer to re-calibrate and be sure that I haven't strayed off the mark.

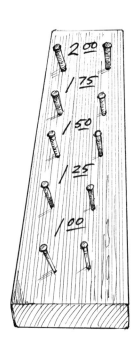
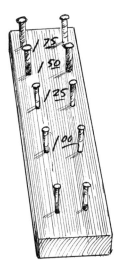

Garlic sizers · Bulbs are passed between pairs of nails in a 2×4 or 2×6. The point at which a bulb can go no further defines its size and price.

Step 12. Store

Once cured, garlic stores best in a cold, dry place with some air circulation. A temperature between 32 and 40°F (0 and 4°C) and relative humidity of 65 to 75 percent will do the job nicely. At higher temperatures — 40 to 60°F (4 to 16°C) — garlic will still keep reasonably well but for a shorter period of time. Above 60°F and at higher levels of humidity, fungal and bacterial organisms are more active and likely to cause trouble.

Some varieties of garlic store longer than others. Under good storage conditions, our 'Rocambole' variety can last 8 or 9 months, which is pretty respectable. Softneck garlics tend to store a bit longer, and so do the hardneck varieties known as 'Porcelain' and 'Silverskin'. Garlic that is harvested in a healthy state, then properly cured, will always have the best chance of a long storage life.

It is best not to store garlic in direct sunlight and to never leave bulbs in a plastic bag or any other container that does not breathe — it will cause them to rot. Some of our customers find it easiest to store garlic in a refrigerator, but we don't recommend this. The moisture inside a refrigerator can cause garlic to sprout, and once bulbs are moved to room temperature, they need to be eaten or they will deteriorate rapidly.

As noted earlier, garlic loses moisture and weight across the length of its storage life, becoming less crisp and juicy than when first pulled from the ground, but the flavor should remain intact.

Step 13. Enjoy the Payoff

I always experience a major sense of accomplishment when our entire garlic crop is harvested, cured, sized, and safely stockpiled in bags or crates in a suitable storage environment. We've come full circle with an amazing plant. And it feels like money in the bank.

CHAPTER 14 RECAP

Quick Guide to Growing Hardneck Garlic

1. Determine when is the best time to plant in your region.
2. Prepare the ground.
3. Divide bulbs into cloves.
4. Plant.
5. Mulch.
6. Weed.
7. Irrigate.
8. Remove scapes.
9. Harvest.
10. Cure and remove tops.
11. Grade and sell.
12. Store.
13. Enjoy the payoff.

Colchester Farm CSA

Manager: Theresa Mycek

Location: Galena, Maryland

Year Started: 2003

Acreage: 10 acres, with 7 acres in cash crops and 3 acres in cover crops

Crops: Mixed vegetables and herbs

WHEN THERESA MYCEK ARRIVED AT OUR FARM in spring 2004, she had already spent a season working at a CSA farm in Pennsylvania and 2 years in the Peace Corps in Guatemala as an agricultural extensionist. From early on, I could see that she was well suited to the farming life. After her season with us, Theresa went to work at Colchester Farm on the eastern shore of Maryland. In her second year, she took on the role of farm manager. She is now in her seventh year at Colchester and has four interns each season to assist her.

Colchester is a nonprofit CSA that operates on 10 acres of a 350-acre farm. The larger farm is conventionally managed and grows corn, wheat, barley, and soybeans for animal feed. The CSA is a separate nonprofit entity set up by the farm's owner. Its mission: To grow pesticide-free food for the surrounding community and offer programs and educational opportunities in sustainable agriculture. The entire 350 acres of Colchester Farm is protected by conservation easements.

Each year, Theresa and her crew grow well over one hundred varieties of 50 distinct vegetables and herbs. Their most popular crops are lettuces and other light salad greens. About 70 percent of what they produce goes to a 150-member CSA that runs for 26 weeks (from May until November). They offer small, medium, and large shares, with prices ranging from $355 to $780 per share. What doesn't go to the CSA is sold at a Saturday farmers' market in nearby Chestertown.

Equipment used at Colchester Farm CSA includes a Kubota L4400 50-horsepower tractor with a front loader, a chisel plow, a 5-foot rototiller, a plastic mulch layer, a water-wheel transplanter, a single-row cultivator, a brush hog mower, and a middle buster for digging potatoes, along with an assortment of wheel hoes, hand hoes, and other garden tools. The farm has a small greenhouse and two high tunnels. It follows organic methods but is not certified.

Colchester Farm CSA makes some of its own compost and gets occasional deliveries of horse or other animal manures. It also uses some granular organic fertilizer, but the farm relies most heavily on cover crops and green manures to maintain and build soil fertility. The preferred summer covers are sweet, white, and crimson clovers; cowpeas; buckwheat; and sorghum sudangrass. For winter cover, Theresa uses winter rye, hairy vetch, winter peas, and oilseed radishes. She especially likes the oilseed radishes because they grow rapidly, block weeds, and have deep taproots that break up compacted soil and scavenge for nitrogen below the root zones of most plants. In Maryland, the radishes die back in January and are pretty much broken down by March. This often makes it possible to plant a cash crop in spring without tilling the soil. At Colchester, the oilseed radishes are usually seeded in late August or September.

Theresa plans to stay on as manager of Colchester Farm CSA for at least the next few years. She enjoys planning for the season ahead, then making sure her plans unfold in the best possible way, given the uncertainty of nature and the elements. She likes growing healthy food on a relatively small scale and in a sustainable manner. She likes the physical activity of farming and enjoys the positive interaction with customers. Theresa hopes to run her own farm some day, but feels that the high land prices in Maryland make this a lofty goal.

Chapter 15

....

Let's Not Forget Herbs

INTEREST IN HERBS HAS GROWN EXPONENTIALLY over the last few decades. Just think of all of the herbal teas and supplements on the market today, and the creams and cosmetic products that boast one or more herbal ingredients. In our disturbingly chemical and synthetic world, herbs represent the natural, the wholesome, and the traditional. They have been around, and used by humans, for a very long time.

There's something reassuring about this, and it might partly explain the resurgent interest in all things herbal. But of course, there's another more obvious reason herbs are so popular — when used knowledgeably, they can greatly enhance the flavor and appeal of almost any dish. An ever-increasing number of chefs and cooks understand this.

After a period of neglect in the English-speaking world, many culinary herbs have become mainstream. Surprisingly, though, not many small farmers have ventured forth to grow them, or at least to grow them in a serious way. This means that opportunities still abound.

We've been growing herbs on our farm from day one. My wife, who, as I've already said, is of Italian ancestry, insisted that I plant basil to go with the tomatoes that were an obvious choice for the first year's crop plan. That was a good move: At our markets, these two have always sold well together. I also planted parsley, since it was a staple herb of my youth. It too proved to be a steady seller. The next year, again at my wife's urging, we added sage, thyme, and rosemary. Before long, others followed.

Today, we're growing and marketing 20 different herbs. Some, such as catnip, summer savory, and anise hyssop, are grown and sold in quite small quantities; others, such as the basil, parsley, sage, rosemary, and thyme mentioned above, have become a significant and lucrative part of our farm business. In this chapter, I'll share what I've learned thus far, starting with our best sellers.

Seven Reasons to Grow Herbs

Over the years, I've developed quite an attachment to our herbs. Each shines in its own way, with unique flavor, aroma, and culinary qualities. Many are closer to being wild plants than most of the vegetables we grow, and this tends to make them hardier; more resilient; and, ultimately, more interesting. They don't occupy a lot of space — probably less than 5 percent of the growing area of our farm, though they make up at least 10 percent of sales. If garlic were included in the picture, at least 30 percent of sales would comprise herbs.

I strongly encourage any beginning (or even established) organic grower to consider including some culinary herbs in his or her crop plan. Beyond this general endorsement, here are seven compelling reasons to grow herbs.

1. Most Herbs Are Easy to Grow

Despite some popular opinion to the contrary, the majority of herbs are not difficult to grow. Usually, all they ask for is a healthy, reasonably well-drained soil with an adequate amount of organic matter and full sun. A mid-range pH — say, between 6.3 and 6.8 — will keep them happy. As virtually everything else you might plant, herbs also need water and adequate fertility. Across the broad spectrum of herbs, water needs might vary quite a bit, but we'll touch on that later.

Certainly, there are a few tricks to learn, depending on the herbs you are growing, but the same could be said of most vegetable crops. Fortunately, there are numerous good books on herb growing, and I'll provide some basic advice in the section, "Tips on Growing and Selling our Favorite Herbs."

Many herbs, especially the perennials, do take their sweet time getting started, so you need patience to be an herb grower. But once they've gained a good footing — or one might say, a rooting — many herbs are tough and tenacious survivors. Some might even outlast you.

2. Most Herbs Are Relatively Pest Free

Humans the world over have found many uses for herbs, but it seems that our species is the exception. The wild animals commonly found on farms — I'm thinking rabbits, chipmunks, deer, and groundhogs — have little interest in most of the herbs that we grow. This means we don't have to erect fences or plant herbs in protected areas. In fact, we often place them along the edges of distant fields or near hedgerows where we suspect our four-legged competitors hang out. Sometimes, we even use herbs as a buffer to slow down or deter freeloading critters from reaching other, more palatable crops.

Neither are most herbs of any interest to birds or insects, but I should mention a few exceptions. Japanese beetles enjoy basil and can eat a surprising amount; however, if they show up in large numbers and start causing significant damage, which they rarely do, they can be easily removed by hand and disposed of. The parsley caterpillar (the larval form of the beautiful swallowtail butterfly) has a fondness for parsley, as its name implies. But these butterflies and their caterpillars are uncommon visitors to our farm and have never damaged more than a handful of plants. We generally let them be. Rabbits, groundhogs, and deer also eat parsley, especially when other food is scarce; therefore, parsley is one herb you should protect, either with fencing or netting or by choosing a location that you deem to be safe — such as close to where your dog patrols at night.

Otherwise, the herbs we grow are, for the most part, pest free. This is a lot more than can be said for many of the vegetable crops on our farm.

3. Herbs Don't Take Up a Lot of Space

This can be said three times. Most herbs require only modest amounts of space in the field, on the truck, and in a retail display. Less land devoted to herbs means more land for other crops. It also means a smaller area to weed, irrigate, and generally maintain. This is especially relevant to growers with limited acreage, as well as to those using a pickup or small truck for hauling crops to market.

There is usually a limited supply of display space at a farm stand, especially frontage. Consider that you might display 20 bunches each of rosemary, sage, and thyme in three small baskets that take up roughly the same amount of space as you would need for a dozen heads of just one variety of lettuce. Though the herb bunches sell for a little more than half the price of the lettuce, they can bring in three times the amount of money. Anyone with a good nose for business will pick up on this very quickly. Bottom line: The financial return on herbs relative to their space requirements — in the field, on the truck, and in the stand — can be quite good.

4. Herbs Don't Weigh Much

You'll appreciate the lighter weight of herbs after dealing with such crops as potatoes and winter squash, which are likely to weigh 50 pounds or more per lug or crate. This can add up to a lot of weight, especially when you consider that every potato or butternut squash might be lifted several times before it passes from your hands into a consumer's. Let me elaborate:

On our farm, we use lopping shears to snip winter squash at the stem, then place them in crates in the field. These crates are loaded onto a pickup truck and taken to a dry basement, where they are unloaded and stored. When needed for market, these same crates of squash are carried out of the basement and loaded onto our larger market truck — a Mitsubishi with a 16-foot box. At market, of course, they are unloaded and displayed. At the end of the day, any squash not sold must be reloaded and brought back to the farm, where they will again be placed in storage.

Over and above all this lifting, consider that a 50-plus-pound crate of squash might sell for $60 or $70, whereas a cooler containing 50 bunches of basil with a net weight of just 10 or 15 pounds will fetch more than $100 — with a lot less muscular exertion and strain on your back. In other words, when compared with squash or potatoes, herbs exact a minimal toll on your body and yield a better return.

5. Herbs Smell Good

Frequently, customers compliment us on the fragrances in and around our farmers' market stand. Often, they're not quite sure what their noses are picking up, but they know they like it. Most vegetables have a certain aroma. Freshly dug potatoes and carrots, for example, have a nice earthy smell if you put them right under your nose. Tomatoes, peppers, and lettuces have distinctive aromas but only when you get very close to them. Invariably, it is the smell of fresh basil, rosemary, sage, thyme, or mint, or some tantalizing combination of these, that captivates passersby and draws them to our stand.

The enticing fragrances of many herbs make them a real pleasure to work with. Harvesting aromatic herbs toward the end of a pick day is a good way to unwind and enjoy some of the rewards that a life on the land can provide.

Our Strongest- and Weakest-Selling Herbs

Here's our basic workhorse herb list, arranged in order of popularity and sales.

1. Italian parsley
2. Basil
3. Thyme
4. Rosemary
5. Sage
6. Kentucky Colonel mint
7. French tarragon
8. Lovage
9. Curly parsley
10. Oregano
11. Marjoram
12. Lavender
13. Cilantro
14. Chocolate mint
15. Orange mint
16. Winter savory
17. Summer savory
18. Anise hyssop
19. Chives
20. Catnip

Don't read too much into the order of popularity in this list. It's simply what works for us at our farmers' market in New York City. Demand for specific herbs will vary widely depending on the nature and location of your market. Let me elaborate with a couple of examples.

Lovage. Unless you happen to be from Romania or another Eastern European country, you have probably never used lovage. It turns out, though, that a fair number of Romanians have found their way to New York City and ultimately to our stand, where they buy multiple bunches of lovage with considerable enthusiasm. We've gone so far as to label lovage with its Romanian name, *leustean,* much to our ethnic customers' approval. Apparently, it makes just about everything taste better to them. If you're selling in other parts of the country, though, where Romanians are few and far between, well, lovage might not be such a big hit.

Parsley. In New York City, Italian or flat parsley is the parsley of choice. Our customers prefer it by a ratio of about five to one, but in some parts of North

America, curly parsley will make a good showing, and it is widely viewed to be a better garnish.

In other words, you need to know your market before you embark on a major planting of any particular herb. It makes good sense to start with modest trial plantings of several different herbs and see how well they sell. Your customers will lead you in the direction you need to go.

To at least some extent, the list at left is also a reflection of the ease (or lack of) with which we can grow various herbs in our soils and climate. Obviously, we are going to favor herbs that perform well on our farm, as long as they are good sellers. It is also a list of the herbs we have learned to grow and for which we have gained some reputation.

Reputation counts for quite a lot. People know they can come to our stand for generous bunches of parsley or basil or for any of the Mediterranean herbs such as sage, thyme, and rosemary, but they tend not to seek us out for cilantro, which we grow much less of. Perhaps it's because our cilantro-growing skills are not so well developed. A second reason might be that cilantro is a rather flimsy plant that wilts quickly after it is cut and displayed. This has always seemed to be a drawback to me; however, another farmer at our market has mastered the art of cilantro growing and marketing, and he sells masses of the stuff.

So, I emphasize, the list at left is definitely not intended as a one-size-fits-all plan. There are other paths to take. North America is a large and diverse land. Depending on the taste buds of residents in your area or the presence of ethnic enclaves, you might explore more exotic terrain. For example, herbs used commonly in South American cooking, or perhaps Thai, Cambodian, or other Asian cuisines, might perform very well where you live.

6. Many Herbs Are Perennials

More than 50 percent of the herbs we grow are perennials; once established in the field, they will provide a harvest for several years. This doesn't mean they are maintenance free. They will almost certainly require periodic weeding, irrigation, and applications of compost or other amendments to keep producing well. It is also important to cut back woody stems and dead flower stalks to stimulate new, marketable growth.

And just because an herb is a perennial doesn't mean it can survive and prosper in all climate zones. Rosemary, for example, is a tender perennial. You'll find it used as a permanent ground cover in California, but in New York State (or any parts of North America that are in Zones 1 through 6), it will usually die in winter. We've found a way to keep rosemary plants alive and productive for several years, but I'll tell you about that a little later.

Sage, French tarragon, and mints are some of the hardiest of perennials. Numerous specimens of each of these have been growing on our farm for 10 years or more. Catnip and anise hyssop are also long-term survivors. Thyme and winter savory are not quite so hardy — some winters are too cold for them.

Then, of course, there are the annuals, such as basil, and the biennials, such as parsley. You have to plant these each year, but they make up for this extra effort by growing rapidly and reaching a marketable size much sooner than the perennials.

7. Herbs Are a Great Add-On Item

Good cooks know that the right herb or herbs can transform an otherwise okay meal into one that is really exceptional. This is why herbs are the perfect add-on item at the cash register. We make most of our money selling fresh staple vegetables — the kind of vegetables you would expect to find on a healthy dinner plate. Among the more popular items are potatoes, carrots, onions, beans, peas, squash, garlic, and a wide variety of greens. It's a rare individual who will design a meal around herbs, but many customers at our stand will readily throw in a bunch of parsley, thyme, or sage, or any of our other herbs, after they've assembled their primary foods.

We keep small quantities of our most popular herbs right on the table next to our scale and cash box. They are excellent candidates for impulse purchases. While the staples are being weighed and tallied up, it's usually just a matter of time before the average customer reaches for the appropriate herb for the dish that he or she is contemplating. It may be only an extra dollar or two, but multiply this by a few hundred customers per day and you can end up with some real money.

Parsley

This widely used herb is a good seller for us. Parsley is a slow starter, but once established, it becomes a strong producer that can last well into fall and provide numerous cuttings. We've sold parsley as late as Thanksgiving, but by then the plants are usually too short to bunch so we simply cut off the tops and throw them into baggies. For the rest of the season, our parsley is bunched with rubber bands and displayed in shallow tubs with about a half inch of water in the bottom. This keeps the bunches looking spritely without the need of a cooler.

Propagating. Under normal seeding conditions, parsley takes a long time to germinate — as many as 25 days. Fortunately, you can cut this time in half by soaking the seeds in water for 2 or 3 days just before you plan to use them. Change the water each day, then let the seeds air-dry on a paper towel. This helps to break their dormancy. We start most of our parsley in a greenhouse in early March and move it to the field in mid- or late April, but we're not likely to cut a bunch for market until late June.

Transplanting. When moved to the field, our parsley is transplanted into 4-foot-wide strips of black plastic — three rows per strip and 12 inches between plants — with two drip irrigation lines underneath. Parsley does best in a soil with high fertility, especially if you plan to take multiple cuttings over an extended period of time. Before laying plastic, we put down a generous amount of compost or other organically approved amendment. By mid-season, if the plants are looking a little pale, we might fertilize them with fish emulsion, either as a foliar spray or as a liquid poured right into the hole in the plastic where the parsley is growing.

Direct-seeding. About 75 percent of the parsley we grow (out of a total of some three thousand plants) is started in our greenhouse in early March. The remaining 25 percent, after a few days of soaking in water, is direct-seeded in the field, in bare soil, in May. Direct-seeding saves us a few steps, but we can spend a lot of time weeding. Without the black plastic to block them, annual weeds will appear several days before the parsley. The weeds get off to a roaring start, while the parsley, even after it has emerged, will remain tiny for a couple of weeks. We need to weed on hands and knees, with careful fingers and a good pair of eyes, to free the little parsleys from the competition, which will otherwise engulf and overwhelm them.

Italian or flat parsley (left) and curly parsley (right)

Flame Weeding

There is one way you can reduce hand weeding in direct-seeded parsley — that is, if you're willing to take a little risk. Seven to 10 days after seeding and definitely before the parsley germinates, you can use a flame weeder — as you would for carrots (see page 308) — to burn off any weeds that have already emerged. A quick pass is all that's necessary. When the parsley finally shows up, it shouldn't have to deal with so much competition, nor should you.

The only requirement is that you have good timing. If you flame-weed too early, additional weeds will come in before the parsley and you won't have gained much advantage. If you flame-weed too late — either just before or after the parsley emerges — you'll destroy your crop and have to seed it again. Before you bring out the flamer, you might try digging up a few parsley seeds to see how far along they are on the road to germination.

Direct-seeded parsley does take more work, but once established, it can perform very well and sometimes better than its transplanted counterpart. As the plants get bigger, it's a good idea to mulch them with straw or some other organic mulch. This will suppress future weeds, help retain moisture in the soil, and keep it cool.

As mentioned above, parsley can be attractive to certain four-legged critters, so some protective measures may be necessary.

Basil

Basil would be at the top of our list, in terms of sales, were it not for the fact that it is the most tender of annuals and will succumb to the first light frost, if not before. This means it has a shorter growing season than many other herbs. Even night temperatures below 45°F (7°C) will negatively affect its appearance and slow its growth. But during summer, when tomatoes are in full swing, basil is an herb that's hard to beat. At our Saturday market in August or September, we have no trouble selling 150 bunches.

Choosing and propagating. Seed catalogs offer many exotic varieties of basil with an assortment of leaf colors, aromas, and flavors. Cinnamon, lemon, lime, purple, opal, and Thai basil are just some of them. Each is distinct in its own way. Over the years I've grown several of these basils, but demand for them has never been high. Our customers have regarded them with interest and curiosity but usually ended up buying the traditional Genovese or Italian large-leaf varieties that are the best for making pesto and using with tomatoes. Today these are the only basils we grow.

Basil is easy to start from seed, as long as the soil, or potting medium, is warm enough — about 70°F (21°C) is suitable for basil and, indeed, most other herbs started from seed. We seed our first basil in a heated greenhouse in late March.

Growing in high tunnels. About 5 weeks after seeding, we transplant our first basil into a high tunnel in 6-foot beds covered with landscape fabric — four rows per bed with two drip lines undernearth. On chilly nights, we'll protect the seedlings with floating row cover, using wire hoops to prevent the cover from touching the basil. This keeps the plants doubly protected from the cold. Our early basil is ready for market by mid- or late June, when fresh, local basil is relatively scarce in our region. It's an easy sell.

Growing in the field. Over the course of a season, we plant basil at least five more times, usually three hundred to four hundred plants at a time, at approximately 2-week intervals. We start all of it in the greenhouse, then transplant it into the open field, in 4-foot-wide strips of black plastic with two drip irrigation lines underneath. Basil likes heat, and the black plastic warms the soil. It also keeps the weeds in check and helps retain moisture.

Our spacing is fairly tight: In each strip of plastic, we plant three rows, 1½ inches apart, with 12 to 15 inches between plants within the rows. Basil likes water as much as it likes heat, so we give it the equivalent of 1½ inches of rainfall every week, through the drip irrigation lines. If it rains a lot, we reduce or withhold irrigation.

Harvesting. The first cutting from basil plants is usually the best, offering broad leaves of a tender green color, but if you look after your crop, you can take multiple cuttings. All you need to do is keep the water flowing, pull any weeds that emerge from the holes where the basil is planted, and pinch off all flowering tops as soon as they appear.

It is especially important that you remove the flowering tops. Most discerning customers will take a pass on flowering or seedy basil. If you're a bit late getting to the flowering tops, it might be better to cut the plants about 4 or 5 inches above the soil with a pair of hedge clippers, throw these cuttings away, and wait for new growth to appear.

When harvesting basil and most other herbs from which you hope to get multiple and sequential bunches, be careful not to cut too low — always stay an inch or so above the branching point because this is where new growth emerges.

Displaying. Like cilantro, basil will wilt quickly on a hot day at market, so you don't want to put too much on display at one time unless business is really booming. It's also a good idea to always keep it out of the sun. We hold the bulk of our basil, and most other herbs, in coolers with cold packs, and display only a limited number of bunches as needed. The temperature inside the coolers should not be too low. Somewhere between 40 and 50°F (4 and 10°C) is cold enough. Temperatures below 40°F can damage basil's tender leaves.

We use 48-quart coolers — the type you might find at a camping supply store and take to the beach. They make it possible for us to display fresh offerings throughout the day. Many people, but especially women, will buy basil on their way home from work. It makes sense — they know it is highly perishable; therefore, they don't want to have to worry about keeping it cool all day. Good-looking and good-smelling basil at 5:00 P.M. is hard to pass up.

Rosemary

Rosemary is a highly aromatic Mediterranean herb that prefers full sun and a well-drained soil. It is perennial in mild climates but seldom survives outdoors in a harsh northern winter. We've overcome this limitation here in southern New York State by growing our rosemary in a high tunnel.

Starting. Rosemary is one of the more challenging herbs to start from seed. It requires a well-drained potting mix that remains moist but not overly wet and stays at just the right temperature (about 70°F [21°C]). Even then, it's a poor germinator and a slow starter — success is by no means guaranteed. A wiser course of action might be to buy rosemary transplants from an established greenhouse with advanced propagation equipment and more experience than you have.

Or you could try propagating rosemary by taking fresh, green cuttings from an existing plant and rooting them in water or a moist growing medium. This requires some knowledge and patience — check out "A Little More on Propagation" (see page 382). Or if you want more details from a real professional, you might consult the book *Growing Herbs — From Seed, Cutting & Root* by Thomas DeBaggio. It's an excellent source for information of this sort.

Growing in high tunnels. If you have a tunnel or some other covered space, I highly recommend growing rosemary in it. If you live in northern climates, the alternative is to plant new rosemary in the field every year. We did this for several years until a fellow farmer and friend — Anne Salomon of Tweefontein Farm — tipped me off to the tunnel method. Sadly, Anne is no longer with us, but I am often reminded of her wonderful smile and generous spirit when I look at our huge rosemary bushes.

Most of the two hundred or more rosemary plants we have in our tunnel are more than 5 years old. They grow in the ground (not in pots) and have turned into substantial bushes that are 15 to 20 inches in diameter and up to 3 feet tall. Each plant is able to provide us with 20 or more bunches of rosemary every year. Because it grows so large, tunnel-planted rosemary needs plenty of room. A spacing of 24 inches between plants should be sufficient.

In late December, we mulch the tunnel rosemary with a couple of inches of straw or aged horse bedding and drape blankets of row cover over them for additional warmth. I think what keeps the rosemary going in winter is the drier soil and lower incidence of freeze and thaw than would occur in the open field.

Growing in the field. The main problem with field-grown rosemary is that it is slow to put on size. In the Northeast, you dare not cut it until August or September for fear of weakening the small plants. Even then, you're not likely to get more than a few bunches before winter comes along and deals its deathly blow.

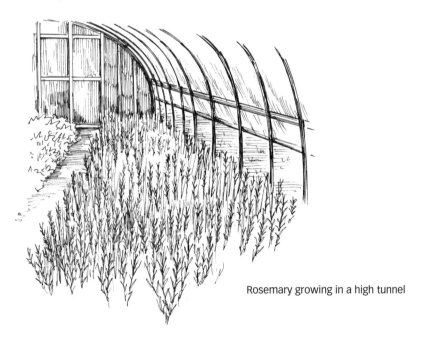

Rosemary growing in a high tunnel

Sage and Thyme

These two robust Mediterranean herbs will prosper in warm, dry weather and a well-drained soil. When I think of them, along with rosemary and parsley, I'm pleasantly reminded of the Simon and Garfunkel song "Scarborough Fair," with its oft-repeated refrain: *Parsley, sage, rosemary, and thyme.* The lilting, nostalgic ring of these words takes me back to a more youthful and carefree time. Now, with experience on my side, I've grown fond of the herbs themselves and appreciate them as much as the music of the words.

Sage is a true hardy perennial that can last for a dozen or more years in the field, provide many cuttings, and come back each spring looking as good as ever. Because sage plants can grow quite large, we space them 2 feet apart.

Thyme is a very popular and versatile perennial herb. It goes well with many dishes — from scrambled eggs to most meats and vegetables. It is hardier than rosemary but not as hardy as sage. On our Northeastern farm, our heavily harvested thyme plants do not always survive winter. Most years, they come through just fine, but some years as many as half the plants are lost.

Starting. Sage and thyme are both relatively easy to start from seed indoors or in a heated greenhouse. Just keep the potting soil moist (but not too wet) and warm — around 70°F (21°C) is ideal. Thyme seeds are very small and should be left uncovered in shallow furrows on the soil surface. Sage seeds, which are much bigger, should be covered with about ¼ inch of soil.

Before germination occurs, it's better to mist the flats rather than water them. After germination, when the little seedlings have reached a height of a couple of inches, we pot them on to roomier accommodations. It's an extra step, but it speeds up the growing process and makes a big difference on the road to a healthy, vigorous plant.

Sage and thyme can also be started from rooted cuttings. One advantage of doing this is that you'll end up with offspring identical to the parent plant. Just remember, it's always best to take cuttings from tender, young growth.

Growing in a tunnel. In the last several years, we've taken to growing some of our thyme in a high tunnel. In this protected environment, we can be sure the plants will survive winter and produce a healthy,

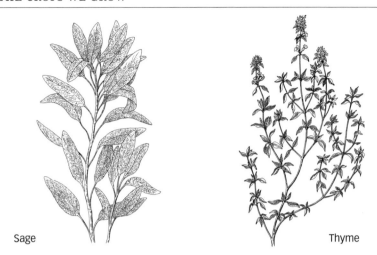

Sage   Thyme

harvestable crop by early summer. In the tunnel, we space thyme plants 24 inches apart; in the field, 15 inches is enough. Numerous varieties of thyme are available; we've settled on German thyme.

Selling. The demand for sage and thyme, as well as for rosemary, grows as the season progresses. These herbs are best with hot, cooked food. In the sizzling days of summer, our customers don't relish the thought of spending hours in the kitchen. They have their sights set on fresh salad greens, vine-ripened tomatoes, basil, parsley, green beans, edible-pod peas, and the like. But when autumn arrives and people sense a nip in the air, their thoughts turn to warm kitchens, simmering soups, and hearty dishes of hot food.

This is when sage, thyme, and rosemary come into their own. On a market day in July, we might sell 20 or 30 bunches of each of these three herbs, but in October, the number can shoot up to 80 or 90. And right before Thanksgiving, when most Americans have turkeys and turkey stuffing on their minds, we've sold as many as 500 of each of these three herbs in a single day. That's serious business, but it doesn't happen out of nowhere. The modest sales of summer establish our reputation and set us up for the big sales of fall.

Mints

Unlike the Mediterranean herbs, which enjoy full sun and moderately dry conditions, most mints prefer lower temperatures and plenty of moisture. They will also tolerate partial shade. They can be started from seed, but because mint often does not grow true to type, the results are unpredictable.

Choosing varieties. There are many interesting mints available; the trick is to find the varieties that you and your customers like. In past years, we've tried quite a few of them, but to avoid being overwhelmed by this aggressive plant, we have narrowed our selection to just three types: 'Kentucky Colonel', 'Chocolate', and 'Orange'. Of these, the most vigorous grower and the best seller is 'Kentucky Colonel'. It has an excellent, all-around mint flavor, which comes through well in teas, cold drinks, and any dish that mint can complement.

For those with more exotic tastes, the 'Chocolate' and 'Orange' mints are always available. The 'Chocolate', in particular, has a most enticing smell, which is somewhat reminiscent of a real after-dinner chocolate mint.

Propagating and growing. It's best to buy mint transplants from a commercial greenhouse or ask a friend if you can dig some mint from his or her garden and replant it. Grow this mint for a season, and try it a few times. If you like the taste and the smell, dig it up the following spring, divide it, and replant it.

If you do this 2 or 3 years in a row, you will have an impressive mint patch. Place the divided sections about 12 inches apart, and water them well. Even without your help, mint will expand its range by spreading specialized stems, called stolons, below the surface of the soil. Whenever you establish a new mint patch, it's a good idea to mulch the plants with straw or some other organic material.

Mints grow well in low spots, where there is more moisture, and along tree-lined hedgerows, where there is partial shade. At market, they sell well on hot summer days, when people are looking for some relief from the heat. On such days, the smell of fresh mint is always a draw. Some growers I know do good business selling chilled mint teas, sweetened with a little honey or maple syrup.

French Tarragon

We started our first tarragon from seed. It grew rapidly and sold well, but only for a few markets. No one came back asking for more. The reason: It was Russian tarragon, which has almost no flavor and no culinary value. At the time, I was not sufficiently familiar with the plant to know how it should look and taste. Soon enough, a few customers informed me of my error, and I withdrew the herb from our stand. If you're going to grow tarragon, and it is an herb worth growing, it's imperative that you get the French variety. Unfortunately, it's not so easy to find, and unlike Russian tarragon, it cannot be started from seed.

Propagating and growing. You can only propagate French tarragon by division or by rooted cuttings, so the first thing you have to do is find a source for the actual plants. Reputable nurseries and herb growers are two options. Once you've got the real McCoy (the leaves of true French tarragon have a mildly biting, licorice-like flavor), put your plants in a light, well-drained soil with full sun. Let them grow and build up their reserves for the first year without taking more than a minimal harvest.

Dividing. Once established, French tarragon is a hardy perennial that will serve you well for a long time. Every few years, the plants should be dug up, sectioned into three or four parts, and replanted in an area where tarragon has not recently been grown. Try to do this on an overcast and damp day in spring, and keep the roots covered with wet soil until they are situated in new ground. Generously water the divided plants, and follow this up with additional waterings once a week for a few weeks. For more detail on dividing tarragon and other plants, see the section below titled "A Little More on Propagation."

French tarragon is not a common herb at most farmers' markets; therefore, it is a good item to have. Once you become a known purveyor of good tarragon, people looking for it will come to your stand. When they arrive, there's a good chance they'll buy more than just tarragon.

Lovage

Lovage, as mentioned earlier, is popular among Romanians but few others, so you should definitely assess your customer base before putting much effort toward this unfamiliar herb. The leaves of lovage resemble those of celery, but lovage leaves have a stronger flavor and perfume-like scent. They are often used in hearty, cold-weather soups and stews with an assortment of vegetables and perhaps meats. Lovage apparently freezes well. Our Romanian customers, and some chefs, often buy it in bulk and freeze it for later use.

Propagating and growing. Lovage is a hardy, strong-growing perennial that produces well for a long time. It likes plenty of moisture and does well when mulched with straw or wood shavings. Its unusual seeds are reasonably easy to start but slow to put on size.

Even if you don't have many Romanians in your area, it might be worth maintaining a small patch of lovage. It's a distinctive herb with an acquired taste, but one that may well gain in popularity as more people learn how to use it. Once you've established a lovage patch, it doesn't take much work to maintain it.

Oregano and Marjoram

I've grouped these two herbs together because they are in the same genus — *Oreganum.* My friend Anne Salomon always referred to marjoram as the queen of oreganos, and I think she was right. Marjoram is sweeter and more delicate than oregano and has a delightful, perfumed, almost balsamlike aroma. It has numerous culinary uses but is particularly good with carrots. It also makes a pleasant and soothing tea. Oregano, of course, is indispensable in much Mediterranean cooking.

Propagating and growing. Oregano and marjoram are perennials, but while oregano can survive in a very cold winter (to about –20°F [–29°C]), marjoram will most likely expire once the temperature drops below a mere 20°F (–7°C); therefore, on our farm marjoram is treated as an annual and planted each spring.

Both oregano and marjoram can be started from seed or rooted cuttings. Mature oregano plants can also be divided in the field. It's relatively easy to start these herbs from seed. Just remember, the seeds are tiny and should be kept moist and uncovered in shallow furrows about ⅛ inch deep. A week or two after they emerge, the little seedlings should be thinned to about one plant every ½ inch. When they reach a height of 1 or 2 inches, pot them on individually to larger cells. Oregano and marjoram will grow best in a light, well-drained soil with full sun.

Dealing with variability. As already noted, one problem with starting these, and some other herbs, from seed is that you can't always be sure of exactly what you're going to get. Some strains of oregano, for example, are stronger and more flavorful than others. These are the ones you want to have. If you already have plants in the field that have good flavor and other qualities that you appreciate, you might want to propagate these individuals vegetatively (by division, cuttings, or possibly layering) rather than venture into unknown territory with new seeds.

Purchased seeds will often give you good-looking plants, but the flavor and aroma can be disappointing. With vegetative propagation, the qualities of the parents always carry over to the offspring.

Oregano

Marjoram

Lavender

Lavender is an appealing herb with stunning blue flowers and a wonderful, lasting fragrance, but limited culinary use. Some of our customers make lavender sorbets and others, lavender tea, which is said to be a soothing remedy for headaches and agitated nerves. Still others sew dried lavender leaves into sachets and enjoy the lingering aroma. Whatever uses our lavender is put to, I'm happy to keep growing it. It's such a nice plant to have around and, while in flower, is much appreciated by bees.

There are several different types of lavenders to choose from, but 'English Munstead' generally has the best fragrance and a good, upright growth habit, which makes it easier to bunch. Most lavenders are moderately hardy perennials that will usually survive a Zone 6 winter, but not much colder unless protected in a high tunnel or perhaps with mulch and row cover.

Propagating and growing. Lavender likes well-drained, gravelly soil and full sun. It can be started from seed, but it has a low germination rate and very slow growth. A week before sowing, you can place seeds in a sealed plastic bag in the freezer to help germination.

If you can obtain mature plants, you might find it easier to propagate lavender by taking fresh, green stem cuttings and placing them in a jar of water. When they develop their own roots, move them to a container of potting mix. An existing lavender plant can also be propagated by layering (see page 382).

Cilantro

With cilantro, if you let the plant go to seed, you get two herbs in one: The leaves, which are properly named cilantro, and the seeds, which are known as coriander. Coriander is commonly used in Indian and Asian cuisine, while cilantro is a favorite herb in Mexican cooking, though both versions of the plant are gaining a much wider audience, especially in the United States and Canada.

Cilantro is a member of the parsley family, which will come as no surprise when you inspect its leaves — they do closely resemble a glossy Italian parsley, though the fragrance is quite different.

Propagating and growing. You can readily start cilantro from seed in a greenhouse or by direct-seeding in the field. A potting mix or soil temperature of 70°F (21°C) will hasten germination of the rather large seeds. Cilantro is an herb that is quick to bolt and go to seed. Direct-seeding, rather than transplanting, will slow down this process.

Because it is a short-lived annual, cilantro doesn't need a lot of space — 8 to 12 inches between plants should be enough. The plants can survive a few light frosts but not a hard one. During their active growth phase, they appreciate plenty of water.

In the field, too much heat or direct sun can make cilantro go to seed faster than you might want it to. In the hot days of summer, it will do best in an area with a little shade, rather than constant overhead sun.

To provide your customers with cilantro throughout summer and early fall, you will need to plant this herb at least once a month until 50 or 60 days before your first fall frost. Unless you want the seeds for coriander, cut cilantro back as soon as it starts to flower. If you do want coriander, you'll need to give the plant at least 100 growing days and, of course, leave the flower stalks in place.

Storing and marketing. As noted earlier, cilantro wilts quickly after it is cut, especially if left in the sun. It helps to store it in a cool place (such as a cooler!) and display only modest amounts at any one time.

The Savories

Though not in high demand, summer and winter savory have managed to retain their places on our herb list. Both go well with bean dishes, especially dried beans, and are said to reduce the incidence of flatulence that can occur after excessive bean eating — information I seldom fail to impart when asked what these herbs are good for. For those few customers who want to know more, I'll add that the savories are also reputed

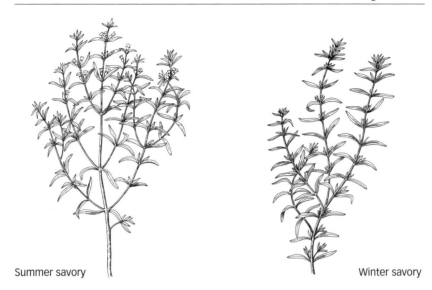

Summer savory

Winter savory

to be aphrodisiacs. Usually, that puts an end to any further questioning. Both savories taste a little like peppery thyme, though the winter variety is the stronger and more biting of the two.

Propagating and growing. Summer savory is a fast-growing, upright annual that is easy to start from seed. It has a short season and goes to flower in early fall. Winter savory is shorter in stature and is a relatively hardy but slow-growing perennial that generally doesn't last for more than 3 or 4 years. It also can be started from seed or a rooted cutting. Both do well in full sun and a well-drained soil.

A Few Others on Our List

The herbs described above are the ones toward which we direct most of our attention. They grow well for us and they sell well. As long as they are in season, our customers can expect to find them at our stand. A few others, though, are catching on and worthy of mention.

Chives. Members of the onion family, chives are very hardy perennials that can survive temperatures as low as –40°F (–40°C). They like fertile soil and full sun but will tolerate some shade. They can be started from seed but, like most alliums, grow slowly in their early days and do not compete well with weeds; however, once established as full-grown plants, chives can hold their own.

As soon as you have a healthy patch of chives, you can easily expand it by digging up the plants, dividing, and replanting them about 10 inches apart. As with most plant division, this is best undertaken in spring under overcast and moist conditions. It is easier and faster than starting new plants from seed.

When harvesting, cut chives close to the ground and lightly bunch with a rubber band. Among herbs, chives can be a good seller. We probably should be growing more of them.

Catnip. This is a hardy perennial in the mint family. Once established, catnip will grow like a weed. It is not, strictly speaking, a culinary herb, but it does make a soothing tea that is reputed to ease menstrual cramps. And then, of course, there are all the cats in the world, many of whom think very highly of catnip. They themselves seldom show up at our stand, but their owners do, and at least a few of them like to bring home a little treat for kitty once in a while. Bottom line: Catnip is an easy-to-grow perennial that can be started from seed. Sales are modest, but so is the effort required to produce it.

Anise hyssop. Also in the mint family, anise hyssop is relatively easy to start from seed. It is a perennial that can survive winters on our farm (Zone 6) but probably not in areas that are much colder. It has broad leaves and a beautiful bluish-purple flower. Its aroma and flavor, which are reminiscent of both licorice and mint, are quite wonderful in a hot tea.

A Little More on Propagation

Most herbs can be started from seed — some quite easily, some less so. For the latter category, there are other methods; depending on the herb, you might take stem cuttings, perform the surgery of division, or try an approach called layering.

Each of these methods requires that you have access to one or more existing plants, and each calls for a little skill and knowledge. All will initially be more time consuming than starting plants from seed; however, once you know what you're doing, they will often provide you with mature plants more quickly and reliably. The "How to Start" column in the table on pages 386 and 387 gives the different possibilities for each of the herbs listed, and the "Tips on Growing" section of this chapter provides further suggestions on which methods to use. Here's a little more detail on the methods themselves.

Stem Cuttings

Spring is the best time to take stem cuttings, but you can have success at any point during the growing season. Among the herbs we grow, mint, rosemary, and lavender are good candidates for this method. Basil can also be started from stem cuttings, but if you're growing more than a few plants, it's easier to start this one from seed.

Cutting and rooting. Use a sharp knife or a pair of scissors to cut a 3- to 4-inch stem of new growth from a healthy plant—preferably one that is growing outdoors. Remove any leaves from the bottom 1 to 2 inches, then place the stem in a shallow cup or jar of water. The water level should be a little below the remaining leaves. You can easily root several cuttings in the same container of water.

Set the jar in a spot with plenty of daylight but not full sun. The sill of a north-facing window works well. Change the water daily, and wait for roots to appear. Depending on the herb you are working with, this could take several days or 2 to 3 weeks.

Rosemary cuttings in a jar of water

Potting. Once the roots are ¼ to ½ inch long, move the cuttings into a moist, but well-drained potting mix. (For stem cuttings, some growers use a lighter-weight potting mix, which usually contains perlite rather than sand to promote drainage; however, we've found that a standard mix works just as well.) Place the newly potted cuttings in a warm spot with full sun. Try to maintain a soil temperature between 70 and 80°F (21 and 27°C). The air temperature can be a bit lower.

At this stage, treat the rooted cuttings as you would any young plant started from seed: Water as needed, but allow the potting medium to dry a little toward the end of the day and at night. If the aboveground portion of your cuttings seem overly dry, you can also mist the plants. Depending on the herb in question, you'll need to wait another 3 to 6 weeks or more before it's time to transplant them into the field.

Best Uses for Herbs

After talking with my wife and others and conducting a little research on my own, I put together a list of the herbs we grow and the foods they might be paired with. A laminated version of this list comes in handy at our farmers' market when we're asked to recommend an herb for a particular dish; however, increasingly, our customers know what they want and have little need of our guidance, which for me is a relief, since my culinary skills are not well developed.

Anise Hyssop: Highly aromatic with a flavor reminiscent of both licorice and mint. Use leaves and flowers in salads or as garnishes. Makes a delicious tea.

Basil (standard Italian pesto type): Has a sweet fragrance and a mild peppery flavor with a hint of clove and mint. A must in pesto and tomato sauce. Excellent with fresh tomatoes and many other vegetables and meats. Basil goes well with garlic, thyme, and lemon. There are many varieties of basil other than the Italian pesto type. Some have strong and distinctive flavors. Experiment with caution.

Catnip: In the mint family. Often grows as a weed. Catnip tea aids digestion, relieves menstrual cramps, and is a good nightcap before bed. Can also be chopped up and added to a salad. Highly desirable to many cats.

Chives: Taste like mild sweet onions. Go well with potatoes, tomatoes, corn, peas, carrots, and many other vegetables. Add to soups, stews, and omelets at the last minute. Add chive flowers to salads.

Cilantro: Has a sage-like citrus flavor. Popular in Southeast Asian, Chinese, and South and Central American dishes. Goes nicely with tomato and pasta salads. Also good with potatoes, beets, onions, and lentils and in curries. A common ingredient in salsa. The seeds of this herb, known as coriander, are used in salad dressings and marinades. They are also good with eggs and in chili sauce and guacamole.

Lavender: Has a wonderful, clean, perfume-like fragrance, which it retains for a long time when dried. Often sewn into sachets. Is used in flavored vinegars, jellies, and sorbets. Add to bath water to stimulate and cleanse the skin.

Lovage: Has a strong, celery-like flavor. Good in soups and stews or chopped up in a salad. Also goes well with potatoes, tomatoes, chicken, fish, rice, and steamed vegetables. Popular in Romanian cooking.

Marjoram: A gently perfumed, calming herb, like a mild oregano. Good with many meats and vegetables. Excellent with carrots. Complements garlic, onions, thyme, and basil.

Mint: Flavor and scent vary depending on variety. Goes well with many meat, fish, and vegetable dishes. Especially good with lamb, peas, carrots, string beans, and potatoes. Makes a pleasant mild tea that aids digestion.

Oregano: Has a hot peppery flavor. Important in Italian cooking. Excellent in tomato sauces. Can be used with most meats, fish, and vegetables. Good with eggs and cheeses.

Parsley: Has a gentle, fresh taste that brings the flavors of other foods together. It is high in vitamins A and C, several B vitamins, calcium, and iron. Goes well with most foods and should not be limited to use as a garnish. Italian (flat leaf) parsley has a somewhat stronger flavor than curly parsley and is preferable for cooking.

Rosemary: Has a piney, mint-like flavor with a hint of ginger. Good with meats (especially roasted meats), tomatoes, peas, mushrooms, squash, cheese, and eggs. Complements chives, thyme, and parsley.

Sage: Has a lemony, camphor-like flavor that goes well with poultry and most meats and vegetables. Essential in stuffings. Makes a pleasant tea (when combined with a little honey) that is especially good for sore throats.

Summer and Winter Savory: Summer savory tastes like a peppery thyme. Blends well with other flavors and helps bring them together. Good with dried beans, peas, lentils, and many other vegetables. Also good in chicken soup, beef soup, and with eggs. Used in herb butters and flavored vinegars and teas. Winter savory has a stronger, more piney taste than summer savory. Excellent with strong game meats and in pâtés. Both savories are reputed to be anti-flatulent!

French Tarragon: Has a strong aniselike flavor. Enhances most meats and some vegetables. Good in salad dressings and cream sauces. Use sparingly; tarragon can easily overtake a dish. Avoid Russian tarragon; it has no culinary value.

Thyme: A wonderful and widely used aromatic herb with a faint, clovelike taste. Good with most meats and vegetables and in soups.

Summary of Growing Information for Keith's Farm Herbs

Plants are listed in order of popularity.

Common Name	Botanical Name	Type	How to Start
Parsley	*Petroselinum crispum*	Biennial	Seed (covered)
Basil	*Ocimum basilicum*	Annual	Seed (covered)
Thyme	*Thymus vulgaris*	Perennial	Seed (uncovered); cutting; layering
Rosemary	*Rosmarinus officinalis*	Perennial	Seed (uncovered); cutting; layering
Sage	*Salvia officinalis*	Perennial	Seed (covered); cutting; layering
Mint	*Mentha* species	Perennial	Seed (uncovered); cutting; division
French Tarragon	*Artemisia dracunculus* var. *sativa*	Perennial	Cutting; division
Lovage	*Levisticum officinale*	Perennial	Seed (covered); division
Oregano	*Origanum vulgare* ssp. *Hirtum*	Perennial	Seed (uncovered); cutting; layering
Marjoram	*Origanum marjorana*	Perennial	Seed (uncovered); cutting
Lavender	*Lavandula angustifolia*	Perennial	Seed (uncovered); cutting; layering
Cilantro	*Coriandrum sativum*	Annual	Seed
Winter Savory	*Satureja montana*	Perennial	Seed (uncovered); cutting; layering
Summer Savory	*Satureja hortensis*	Annual	Seed (uncovered)
Chives	*Allium schoenoprasum*	Perennial	Seed (covered); division
Catnip	*Nepeta cataria*	Perennial	Seed (covered); cutting; division
Anise Hyssop	*Agastache foeniculum*	Perennial	Seed

Root Division

A damp, overcast day in spring is always the best time to divide plants. If there's a little drizzle or light rain in the air, so much the better. On our farm, French tarragon and mint are the main candidates for division. We do them every spring.

Tools. You will need a wheelbarrow with a few inches of damp compost in it, a large garden fork or shovel, a small hand fork or trowel, a sharp knife, a pair of pruning clippers, and a hose or watering can.

Digging. Select the healthiest, most vigorous specimens for division. Using the shovel, dig a wide and deep circle around the first candidate — a circle that is 10 to 15 inches deep and about 18 inches in diameter

Distance Between Plants (in.)	Days to Harvest (from Seed)	Cold Tolerance	Light Preference
8–12	75	to 15°F (−9°C)	Full or partial sun
12–15	75	to 35°F (2°C)	Full sun
10–15	100	to −20°F (−29°C)	Full or partial sun
10–15	150	to 10°F (−12°C)	Full or partial sun
12–18	130	to −20°F (−29°C)	Full sun
12–15	60	to −20°F (−29°C)	Full or partial sun
18–24	n/a	to −20°F (−29°C)	Full sun
24–30	120	to −30°F (−34°C)	Full or partial sun
12–15	100	to −20°F (−29°C)	Full sun
10–12	100	to 20°F (−7°C)	Full sun
18–24	130	to 5°F (−15°C)	Full sun
8–12	55	to 25°F (−4°C)	Full or partial sun
10–15	100	to −10°F (−23°C)	Full sun
10–12	65	to 35°F (2°C)	Full sun
8–12	80	to −30°F (−34°C)	Full sun
15–20	80	to −40°F (−40°C)	Full or partial sun
8–12	80	to −15°F (−26°C)	Full or partial sun

should be fine, as long as you are not encroaching too far into the root zone of a neighbor. Dig up the entire plant, trying to minimize any damage to its root system. As you dig deeper you might find that the garden fork (or small hand fork or trowel) does a better job of loosening the soil around the roots.

Dividing and transplanting. Once you have removed the entire plant, separate it by hand, if possible, into two or more pieces (you might get several), with some healthy root attached. With a little coaxing, some herbs will break apart along natural lines. Others might need to be snipped with the pruning shears or cut cleanly with the knife. Set the several pieces in the wheelbarrow, lay them down at an angle, and cover the roots with damp soil or compost— this is called "heeling in." Proceed to dig up more plants as needed.

The next step is pretty obvious. You need to find a new home for each divided piece. Dig holes big enough to accommodate the roots with minimal bending. Refill the holes with soil, putting the top few inches of soil in the bottom of the hole, along with some of the compost in your wheelbarrow. Tamp down, and water in well — the water should permeate at least as deep as the deepest root. Unless there's plenty of rain in the following few weeks, it's wise to go back and give the divided plants a couple more deep waterings.

For those herbs (and other plants) that will tolerate it, division is straightforward and effective, even if it does appear to be a somewhat violent act. The new pieces take off surprisingly quickly and often can be cut for harvest within a couple of months, but if you do this, don't take everything — be sure to leave some new green growth behind.

Layering

Among the herbs we grow, lavender, rosemary, sage, thyme, and oregano are candidates for layering. Early spring, when the soil begins to warm up, is a good time to undertake this process.

First select one or more supple, low-lying branches of the plant you want to reproduce. Instead of severing these branches, leave them attached. Lightly scrape away a one or two inch strip of bark on the branch, then bend down this portion and press it into a couple of inches of soil adjacent to the mother plant. You will usually need to hold the branch down with a wire staple or a rock.

Keep the soil moist. If all goes well, within several weeks the layered portion of the branch that is in contact with the moist soil will develop its own roots. When this happens, you should feel some resistance if you remove the staple or rock and gently pull on the branch.

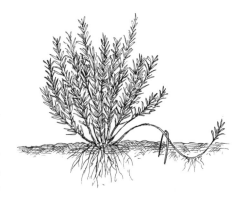

A supple outer branch is lightly stapled to the ground and a portion of its stem is buried in the soil.

The next step is to use a pair of pruning shears to sever the branch from the parent plant. You now have a separate and new plant with exactly the same genes as the parent. You can dig it up and relocate it wherever you wish.

Layering doesn't always require human intervention. Some herbs and other plants, especially those with creeping growth habits, are quite capable of doing it by themselves when conditions are right and one or more of their branches touch moist soil.

A Few More Tips on Harvesting

Most annual herbs can be harvested repeatedly, either by snipping off enough sprigs to make a bunch or by cutting the entire plant a few inches above the ground. By cutting back annuals, you will stimulate new growth and ensure future harvests — that is, assuming the plants are getting enough sunlight, water, and fertility to keep them happy. Of course, when winter approaches, the annual's life cycle comes to an end.

Perennials should be treated a little differently. In their first year, allow them to put on a good amount of growth before going at them with a harvesting knife. When you do harvest perennials, avoid cutting them too close to the ground; leave at least a third of their height intact.

Once they are well established, perennials may be harvested repeatedly, though it's a good idea to leave them with some aboveground, leafy growth as they head into winter. Heavy harvesting at the end of the season will deplete their energy reserves and can jeopardize winter survival. This often presents us with a vexing choice, since demand for sage, thyme, and rosemary always climbs dramatically before Thanksgiving.

Harvest before flowering. When herbs flower, their flavor usually deteriorates — in varying degrees, depending on the herb in question. As a general rule, harvest herbs for culinary use before they flower, and if they do flower, cut them back and wait for regrowth.

Exceptions to this rule include lavender and sometimes sage and thyme. Our lavender sales always go up in midsummer when the plants have their beautiful blue flowers. Bunches of flowering sage can also be attractive; we sell them as miniature floral bouquets, with the option of using the leaves for culinary purposes. We've also sold flowering thyme to New York City chefs who sprinkle the little white flowers over some of their edible creations. Most people, though, prefer their sage and thyme without flowers.

Harvest in the morning if possible. For maximum flavor and fragrance, most serious herb growers recommend harvesting herbs in the early part of the day, when their volatile, aromatic oils are most potent. The plants should be dry but not overly heated by the sun. I'm sure this is good advice, but I have to admit that we usually harvest our herbs in late afternoon.

Pick days on our farm (always the day before a market) are very busy. There's a lot to harvest — much more than we can accomplish in a few morning hours, so we are forced to prioritize. Lettuces and other greens quickly get limp on a hot day, so we bring them in first. Crops such as peas and beans also hold up better when harvested under cooler conditions. During the midday hours, we dig potatoes, pick tomatoes or peppers, or wash mesclun salad greens — this last one is a coveted indoor job on a hot day.

That leaves us with late afternoon for cutting and bunching herbs. It might not be the ideal time for capturing those essential oils at their peak, but for us, at least, cutting herbs as the sun goes down is a peaceful

Medicinal Herbs

Though our farm made an excursion into the world of medicinal herbs in its early days, experimenting briefly with skullcap, feverfew, St.-John's-wort, and a few others, we are now firmly entrenched in the culinary herb camp. This is not to say that many of the herbs we grow don't possess medicinal properties. They do, but we market them for use in the kitchen and leave it to our customers to explore their curative aspects, should they choose to. One good resource is *Rosemary Gladstar's Medicinal Herbs: A Beginner's Guide* (Storey Publishing, 2012).

Some of the strictly medicinal herbs are quite easy to grow, but I soon realized that I was not comfortable explaining to earnest customers how to use feverfew to ease migraine headaches or St.-John's-wort to treat depression. My knowledge of these herbs and their healing properties did not seem equal to that responsibility, and the demand was not great.

and restorative activity. If we were intending to dry our herbs, we would certainly go for a morning harvest.

Use coolers, but be careful. I've already mentioned that we like to transport our herbs to market in coolers. This keeps them fresh and helps retain fragrance. Frequently, we place several different herbs in the same cooler, but we always keep the different varieties separate — each in its own open-mouthed plastic bag. This method works only when herbs are harvested under dry conditions. Wet herbs that are placed in a cooler or in a plastic bag, where there is no air movement, can blacken and become rank, especially as the temperature rises. If you're forced to harvest in the rain, always pack your herbs in open crates.

CHAPTER 15 RECAP

Seven Reasons to Include Herbs in Your Crop Plan

1. Most herbs are easy to grow.
2. Most herbs are relatively pest free.
3. Herbs don't take up a lot of space.
4. Herbs don't weigh much.
5. Herbs smell good.
6. Many herbs are perennials.
7. Herbs are a great add-on item at the cash register.

The Neighborhood Farm

Operators: Kate Canney and Jude Zmolek

Location: Needham, Massachusetts

Year Started: 2008

Acreage: 2.5 acres (11 different sites)

Crops: Mixed vegetables, herbs, and flowers

KATE CANNEY GREW UP IN NEEDHAM, MASSACHUSETTS, a suburb on the outskirts of Boston, and she's back there today. From a very young age, Kate dreamed of being a farmer. Before arriving at our place in 2001, she had studied ceramics in art school, graduated from the University of Massachusetts with a degree in sustainable agriculture, and hiked all 2,160 miles of the Appalachian Trail. Kate moved on to receive a master's degree in education and taught school for 4 years but always hankered to be outside and felt an overriding desire to grow plants. The main obstacle she faced was access to land.

Walking through her suburban neighborhood, Kate kept noticing all the vacant yards and lawns that could be put to productive use, and the idea of the Neighborhood Farm was born. She approached landowners and proposed that they allow her to convert their largely unused front yards or backyards into vegetable gardens. They would receive a portion of the harvest, and Kate would take the rest, and she would do all the work. Within a month, she had more than a dozen land offers.

Not all of the plots that were offered were suitable for vegetable production. Shade trees in the wrong places ruled out some. Shallow soils and ledge rock ruled out others. And because she wanted to grow organically, Kate eliminated any land that had received lawn chemicals within the past 3 years. The best plots were those that were larger in size and relatively level, possessed good soil, and had at least 8 to 10 hours of full sun during the growing season. It was also important that they be clustered in a relatively small area. Traveling long distances between sites was not practical.

In her first year, Kate selected five plots from those that were offered. Together, they totaled just ¼ acre. Four years later, the Neighborhood Farm is growing vegetables on 11 different sites, totaling 2.5 acres. It has evolved into a viable microfarming business that generates enough income to support both Kate and her partner and wife, Jude Zmolek, who joined the business in 2009.

In its second year, the Neighborhood Farm purchased a BCS walk-behind tractor with a rototiller and a rotary plow. The rotary plow is good for breaking up sod before the tiller comes in to prepare planting beds. Kate also notes that the plow does a good job of digging trenches for planting potatoes. Next on the list of implement acquisitions for the BCS is a mower to cut cover crops before incorporating them into the soil.

The Neighborhood Farm makes some of its own compost and buys the rest. In addition, they use leaf mulch, straw, bagged organic fertilizer, and foliar sprays to build fertility and suppress weeds. Crops are watered through drip irrigation that is fed from the landowners' municipal water supply. Each of the sites is subjected to a soil test before planting commences. The Neighborhood Farm's principal crops are tomatoes, garlic, assorted leafy greens, cucurbits, peppers, eggplant, carrots, onions, and herbs.

According to Kate, developing a crop rotation plan for 11 sites, of different shapes and sizes, can be challenging, but the setup also has some advantages. Often, a single plot is used for two or three mutually beneficial crops, such as tomatoes and basil or pole beans and lettuce. Because there is usually plenty of intervening land between plots, pests and diseases seldom have the opportunity to skip from one site to another. Most of the Neighborhood Farm's produce is sold at two farmers' markets in the Boston area.

In exchange for providing the use of their land, the property owners receive credit for vegetables — $75 worth of produce for every 500 square feet of land, up to a maximum value of $350. The landowners mostly use this credit at Kate and Jude's two farmers' market stands. This means they can choose from a range of produce, rather than be limited to only those vegetables grown on their land.

Kate credits some of the Neighborhood Farm's success to building strong relationships with nearby local farms. These connections have enabled Kate and Jude to barter for greenhouse space, mine for advice, and borrow equipment and even workers on occasion.

Kate and Jude are not the only people farming in backyards. Google "SPIN [small, plot-intensive] Farming" and you'll see there's a growing mini-movement in microfarming.

PART V

HARVESTING AND MARKETING

....

Chapter 16

....

Harvesting and Storage

UNLESS YOU RUN A STRICTLY PICK-YOUR-OWN operation, harvesting and preparing crops for market will occupy a fair amount of your time — perhaps as much as one-third; therefore, it is a subject to be taken seriously. Harvesting vegetables and herbs on a small, diversified farm is a largely hands-on affair that will very likely take place more than once a week during the growing season. It is labor intensive and seldom automated, and you will almost certainly be dealing with a changing assortment of crops through the course of the season as well as variable weather conditions. So whatever system you develop needs to allow for flexibility. It will also need to address the matters of efficiency and speed, but I'll have more to say about these two critical elements later in this chapter.

Before you harvest, you will need to develop a pick list, along the lines of what I described in chapter 3 (see page 61). The main point to remember is that you want to bring in the appropriate amount of each crop to meet anticipated demand — whether that be for a farm stand, a CSA, a farmers' market, or any other outlet you've lined up. If you're harvesting for a restaurant, a health food store, or a college cafeteria, they will most likely tell you the quantities of each vegetable or herb they want. For most other venues, the target amounts will be somewhat less precise and may vary, depending on availability in the field or in storage.

The key element is always anticipated demand. It's okay to harvest a little more of the perishable items than you think you can sell (up to about 10 percent) but not less, if you can help it. You have more leeway with storage crops, such as onions and potatoes, because if they don't sell, you can bring them back to market another day.

Best Time of Day to Harvest

Most crops are best harvested in the early part of the day, before the sun gets hot, then quickly moved into a cool storage environment. During summer, however, there usually aren't enough cool morning hours in a day to do all the harvesting that needs to be done, so you have to pick and choose or, should I say, choose and pick. If you have an abundance of crops in your fields, it's unlikely that everything you want can be harvested at the absolute optimum time.

Once again, I use our farm as an example, with the full understanding that others might approach harvesting a little differently, especially in different regions of the country. As always, there are a lot of variables to consider and no single guidebook in this farming endeavor. And who would want one? Part of the satisfaction of being a farmer comes from developing systems over time that work for you — given your land, your crops, your climate, your resources, your skills, your markets, and even your personality. Allow some time for this to happen.

On a typical midsummer pick day on our farm—let's say, a sunny day in mid-July — we begin by cutting mesclun salad greens because these are the most delicate and fragile greens we have, and the most prone to wilting as the sun climbs in the sky. After the mesclun, we move on to lettuces and other bunching greens such as kale, dandelion, and Swiss chard. Once all the greens are out of the way, we go for peas (if there are any left in midsummer) and beans, then proceed to summer squash, cucumbers, and zucchini — items that can handle a lot more heat.

By midday, two or three of us begin washing and spin-drying the mesclun greens while others dig potatoes and garlic or pull green onions, shallots, and scallions. After lunch — by which time it is usually as hot as it's going to get—we pick tomatoes, peppers, and eggplant. Tomatoes are a major crop for us. Once they are in full flush (usually not until late July in our area), it can take 2 or 3 hours to pick them, with all hands on deck.

The last stop of the day is herbs, which is somewhat regrettable. As mentioned in the previous chapter, it is generally considered best to pick herbs in the early hours of the day while their essential oils are strongest. We would like to do this but simply don't have the time and feel it is more important to harvest the greens before the day heats up. At least by late afternoon, the sun is low in the sky and the temperature has moderated. We hope that the herbs (and our herb-buying customers) appreciate this small, end-of-day deference we pay to them.

All crops will be sold at market the following day. The garlic may not yet be fully mature, nor the onions and shallots, and the potatoes are probably smaller than they will be a couple of weeks down the road.

Advantages of Selling Direct

Selling directly to the public is one of the advantages of being a small farmer — you don't have to conform to the supermarket model and sell only mature vegetables of a certain grade and size. Young, immature plants can be tender, delicious, and nutritious, especially when eaten within a couple of days of harvest. We offer customers truly fresh, unusual, and often highly perishable products that they would be hard-pressed to find in any other retail food outlet. These atypical offerings make shopping in a farmers' market a uniquely rewarding experience.

Nevertheless, we want to take each of these crops to market because we know they taste good, we know our customers will buy them, and we want our stand to look bountiful.

As the season progresses, the picture changes. At our place, by mid-July it is usually time to harvest mature garlic en masse, and a little later, the same goes for onions and shallots. Harvesting these alliums chews up a fair amount of time. By mid- or late September, when we have less field work to do, we bring in all the remaining spuds. Winter squash and pumpkins come next — usually in early October. Once we're through harvesting these storage crops, on a pick day, we simply retrieve what we think we need from our root cellar, barn, or basement. Later in this chapter, we'll take a closer look at how to harvest large quantities of specific crops, then store them.

Harvesting Tools

Because harvesting crops can use up a lot of your time, it pays to use the right tool for the job. By this I mean the tool that will enable you to get your work done as efficiently and speedily as possible, and with the least amount of bodily effort. That tool will vary depending on the crop being harvested and the grower's personal preferences. Here, we'll talk about some of the different knives and cutting shears that small farmers use. We'll also consider the different types of containers into which you might harvest your crops.

Knives

Our most common harvesting tool is a knife, preferably a sharp one with a stainless steel blade. In a pinch, any serviceable knife will do the job, but there are specially designed harvesting knives that will make your life a little easier. Sooner or later you'll want to get your hands on some of these. If you grow lettuces, it's good to have a lettuce-cutting knife. These knives have a broad tip, which the user presses into the stem of the lettuce just above soil level. A lettuce-cutting knife will also do a good job of cutting cabbages and cauliflower. If kept reasonably sharp, it will slice off a head with one firm, well-directed thrust.

A somewhat larger, heavier, more traditional knife works well on bunching greens such as kale, collards, and chard. Broccoli, with its thick, tough stem, also calls for a sturdier knife, which can be used in a short slashing motion almost as though it were a machete. Some growers favor knives with sickle-shaped blades. They are often serrated and do a good job on most greens, especially those with heavy stems, such as brussels sprouts and broccoli. It takes a little getting used to the hooked cutting motion of these sickle-style knives.

Many seed and garden supply catalogs sell a range of harvesting knives. Try a few different models, decide which you like best, then stock up on them. Once you've made your choice or choices, stick with them for a while. Finger-cutting accidents are less likely to occur when you use a knife that you are familiar with and when you wear a glove on the non-knife-holding hand (see page 521 for more advice on the safe use of knives).

Harvesting knives

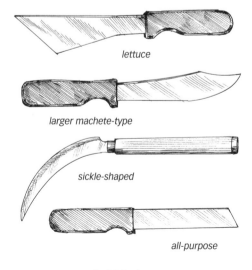

lettuce

larger machete-type

sickle-shaped

all-purpose

A Word about Sharpness

When you're in the field wielding a knife for a few hours, it definitely pays to keep it sharp. A sharp knife will allow you to get the work done more quickly, more safely, and with less expenditure of energy. But you don't need to go overboard and hone the blade to a razor edge that you could shave with. If the angle of the blade is too acute, it will blunt more rapidly, especially when it comes in contact with soil, which is highly abrasive.

If you're really into keeping your knives sharp, you might want to acquire a belt sander. In knowing hands, these do a good job, but they are expensive. Electric knife sharpeners, used by many chefs, are another option. They enable you to guide the blade between two grinding elements at an appropriate angle (usually between 20 and 25 degrees), but most of them can accommodate only thin-bladed knives. If you're drawn to the tools of our ancestors, look for a secondhand revolving whetstone. These relics are low-tech, but once you learn how to maintain the proper angle, they will do a nice job. Or you might try a more readily available stationary whetstone.

There are a number of other sharpeners on the market, but many of them don't yield very satisfactory results or last very long. Some farmers find that a simple file is all they need. Josh Volk, who farms in Oregon and writes a highly informative column for *Growing for Market,* uses a 6-inch single-cut mill bastard file to sharpen most of his harvest knives and hoes. (The word "bastard," here, does not connote any illegitimacy on the part of the file but rather refers to the size of its teeth — somewhere between coarse and fine.) Note: If you're using a file, they cut on the forward stroke only. It is counterproductive for both the knife and the file to drag them back across the blade.

For more information on sharpening or to buy supplies, check out the resources section.

I tell my crew members to select a knife or a couple of knives (for example, a lettuce-cutting knife and a larger knife for other greens) and become familiar with them. I encourage each person to take responsibility for keeping his or her own knife, or knives, sharp (most don't know how to do this, so I give some training) and discourage them from using the knives of coworkers. A piece of tape around the handle, with the individual's initials on it, helps in this regard. This policy reduces the incidence of both knife misplacement and cut fingers.

Shears

Every grower should have one or more sets of pruning shears and lopping shears in the tool shed. They'll come in handy for a variety of jobs around the farm. Here, I'll focus on the merits of each as harvesting tools and mention some of the crops they work best on.

Pruning shears. Although I'm happy to cut most herbs with a sharp knife, some of my workers prefer to use bypass pruning shears. They do have their advantages, especially when you're cutting older, perennial herbs — such as sage, thyme, and rosemary — which can have tough, woody stems that are hard to slice through even with a sharp knife. Pruning shears can also facilitate the harvesting of such fruiting crops as peppers and eggplants and reduce damage to these sometimes less-than-robust plants, which can occur when you simply pull off the fruits with your hands.

High-quality pruners will last a long time but are expensive. They are probably a good investment if you are able to look after them and keep

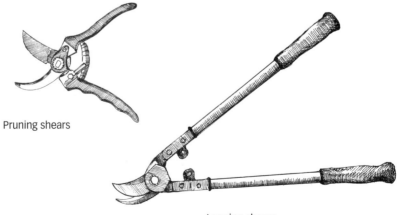

Pruning shears

Lopping shears

track of their whereabouts. But you might opt instead for the less costly models from China—they work okay for a year or two, until they become lost or fall apart. One nice thing about pruners is that, unlike knives, they don't need to be sharpened often.

Lopping shears. These are essentially long-handled, heavy-duty pruners. Lopping shears are the best tool for harvesting winter squash and pumpkins, the stems of which are usually a bit thick and hard for the lighter-weight pruning shears to cut through. It is important to leave some stem on these vegetables; when they break off entirely, the fruits will have a much shorter storage life, and very few people will buy a pumpkin without a stem. We try to leave 1 or 2 inches of stem on winter squash, and 2 to 4 inches on pumpkins. On pumpkins, longer stems look good, but they break off more easily. One good thing about lopping shears is that you don't have to bend over so far to snip the squash and pumpkins lying on the ground.

Harvesting Containers

With the possible exception of large pumpkins (which you might simply stack in the back of a pickup truck because they don't fit well in crates), you need to place crops harvested from the field in containers so they can be easily transported to a storage environment or a market-bound truck. There are numerous choices. We started out using wooden crates but long ago switched to plastic lugs because they are lighter and easier to clean. Whatever containers you choose, make sure that they can be easily stacked on top of each other.

If you sell at a farmers' market or farm stand, you will most likely use your harvest containers to display produce, so you want something that will function well in both settings. If you're delivering to a restaurant or an off-farm CSA, you might prefer to use waxed cardboard cartons with the expectation that you won't see them again, unless you can make an arrangement to have them returned to you.

Plastic lugs. If you buy them new, durable harvesting containers can set you back a fair bit of cash. Look for used ones at vegetable farm auctions or going-out-of-business sales. They usually still have many years of good service in them.

In our area, there are several importers of tulips and other bulbs from Holland. The bulbs come in reasonably sturdy, ventilated plastic lugs that are 23½ inches long, 15 inches wide, and 9 inches deep. Some years, the bulb importers send the empty lugs back to Holland for reuse; other years, they find the freight charges too high and are willing to sell the lugs to us and other farmers for $2 apiece. Over a dozen or more years, we've managed to acquire about eight hundred of these secondhand lugs. If we had purchased new containers of similar dimensions here in the United States, we'd have spent at least five times as much.

You might think that eight hundred plastic lugs is excessive, but the number is not so high when you consider that we use them not only for harvesting crops and taking them to market, but also for storage. On our farm, a good crop of potatoes, garlic, or winter squash can easily occupy a couple hundred lugs each. Add onions, shallots, carrots, and other root vegetables, and you can see why we need a lot of containers. If we run low on lugs, we usually store the alliums in mesh sacks, but these don't stack well.

Plastic tubs. Plastic tubs with lids are good for harvesting mesclun salad greens. They are a bit lighter than the Dutch bulb lugs, and because they don't have any ventilation openings, there is no danger that small leaves will fall out. The lids keep the sun off the delicate greens but will cause them to heat up, so you need to move them to a cool and shaded place as soon as possible.

Shopping baskets. Supermarket shopping baskets come in handy when picking fruits such as tomatoes, peppers, and eggplant. You can carry the basket in one hand and use the other hand for picking. Once a basket is full, we usually transfer its contents to a lug. These same shopping baskets can also boost sales at our farmers' market. Customers use them to accumulate different items, just as one would at a supermarket, before coming to the cash register. The last time we bought shopping baskets, we received a substantial discount from the supplier because we were willing to accept baskets with other businesses' misprinted logos on them. These misprinted baskets do the job just fine, and our customers find them amusing and in keeping with the rough-and-ready ambience of our stand. If you like to cut costs, and don't mind asking questions or waiting for the right opportunity, there are plenty of good deals out there.

Berry baskets. If you grow cherry tomatoes, and I recommend that you do, you'll find you can minimize cracking or splitting of fruits (this often occurs after rain) by harvesting them directly into pint-size berry baskets. We place six empty berry baskets into a shopping basket and go to work. Once they are filled, we set them into a lug, which can hold

15 pints, then load up the shopping basket with another six empties. It's the fastest way we've come up with to bring in the cherries.

These little baskets are made of green pulp fiber or plastic and are the same type you see at supermarkets and grocery stores. The pulp fiber baskets are less expensive and easier on the environment. They are less reusable than the plastic type, but this isn't highly relevant since most buyers of cherry tomatoes prefer to take the baskets with them.

Berry baskets — including half pint and quart sizes — are also good for marketing other small vegetables such as beans, peas, shallots, and small onions and potatoes, though it is usually easier to harvest these crops into larger containers and transfer them to berry baskets later on. Of course, berry baskets are also perfect for small fruits and berries.

Coolers and cold packs. If you don't have a refrigerated truck (we don't), you can keep certain high-value crops cool — both on your truck and while at market — with camping coolers and frozen cold packs. We use them for mesclun, washed arugula, peas, beans, and most herbs. They keep the crops fresh, and as a bonus, they concentrate and hold onto the pleasant aromas. But beware: Direct contact between a frozen cold pack and a vegetable or herb will damage the crop, so always be sure to wrap the cold pack, or separate it in some way from the contents of the cooler.

Because they are insulated, coolers also come in handy in cold weather. If the temperature is forecast to drop below 32°F (0°C), we put potatoes, squash, carrots, cabbage, and the like in coolers when we load our truck the night before market. We then place a small electric heater in the back of the truck overnight. This way, we seldom have a problem with anything freezing.

Harvesting Methods

Just as different growers will have their preferred harvesting tools, they will also have their favorite harvesting methods, including their own approaches to postharvest cooling, washing, and bunching. But there are a few basic dos and don'ts that almost all of us have to abide by. These mostly relate to the weather.

Working around the Weather

Except in regions with extremely stable climates, the vagaries of weather are a fact of life that farmers must constantly take into account. This applies to harvesting as much as it does to most other tasks that need to be done on a farm.

For leafy greens, the ideal harvesting weather is cool, overcast, windless, and a little damp. Warm and dry weather is good for tomatoes, peppers, eggplant, summer squash, and alliums. Cool and dry weather is better for beans and peas and will work well for most herbs.

Root crops, such as potatoes, carrots, turnips, and radishes, are not too fussy, though it is generally easiest to dig them when the soil is a little moist but not wet. You just don't want to leave them lying in the sun for too long. One exception is potatoes dug for storage. Immediately after digging, they will benefit from an hour or two in the sun; this will harden their protective skins and make them store better. Winter squash and pumpkins have few harvest demands, as long as they are dry and mature and have not been exposed to frost.

These are the preferred harvest conditions for many common vegetables grown in temperate regions, but you will seldom find all of them lining up for you in a single day. All you can do is watch the weather and allocate your time in a way that will do the most good and the least harm. Let's take a closer look at two important harvesting considerations: shade and rain.

Look for shade. Once they are harvested, most vegetables, but especially greens, should be moved to a shaded spot as soon as possible. Being exposed to hot sun for even a very short time takes a toll on them. Unless the sun is right overhead, any large object in the field — a tree, a hedgerow, a tractor, the shaded side of a pickup truck — can provide temporary shade to harvested crops. Or you might try building some type of covered wagon to tow behind your tractor or pickup.

Watch out for rain. If significant or prolonged rain is forecast on a pick day, it might be wise to harvest certain crops — such as tomatoes, especially heirlooms and cherry tomatoes — a day earlier, so you can bring them in dry. Too much water, usually in the form of rain but sometimes from excessive irrigation, will cause fruits to crack, which greatly reduces their marketability and storability. Also, handling the foliage of wet tomato plants can spread some of the diseases that beset this high-value crop — another reason to harvest under dry conditions.

Onions, garlic, and shallots are also best harvested when dry. But if they have to come in wet, you can always dry them off indoors with a fan. Green beans are another crop to pay close attention to. After harvest, we like to pack our beans in coolers with frozen cold packs to keep them crisp and fresh; however, putting wet beans into a cooler, which has no air movement, is a big mistake. They will soon develop a fuzzy mold and sometimes brown spots. If we have to bring beans in wet, we always set them out in front of a fan for an hour or two before packing them for market.

The same goes for most herbs. Never pack them wet in a sealed container such as a cooler; they'll blacken and go rank. Either dry them off or take them to market in open containers, which allow air to pass through. Potatoes are another crop to be careful with. On hot days, if you think washing is necessary, always allow the tubers to cool down first. Washing potatoes while they still have field heat in them is a bad idea. The pores are open and will absorb water and surface germs, which will very likely cause the tubers to rot.

It's not good for your soil to harvest crops during or after heavy rain. Even foot traffic on wet soil can cause significant compaction. If you have a permanent bed system, with grassy pathways in between beds, this will not be such a problem. Wet grass or sod can handle foot traffic a lot better. Just try not to step in the growing beds.

Cooling after Harvest

To prevent leafy greens from wilting and to preserve freshness, it's important to get the field heat out of them as soon as possible. A refrigerated room, if you have one, will do this. A faster and less expensive method is hydrocooling, which is a fancy way of saying dunk the crop in cold water for a few seconds, or if you prefer, sprinkle it with cold water for a couple of minutes. Then, store the crop in an air-conditioned or refrigerated room if you have one, or some other cool space.

Not all crops need to be wetted after harvest. All the alliums (onions, shallots, and garlic) are best left dry, and there is seldom much to be gained from wetting crops such as squash, cucumbers, potatoes, tomatoes, peppers, and herbs. Peas and beans may be briefly hydrocooled but should be allowed to dry off well afterward. Leafy greens are the main candidates for hydrocooling or sprinkling with cold water.

Washing — Yes, No, and Maybe

Though both involve water, washing and hydrocooling are not quite the same thing. The purpose of hydrocooling is to lower the temperature of a crop. The purpose of washing is to remove dirt or dust. Certain crops look better and sell better after they have received a quick wash. These include most of the root crops — carrots, turnips, beets, parsnips, radishes, kohlrabi, and celeriac — which usually come out of the ground with some soil on them. Note the omission of potatoes. We choose not to wash potatoes unless they are harvested under wet conditions and have some soil on them. If they are dirty, washing is highly appropriate, once the tubers have cooled down.

Whether to wash leafy greens and crops such as tomatoes and peppers is a matter of choice. Aside from the root crops mentioned above, most of the vegetables harvested on our farm and taken to market are usually sold unwashed. We make an exception when heavy rain has splattered soil onto the leaves of our lettuces and other greens. Otherwise, we leave it to our customers to wash what they buy, to the degree that they see fit.

We feel that a little field dirt on a potato or the bottom of a lettuce, where it had contact with the soil, is not a problem and, in fact, is a good reminder that we are a small, local farm — not a giant, impersonal food corporation. Our customers understand that we grow food without the use of chemicals and sell it pretty much straight out of the ground — no Styrofoam, no shrink wrap, no gas treatment, and none of those infuriating sticky labels that seem to decorate every piece of fruit that the industrial food system has to offer, some fair number of which, presumably, enter the stomachs of unwary consumers daily.

Having just delivered that little tirade, let me now say that there are some crops that we do wash very carefully — principally, mesclun salad mix and arugula. These are washed in large tubs in a cool, clean room in our lower barn, then dried in a commercial salad spinner and put

Potable Water Test

Washing should always, of course, be done with clean water. Water from ponds or streams will not suffice. In fact, organic certification requires an annual potable water test for any water used for washing produce. State and local departments of health can tell you how and where to get a potable water test.

Excesses of Youth

> ## My Story

IN MY FIRST COUPLE OF YEARS OF FARMING, my desire to keep produce looking fresh knew no bounds. Instead of loading the truck the night before market (which is what we do now), I would get up at 2:30 in the morning and do the job then. Since I didn't have a cooler, I would leave crates of greens sitting outside with an oscillating sprinkler misting them through the night. This was a case of hydrocooling taken to the extreme. It definitely kept the greens looking good, but loading the truck so early in the morning added an extra 45 minutes to an already very long (usually 17 hours) day.

But don't let this story dissuade you from considering selling at a farmers' market. Our market in New York City, which starts very early and runs until the end of the day, does not represent the norm. Most farmers' markets don't run so long and usually are closer to the farmers that serve them, which means there is less travel time. In any event, the early-morning truck loading didn't go on very long. I soon got older and wiser.

into clean 48-quart camping coolers with cold packs in the bottom. This way, they stay cool to the point of sale. Our signage clearly indicates that the mesclun and arugula are washed and ready to eat, and our pricing reflects the extra effort we have put into them, which is considerable.

Salad spinner. If you plan to sell mesclun or other greens as washed and ready to eat, I recommend that you invest in an industrial-grade, electric salad spinner — the type that is used in restaurants and salad bars. They cost at least $500, but they'll spin a lot of salad for you, and you won't have to worry about contracting a case of salad spinner's elbow (never a good thing). Some cost-conscious growers have modified old washing machines and clothes dryers for the same purpose.

Wash station. If you are of a mind to wash most of your vegetables, sooner or later you'll probably decide to build a wash station and equip it with sinks, tubs, hoses, and a good supply of water. A simple outdoor wash station might be housed in a pole barn with one or more open sides. It's easier to wash your crops while standing rather than bending over, so factor in comfortable work surfaces a few feet off the ground. You'll also need good drainage and a way to collect dirt and debris, as well as enough room for a couple of people to move around.

Barrel washers and other devices. Depending on how many root crops you grow, you might find it worthwhile to invest in a barrel washer. These rotating barrels, which have multiple drain holes or separated wooden slats, shoot jets of water at such crops as carrots, beets, turnips, parsnips, celeriac, and potatoes (if you choose to wash them). Dick de Graff of Grindstone Farm in Pulaski, New York, makes and sells an excellent barrel washer.

Other washing-device options include brush washers, which pass root crops over rotating brushes whiles spraying them with water, and batch washers, which feature sprinklers, rotating brushes, and a walled-in conveyor belt.

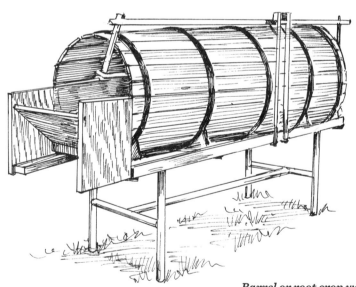

Barrel or root crop washer ·
The pitched, rotating barrel shoots jets of water onto carrots, parsnips, turnips, and other root vegetables.

Bunching

We believe that most vegetables and herbs that you can put a rubber band around should be sold by the bunch, rather than by weight — for the simple reason that it sells more product. For a more in-depth discussion of the virtues of bunching, see page 443.

Most of our greens and herbs are bunched in the field. We use a wide, #32-size rubber band for greens, scallions, turnips, beets, and the like, and a smaller and narrower #12-size band for herbs. It's important not to wrap the rubber band so tightly around the stems of the vegetables or herbs that you leave a permanent crease. Some growers prefer twist ties to rubber bands for this reason. They put less crimping pressure on the stem of the bunch, but we find they slow us down in the field. Experiment with both, and see which you like best.

Bunches of the same vegetable or herb don't have to look exactly the same, but they should be close — in both size and weight. Most of our bunches of greens range between 8 and 12 ounces (we aim for 10 ounces).

Herb bunches should also be fairly consistent, though they will vary in weight and size depending on the herb in question. Bunches of mint, parsley, and sage, for example, are generally bigger than bunches of French tarragon, marjoram, or winter savory. It takes a bit of practice to get bunch sizes and weights to be fairly consistent, and not all harvesters are equally adept in this regard. Until the harvester's skills are honed, you may need to rearrange some bunches later, with the aid of a scale. When training new crew members, we sometimes bring a hanging scale into the field and attach it to the side of a pickup truck, so that people can periodically check their bunch weights.

We don't do all of our bunching in the field. We find it easier and more efficient to bunch scallions and immature garlic, onions, and shallots back at our barn. This gives us a better opportunity to brush dirt off the roots and remove any dry leaves. Crops that look better after a quick wash, such as turnips, beets, and radishes, are also bunched at our barn after washing.

Efficiency and Speed

Because harvesting crops and preparing them for sale can consume a lot of time, it's imperative that these tasks, as so many others on a small, diversified farm, be performed with efficiency and speed. It's not a matter of working harder or pushing yourself to the breaking point, but rather of developing good technique, along with focus, rhythm, and pacing.

You'll neither become a master of the harvest overnight, nor probably over the course of a single season, but you should get better with time. Try to remain open to the possibility of improvement, and keep on the lookout for more efficient and time-saving methods. Every athlete knows there are subtleties to even the simplest actions. Think like an athlete, and try to find those subtle moves.

Each harvester will find his or her preferred way of bringing in the crop, and it's not up to me to say how the job should be done, but I can't resist sharing a few pointers on how to improve harvesting efficiency:

- **If you're working on sloping ground, always harvest on the downhill side of your crop.** This way you won't have to bend over so far.

- **Avoid having the sun in your eyes.**

- **Stay on your feet, with your legs a comfortable 18 to 30 inches apart** (depending on how long they are) and your knees slightly bent, almost as though you were sitting on a horse. This position puts less strain on the back and requires that your legs do more of the work. Never sit down to harvest. It takes too much effort to get up and move to the next plant. In some cases, perhaps when working with herbs, kneeling is okay. If you do like to kneel, try wearing knee pads.

- **When using knives (or clippers) and rubber bands to bunch greens or herbs,** slip the knife into an open sheath while working with a rubber band. You'll find it's very difficult to hold everything (the knife, the bunch, the rubber band) in your hands at the same time. You can set the knife on the ground, but then you have to bend over to pick it up. A third approach, and one that I like, is to tuck the knife under one arm while using the rubber band. Just take care not to stab yourself in the armpit. I haven't done it yet!

- **Try keeping a supply of rubber bands on the handle of your knife,** or around the thumb or fingers of your non-cutting hand.

- **When using a knife, always cut at an angle (30 to 40°),** rather than straight across, and rely on the part of the blade closest to the handle, rather than the tip. The cutting motion should be an easy, focused stroke rather than a hacking action.

- **It goes without saying that your knife should be fairly sharp at all times.** You might consider carrying a small sharpening steel into the field, so you can touch up the blade periodically.

- **Check how you're doing.** Time how long it takes you to cut 10 or 20 bunches of a particular vegetable or herb, then try to beat your record without compromising bunch quality.

- **Work on the same harvesting job with one or two others, and see who is the fastest.** A little friendly competition is not unhealthy. We've found it sometimes helps to keep a running verbal tally of the bunches. If it turns out that someone is unusually slow at harvesting a particular crop, consider directing his or her energies elsewhere.

- **Don't allow your mind to wander too much.** Stay focused. You'll get the job done sooner and be less likely to cut yourself.

The payoff: Using your body effectively and efficiently and employing techniques that work well for you can make the harvesting process surprisingly satisfying. Understanding the ergonomics of efficient harvesting will enable you to get more work done in a given amount of time and feel better while you are doing it.

Short-Term Crop Storage

Most growers differentiate between short-term and long-term storage. Crops that are harvested for sale within 2 or 3 days fall into the short-term category. They are brought in from the field, possibly hydrocooled or washed, then placed in a cooler or some other cool place. They remain in this environment until it is time to transport them, usually by truck, to their destination.

The most common type of short-term storage is the walk-in cooler. On small farms, this is typically a well-insulated box, 6 or 8 feet wide, at least 10 or 12 feet long, and tall enough for a tall person to stand up in. The box should have a solid, insulated door that seals tightly when closed. When opened, it should be wide enough to walk through with a crate of vegetables in your hands. Consistent with its name, a walk-in cooler also needs to have some type of cooling device, traditionally a refrigeration unit consisting of a condenser, compressor, and evaporator. Purchased new, a fully outfitted walk-in cooler could set you back $10,000 or more.

Fortunately, secondhand insulated boxes are not hard to find, and there are low-cost alternatives to traditional refrigeration. Here are a couple of ideas.

Our Cooler

The walk-in cooler on our farm is an insulated box with interior dimensions of 8 by 16 by 7 feet high, with a window air-conditioning unit. This particular box once lived on the back of a snack-food delivery truck that was in an accident that smashed in one of the front corners of the box. The cab and engine were salvaged, but the box had sat around for a while, and no one wanted it. In fact, the owner of the yard where it was parked wanted to be rid of it. I paid him a couple hundred dollars to deliver the damaged box to our farm.

A friend and I then went to work: We cut out most of the mangled metal, straightened the rest with sledgehammers and pry bars, added some reinforcement, patched the outside with metal siding and self-tapping screws, put insulation into the wall space, and covered the interior walls with more galvanized metal siding. We then inserted an air conditioner where the truck's refrigeration unit had been. All told, the whole job cost about $700. It's a solid box with a heavy door that closes tightly. It does a good job for us.

Why only an air conditioner? You may be wondering why we chose an air conditioner over a refrigeration unit. Well, air conditioners are a lot cheaper, for one thing, and they use less electricity, for another. It's true they don't keep things as cold as refrigeration units do — most of them are programmed not to go below 60°F (16°C) — but we've found that this is an acceptable temperature for us.

Even at a constant 60°F (16°C), the greens in our air-conditioned box are usually a good 20 to 30°F cooler than they would be if left outside in the shade on a summer's day. At the end of the day, the greens are loaded onto our market truck, which is neither refrigerated nor insulated. It seems to me that moving vegetables from the field, where it is 85°F (29°C), to a refrigerated box that holds them at 35°F (2°C) for several hours, then moving them back outside to endure whatever the ambient temperature is, both in the truck and at market the next day, does not serve any great purpose. Others may see the situation a little differently.

If our market truck were insulated, I might see things differently myself, and this could come to pass, thanks to the ingenuity of a fellow farmer.

The CoolBot

Ron Khosla is a farmer/inventor. Among his inventions and adaptations is a device called the CoolBot. A CoolBot, when wired to a window air conditioner, effectively overrides the minimum setting and allows the air conditioner to chill down to as low as 34°F (1°C) without freezing up. Thousands of farmers are already using CoolBots in place of refrigeration units in their walk-in coolers, and a growing number are adapting them for use on insulated trucks as well. To learn more about it, go to Ron Khosla's website (see Resources).

Postharvest Treatment of Storage Crops

Fresh produce, picked a day or two before point of sale, is what really excites local food enthusiasts. But this doesn't mean they have little interest in storage crops that were harvested weeks or months before. Far from it. Well-stored alliums, root crops, and winter squash are also in high demand, especially in the colder months of the year, and most farmers are happy to oblige.

After all, the season for many fresh vegetables across much of North America is short. Sugar snap and snow peas are great while you have them, but depending on where you live, that might be for only 6 weeks out of 52. The same can go for tomatoes, if you live in northern climes. For a farmer who has a long marketing season, storage crops can provide income over an extended period of time. The only problem is that you need to have suitable rodent-free environments in which to store them. We'll get to that in a moment. First, I'd like to address the issue of what to do immediately after harvest.

Alliums. Crops such as onions, garlic, and shallots should be harvested once they are mature. Leaving them in the ground beyond maturity will shorten their storage life. After harvest, these crops need to be cured with the tops left on. Garlic is usually hung in bunches in a dry place, out of direct sun, for several weeks, or until the tops are fully dry (see page 354).

Onions and shallots can be spread out on the ground and left to dry in a covered space or outside in an area with partial shade, as long as there is no rain. But beware of hot, direct sun — it can cause a condition called sunscald whenever the plant's roots are removed from the soil, especially immediately after harvest on a hot day. On our farm, when onions and shallots are partially dry, we put them in half-full, open lugs, in a covered space, and blow fans on them for a few weeks. We stack the lugs in a U shape and put 4×8 sheets of plywood behind them and on the sides. This arrangement funnels the air from the fan and forces it up through the lugs, promoting more rapid drying. As soon as the tops of the alliums are fully dry, it's safe to cut them off and consolidate the bulbs in mesh sacks or lugs for long-term storage.

Winter squash and pumpkins. These crops can be harvested for sale as soon as they are mature and have their distinctive ripe color — usually 3 to 4 months after planting, depending on the variety. When harvesting for storage, the fruits should be dry and the rinds hard. Always treat squash and pumpkins gently — any nick or puncture of the rind can cause rapid infection and decay.

After harvest, winter squash and pumpkins can be left in crates, or even in piles, in the field for a few days, to harden off but should be brought into a protected space before the first frost. In a cool, dry environment, with some air movement, they can last for up to 3 months.

Root crops. Potatoes, carrots, celeriac, rutabagas, kohlrabi, and storing turnips can be harvested whenever they reach a satisfactory size, or toward the end of the season, before the ground freezes. They will keep in the soil, but the longer they remain there, the more vulnerable they become to wireworms, voles, and other pests. With the exception of potatoes, it's a good idea to wash these crops and then dry them off for an hour or two before storing them.

Long-Term Storage

Creating ideal, long-term storage conditions for each of the above categories of crops can be quite a challenge and one that, I expect, many growers are not able to fully meet. Unfortunately, there is no single environment that works well for all categories. To cover all bases effectively, you would need four separate storage environments. But remember, these are for optimum storage. Under less-than-perfect conditions, and with some ingenuity, you can still do a respectable job and hold many crops well into winter.

If you do have controlled storage facilities, you'll want to purchase one or more max-min thermometers (see box below) and hygrometers, so you can monitor both temperature and humidity in different storage areas and make adjustments when necessary.

All storage crops should be handled with care. During harvest, avoid any rough treatment that might cause bruising. You might not detect the damage caused by a bruise at the time it occurs, but you can be reasonably sure you or someone else will at a later point. Also cull vegetables with nicks, cuts, and exposed flesh. Injuries of this sort are entry points

Four Distinct Long-Term Storage Environments

Storage Environment	Temperature Range	Percent Humidity	Crops
Cold and dry	32–40°F (0–4°C)	65–75	Onions, shallots, garlic
Cold and damp	32–40°F (0–4°C)	85–95	Beets, carrots, turnips, rutabagas, kohlrabi, celeriac
Cool and dry	50–60°F (10–16°C)	55–65	Winter squash, pumpkins
Cool and damp	40–50°F (4–10°C)	85–95	Potatoes

What's a Max-Min Thermometer?

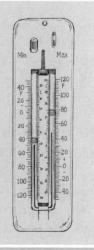

A max-min thermometer will give you the current ambient temperature. It will also record and save the lowest and the highest temperatures that occurred over a period of time, or until you reset it. This is very handy, because you might not be around during the warmest and coldest times of the day or night. There are digital versions of max-min thermometers and old-style mercury versions, which you reset using a small magnet. Either will do the job.

Storage Life of Common Storage Crops (under optimal conditions)

Crop	Months to Store	Crop	Months to Store
Beets (without tops)	4–6	Parsnips	4–6
Carrots (without tops)	6–8	Potatoes (early)	4–6
Celeriac	6–8	Potatoes (late)	6–10
Garlic	6–9	Rutabagas	4–6
Kohlrabi	2–3	Shallots	6–9
Leeks	2–3	Turnips (storage type)	3–5
Onions (salad)	2–3	Winter squash and pumpkins	2–3
Onions (storage)	6–8		

for bacterial and fungal disease organisms, which will definitely shorten the storage life of the individual pieces involved and may compromise others that are touching them.

Avoid storing fruits and vegetables that are overripe or underripe. You'll have the best success with mature, high-quality produce. If you must hold on to less-than-perfect specimens, keep them together in their own crate or lug, in a highly visible spot, so that you can check on them from time to time.

More Postharvest Information

The USDA's Agriculture Research Service has compiled a wealth of information and advice on the postharvest handling of many different fruits and vegetables. This material is contained in Agriculture Handbook Number 66, or can be read online (see Resources).

Long-Term Storage on Our Farm

I should begin this section by telling you that, at this stage in our evolution as a farm, we don't market our produce beyond late December, so we've been able to get by with less-than-perfect storage facilities. If we wanted to sell year round, we would have to enlarge some of our storage space and look into installing better temperature and humidity controls. In the meantime, this is what we have, and it has served us reasonably well.

The root cellar. As the name suggests, root cellars are good for storing root crops, such as potatoes, carrots, beets, and turnips. In the days before refrigeration, most farmsteads had one. Root cellars are usually built several feet into the ground or into the side of a hill. They rely on the fairly constant temperature of the soil, once you get down a few feet, to keep crops cool in summer and prevent them from freezing in winter. They are generally humid places.

There are several different versions of a root cellar, ranging from a simple 55-gallon drum or old freezer buried in a mound of earth to fully underground, compartmentalized, and ventilated walk-in rooms. If you don't have a root cellar and would like to build one, do a little research online or look for a book that presents different options and construction information. You'll want a cellar that is especially well suited to the crops you're planning to store. The book *Root Cellaring* by Mike and Nancy Bubel is a good resource. Or type in "root cellar construction" on your search engine and you'll find plenty of sites with advice and instructions on how to build various types of root cellars.

Because we live on an old dairy farm, we use a barn that once housed milking cows and hay to store all sorts of things. Fortunately for us, one entire side of the barn is built roughly 7 feet into the ground. In the corner that is most deeply buried in the earth, we created our own version of a root cellar. We walled off a room that is 10 feet long by 6 feet wide and installed several inches of foam insulation. We built a wide, insulated door and lined the room with sheet metal (to keep rodents out). Using 3-inch PVC pipe, we created ventilation lines that enable us to bring in air from the outside when we want to, giving us some control over both temperature and humidity.

In summer, the temperature in this room is about 10°F (6°C) lower than the rest of the lower barn, where the cows used to be milked (the upper barn, which was used to store hay, is warmer and drier in summer). In winter, our root cellar's temperature ranges between 32 and 45°F (0 and 7°C). In very cold weather, we use a small, thermostatically controlled electric heater to keep the temperature a few degrees above freezing. This room does a good job of storing any root crops we have left at the end of our season. Although it may be cooler than ideal for potatoes, we routinely store one thousand or more pounds of this crop for personal consumption over winter and to use as planting stock the following spring.

Good Agricultural Practices (GAPs)

GAPs are a set of recommendations designed to ensure the safety of food at all production stages, from field to market, with the goal of reducing the incidence of food-borne illnesses. Not long ago, small, diversified produce farms were pretty much under the radar in terms of food safety regulations, but that appears to be changing.

Even though, to my knowledge, all the recent food-borne illnesses have been traced back to large industrial farming, packaging, and processing facilities, small organic growers, as their numbers increase, are finding themselves lumped in with the big guys. It's looking more and more like a "one-size-fits-all" model will emerge. Today, all commercial growers, regardless of scale and status, should be well informed on GAPs. Federal or state legislation is most likely on the way. But regardless of what the future holds, it behooves small growers to maintain high production and cleanliness standards — for the good of our customers and for our own good as well.

The best way to learn about and keep abreast of GAPs as they unfold is to visit the National GAP Program website and various state websites. I also encourage growers to attend a GAP training course. The subject is an evolving one.

Here are some basic GAP requirements:

- Handle all manure and nonregulation compost according to organic "days to harvest" guidelines.
- Use clean, potable water for washing produce.
- Remove any animal feces from production areas.
- Keep hands clean while harvesting, washing, and packing produce.
- Make sure surfaces and containers that have contact with produce are clean.
- At market, keep lugs and crates that contain produce at least 9 inches off the ground.
- Encourage customers to wash produce before eating it.
- Discard produce that shows even minimal signs of disease or rot.

Most of the GAP requirements are basic common sense, but it's common sense that you need to keep front and center.

House basement. About 10 years ago, we purchased a small modular home to house farmworkers. At the time, we could have set the house on a concrete slab but chose, instead, to build a full-size concrete block basement. It added about $10,000 to the cost of the house, but we've always been glad we did it. Within the basement, which can be accessed from the outside, we built a 150-square-foot room to house a hot-water heater and furnace, which heat the living quarters above. This room stays warmer than the rest of the basement—it seldom drops below 65°F (18°C). It's a good spot for drying herbs, ripening green tomatoes at the end of the season, and green-sprouting potatoes (see box on page 420) in spring.

In winter, the remainder of our basement stays at about 60°F (16°C) and reasonably dry—okay for storing winter squash but a bit too warm for onions and garlic. Any alliums we have left over are better stored in a cool, dry pantry in our house. We've thought of walling off another room in the modular basement that would allow us to control the passage of cold air from the outside so that we could store more alliums over a longer period of time. It's a project we haven't taken on yet because we always manage to sell out of alliums before the end of the season.

With a good-size root cellar, a fairly dry basement, and a little ingenuity, you can create environments that will keep the major storage crops reasonably well. Supplement these two storage environments with a walk-in cooler, and you'll have covered most of your bases. Just don't feel you have to have all these facilities in place on day one. These are the sort of projects that will fit nicely into your long-range plan.

One final point about storage facilities: Always make sure they are insect and rodent proof. A family of mice (or rats) will be more than happy to cohabit with your winter squash but will cause you much grief and revenue loss.

Green Sprouting

Over winter, store potatoes in an environment that is cool, dark, and moist. Before planting, though, it's a good idea to gradually expose the tubers to warmth and light. This is known as green sprouting (also called *chitting* or *pipping*). It promotes early growth, early tuber set, and increased yields.

To start the process, move the spuds to a warmer spot — 60 to 70°F (16 to 21°C) is good — about 4 weeks before planting. Leave them in the dark, at this warmer temperature, for a couple of weeks. Then spread them out, with most of the eyes facing up, in an area where they are exposed to moderate light — either indirect sun or incandescent light — for another week or two. Continue to keep them warm. This encourages the development of short, stubby sprouts that don't break off when you plant them in the field.

Chapter 17

Marketing What You Grow

EVELOPING AND MAINTAINING A MARKET for the crops you grow is almost as important as learning how to grow them. A successful harvest will not do you much good if you don't have a ready outlet that will bring you a good return. Usually, this excludes intermediaries and wholesale arrangements. Like it or not, for most small organic farmers, active, direct marketing is part of the job description.

In the 1990s, the public's interest in fresh, local, and nutritious food was awakened from a long slumber. Ever since, that interest has been gaining momentum at an impressive rate, and the trend shows no sign of abating. This a good thing for farmers because it has opened up and expanded numerous marketing opportunities, many of which were barely on the map 25 years ago.

Direct-Marketing Venues

In this chapter, we'll start by examining the direct marketing that occurs through farmers' markets, CSAs, and farm and roadside stands. These venues provide the lifeblood that keeps many small, diversified farms in business. They are usually direct and personal. The buyer is almost always the eater, or at least one of the eaters. And the buyer and seller often get to know each other well.

Farmers' Market

Since 1994, the trend has been clear: Each year, there are more farmers' markets than there were the year before, and many of them are bigger and better attended. In 1994, the year the USDA began counting, there were 1,755 farmers' markets in the United States. In 2012, the number had jumped to 7,864. That's a serious rate of growth — nearly 450 percent in just 19 years.

Signing up to participate in a farmers' market is a good way to get a small farm business off the ground, and it can be a surprising amount of fun. At a farmers' market, you can learn on the job, to some extent, and probably no one will hold this against you. After all, people are under no obligation to buy what you have to sell. If your offerings are scant, uneven, and sometimes of poor quality, you're the one who will suffer. When you operate a CSA or sell to restaurants or food co-ops, you need to be a little more on top of your game. Don't infer from this that you can get away with doing a shoddy job at a farmers' market. You can't. It's just that you have a little more time to get things right.

In the more heavily populated regions of North America, the great majority of small farmers now live within an hour's drive or less of a farmers' market. In other areas, the drive may be longer, but the journey still worth it. Check out the Local Harvest website (see Resources) for a list of farmers' markets and small farms near you.

In some places, so many new markets are popping up that farmers have begun complaining that customers are spread too thinly. Perhaps it wouldn't hurt to see the growth rate in markets slow down a bit. And certainly, it would be smart to get an idea of the average gross sales of farmers at the market you are thinking about attending before signing up. In the long run, the public's growing appetite for fresh, local food should cancel out these temporary instances of supply exceeding demand. Farmers' markets, I believe, will remain a great way to sell what you grow for a long time to come. In the next chapter, I'll cover in more detail what you need to get started at a farmers' market.

CSA (Community Supported Agriculture)

When I started farming in 1987, very few people, myself included, would have been able to tell you what the letters CSA stood for. Now, if you have any connection at all to the world of small farms and local food, you'll have no trouble with this acronym. From the humble but enlightened beginnings of a few pioneers, the CSA has blossomed into a major force in sustainable agriculture. It is a rather brilliant and socially progressive marketing model that works well for many diversified organic farms. The 2007 U.S. Census of Agriculture counted more than 12,500 farms that market some or all of their products through a CSA. No doubt, the number has grown since then.

The CSA Contract

A CSA is basically a contract between a farmer and a group of consumers who are usually called shareholders or members. Before the season begins, the farmer invites interested buyers to purchase shares of the coming harvest. Vegetables to be grown and anticipated quantities are spelled out in general terms. The cash the farmer receives in advance enables him or her to buy seeds and supplies and draw up a crop plan based on the number of shares sold.

Share prices vary widely depending on the number of weeks that the CSA will provide its members with food and the variety and quantity of the food in each weekly share. Twenty weeks would be a relatively short-season CSA and 40 weeks a long one. Farms in more temperate climates with mild winters obviously will be able to keep going longer. Growers who dedicate a lot of time and space to storage crops and use season-extension devices such as high tunnels might be able to provide shares every week of the year.

The majority of CSAs have fewer than two hundred members, but much larger operations with five hundred shares or more are not uncommon. Once harvesting begins, shares can be picked up at the farm where the produce is grown or at other central distribution points. In some instances, packages are delivered directly to members.

The CSA contract usually states that members will share in the as-yet-unrealized risks and rewards of the season. In a good year, all members receive generous shares of fresh, local produce. In bad years, they are likely to get a less bounteous quota.

Clearly, a number of bad years, whether from natural causes or inexperience on the part of the farmer, will put a severe strain on the farmer-shareholder relationship and will likely result in members discontinuing their contracts. For this reason, new farmers should approach the CSA model cautiously. It is far better to offer fewer shares and be able to fill them generously, with a good variety of vegetables, than to overestimate production and disappoint shareholders.

Most CSA growers start out small and increase their offering of shares as they become confident in their ability to meet expectations in all but very bad years. Some growers operate a CSA and participate in a farmers' market at the same time. They meet their CSA obligations first and sell surplus produce at the farmers' market.

Nash's Organic Produce

Proprietor: Nash Huber

Location: Dungeness, Washington

Year Started: Late 1970s

Farm Size: 450 acres, with 120 acres in production and the remainder in cover crops, fallow, or livestock

Crops: Mixed vegetables, berries, grains, seeds, hogs

NASH HUBER GREW UP ON A DIVERSIFIED FARM in Illinois in the 1940s and '50s. He studied chemistry in college, then worked for a large corn and soy processing company. As the 1960s unfolded, Nash was restless; the land beckoned him. Before long, he found himself in Washington State with a hay-growing business. In the 1970s, he moved to Dungeness, Washington, and started growing an acre of organic vegetables. Nash enjoyed what he was doing, and the market for organic produce was gaining traction. Fast-forward to 2012: Nash is operating a 450-acre farm on the North Olympic Peninsula and growing 120 acres of certified organic vegetables every year. He didn't expect to get this big.

Most of the land Nash farms is under 20- to 30-year lease arrangements, and about half of it is protected by conservation easements that Nash himself has worked hard to obtain. He and his wife, Patty, own about 10 acres. Nash is the recipient of several awards that recognize his land stewardship and farmland preservation efforts, including the American Farmland Trust's 2008 Steward of the Land Award, and the Vim Wright Foundation's 2006 Farming in the Environment Award.

The North Olympic Peninsula's alluvial soils and cool summers (80°F [27°C] is a hot day) are ideal for carrots, potatoes, beets, broccoli, brussels sprouts, cauliflower, and many leafy greens. Nash's farm grows all of these in abundance, along with numerous other vegetables and herbs. The farm employs between 30 and 35 people year round and has an experienced management team. Marketing outlets include an on-farm store, a box-type CSA, a debit-style CSA, wholesale venues in the Puget Sound region, and as many as eight individual farmers' markets in summer and four in winter. Nash's Organic Produce is famous for its 'Nantes' carrots and several of its brassica crops. It is also known for its red wheat berries, strawberries, and pork.

Equipment at Nash's farm includes four large tractors, a chisel plow, a seldom-used moldboard plow, a large disc, and some 20-odd small cultivating tractors. The major cover crops are annual rye grain and common vetch. The farm makes a large amount of compost and has its own compost windrow turner. Ingredients in the mix include wood chips, manure from a raw-milk dairy, fish parts from a nearby seafood processing plant, and early-season green-chopped hay. Crops are irrigated with overhead sprinklers that pump water from a valley-wide irrigation system.

Seed Production

In addition to growing vegetables for market, Nash's Organic Produce operates its own seed-producing enterprise. The farm puts significant resources into maintaining and improving open-pollinated vegetable varieties — both for its own use and for sale on the national seed market. Land is set aside for seed production of preferred varieties of open-pollinated cabbages, carrots, cauliflower, beets, kale, and other crops.

Nash explains that the typical F1 hybrid that the seed companies sell takes plenty of dollars and 5 to 10 years to develop. Such varieties derive from two in-bred parental lines that must be carefully maintained. The parents are chosen for specific traits, such as uniformity of shape and size, long shelf life, rapid growth, transportability, and exotic color. The resulting offspring are highly selected and generally do not adapt well to variable growing conditions or produce usable seed. They are created for what Nash calls "spoon-fed" production, by which he means a stable climate and soil and a side dressing of chemical fertilizer.

Nash thinks it is vital that organic growers the world over hold on to and nurture open-pollinated vegetable varieties that have been carefully selected, saved, and handed down by our forebears. It is a mistake, he believes, to leave seed production solely in the hands of the big international seed companies, which are primarily interested in hybridization and proprietary products that they have full control over.

Nash worries that many of the F1 hybrids will not perform well under the erratic and extreme weather events (droughts, floods, extreme heat and cold) that the world now seems to be experiencing with increasing frequency. He fears that there might not be enough time to develop new varieties in response to the new world we have created. "I want to hold on to plants that have the ability to adapt to changing conditions," he says, "and can feed themselves."

Different CSA Models

CSAs come in many different forms. In fact, they are almost as individualistic as the farmers who run them. Some offer working shares at a discounted rate, in which case members are asked to come to the farm each week, or every other week, and help out in some way — perhaps with weeding, harvesting, or preparing shareholder packages. Work shares are a good way to involve CSA members in farm activities, but they do require some additional management, such as keeping track of hours and making sure a job is done properly — something you cannot necessarily assume with an untrained worker.

Some CSAs cater to local residents within the vicinity of the farm, while others truck their wares to pickup points in urban centers, and in these cases members may have little or no contact with the farm itself. Many CSAs provide a weekly, biweekly, or monthly newsletter to their members, in either paper or electronic form. These newsletters typically tell members what vegetables and fruits they can expect in the coming weeks and often include a relevant recipe or two and perhaps a short vignette about farm life.

Many CSAs offer half shares for singles, senior shares, and large-family shares at different prices. Some determine exactly what will go in each member's package. Others allow members to choose from a variety of options (for example, take five items from list A, three from list B, two from list C) to create their own packages. Still others have a "trade table" where members can leave behind produce they don't want and take more of what they do.

Some CSA farms have expanded to also provide meats, cheeses, seasonal fruits, and even grains. If you want to start a CSA, it would be wise to talk with several experienced CSA farmers and try to find out how well each of the various arrangements work for them.

A farmers' market CSA combines the traditional CSA concept with a farmers' market or farm stand venue. Participants pay the farmer a membership fee before the season begins. Instead of picking up a weekly package of food at the farm or some other distribution point, members come to the farmers' market and take produce from the farm stand on a debit-style basis. Because they have paid in advance, they receive a discount (usually 10 to 15 percent) off the stand price that is charged to non-member customers.

Market CSA members get to choose exactly what they want each week and even skip a week or two, without penalty, if they go on vacation. They spend down their share at a pace that suits them. Farmers don't have the bother of assembling identical packages each week, but they do have to develop a bookkeeping system to track each CSA customer's account balance. Some farmers might allow members to add to their balance periodically and continue to receive produce at the discounted rate. Traditional CSAs and farmers' market CSAs are not mutually exclusive. Both can function at the same time, as long as the farmer attends a farmers' market.

CSA Management Systems

As CSAs grow in size, they can become increasingly complex to manage. Many CSA farmers are turning to software systems, such as Farmigo, Member Assembler, and CSAware, which are designed to help them stay on top of the details and keep customers well informed. These web-based packages have lots of nifty features. They can sign up members, track balances, take credit card payments online, announce share options and pickup locations, run newsletters, keep the records straight, and more.

Farm and Roadside Stands

Farm stands have been around for a long time and no doubt will continue to be a good way for some farmers to market what they grow. To be successful, a farm stand needs to have a steady flow of customers. In our age of supermarkets, shopping malls, and a generally distracted electorate, this is not always so easy to come by.

Location and signage. Location is critical. A farm stand on a well-traveled highway, close to an urban center, will draw more customers than one that is tucked away on a backcountry road — that is, unless the latter has earned a winning reputation. Easy vehicular access and adequate parking and turnaround space are also very important. Ditto for eye-catching signs. These should be located at the stand itself and, if possible, a half-mile away on either side of the stand to give potential customers advance notice.

Stands that are situated on or adjacent to farms will have more appeal and authenticity than those that are located off-site, though some entrepreneurial growers have done quite well renting roadside space in high-traffic locations away from their farms. Areas that have a large population of vacationers with summer homes or rentals can be fertile ground for a farm stand.

Developing a customer base. As in any marketing situation, high-quality produce, good service, good presentation, and fair prices will win you happy customers, and happy customers will come back and will tell their friends about you. If you're planning to build a farm stand where no stand has been before, give yourself time to develop a customer base and look for creative ways to bring in those first buyers. People who stop at farm stands are usually looking for their favorite seasonal crops, such as tomatoes, basil, corn, squash, and bedding plants in spring. They are less likely to buy unusual herbs and exotic Asian greens. Save these for the more sophisticated urban farmers' market shoppers.

Less setup time. One nice thing about a farm stand is its relative permanence. Unlike a farmers' market stand, you don't have to set up your shelter and display each morning, then break it all down and pack it up at the end of the day.

Business and Institutional Customers

We've looked at direct marketing through farmers' markets, CSAs, and farm stands. These are good examples of a local food system functioning at the grassroots level. Now let's consider other, somewhat less direct marketing opportunities that are generally better suited to larger and more experienced growers. Included in this group are high schools, college campuses, nursing homes, and any other institutions that feed large numbers of people. We'll also take a look at smaller business customers, such as restaurants, health food stores, and food co-ops.

College Campuses, Schools, and Other Institutions

Across North America, and in many other parts of the world, young people are a principal driving force behind the local food movement and the push for global sustainability. Nowhere is this more evident than on college campuses, where eco-friendliness and going green are common sentiments. Many students are demanding fresher, better-tasting, and more nutritious food in their dining halls, and these demands usually translate into food that is grown organically and sourced locally.

Student Farms

Some college students are so enthusiastic about local food that they have established their own farms on or near their campuses. The book *Fields of Learning: The Student Farm Movement in North America,* coedited by Sean Clark and Laura Sayre, looks at about one hundred campus farms across North America and tells the story of the burgeoning student farm movement. One such example in my neck of the woods is the Poughkeepsie Farm Project at Vassar College in Dutchess County, New York. Students work on a 10-acre farm that offers CSA shares and operates a food-share program that focuses on the issues of hunger and affordability at the local level. Programs like this are excellent incubators and training grounds for future farmers.

At the same time, college administrators have begun, somewhat belatedly, to recognize that superior local food can attract more and better applicants to their programs. Such food may cost a little more, but it can pay off in higher enrollments. This recognition is leading to a shift in food procurement policies: Many colleges now include at least some local and organic food on their menus. There are 15 million college students in the United States, and the college food industry accounts for sales of over $4.5 billion annually, so there is a big market here, and it is likely to get bigger.

The same thing is going on in schools across the country, as parents push for more nutritious food in school cafeterias. Various farm-to-school, farm-to-college, and homegrown lunch programs help build links between schools, colleges, and nearby farms. The Community Food Security Coalition is one group that partners schools and local farms and provides information and training on how to build a food system that promotes health and sustainability, rather than merely profits.

To win over institutional clients, growers must offer high levels of consistency and volume and be able to keep this up over an extended period of time. Offering nice lettuces and basil one week but not the next will not impress these large buyers. It would be an ambitious, and probably unwise, move for a beginning farmer to attempt to break into this type of market, but it's something to think about for a few years down the road.

The institutional market, which might be expanded to include hospitals, nursing homes, and the like, can be rewarding for experienced growers who have a high production capacity and can sell at a discounted rate. As interest in fresh and healthy food grows, even emergency feeding programs such as soup kitchens and food banks might be interested in purchasing local produce at reasonable prices.

Restaurants

Many of the best chefs in North America have taken up the cause of local food. They understand that locally grown, fresh produce is easier to work with, requires little if any doctoring up, and almost always tastes better. When they go local, or at least go local in part, they see their clientele respond positively and their reputations enhanced. Some restaurants have taken to stating on their menus that the ingredients of certain dishes come from local farmers and ranchers, and they'll often go so far as to identify the farmer or rancher. Sweet, if you're the farmer.

The farmers' market we sell at in New York City is regularly patronized by well-known chefs who often show up personally, with a team of porters in tow, to bear away the haul of the day, which is destined to appear on that evening's menu. Not all of these chefs and restaurants are persuaded of the need to buy organic — especially when there is a price difference between it and local, conventionally grown food — but an increasing number are. More and more, it comes down to quality. If they think the food tastes better, they are willing to pay a little more for it.

Delivery to restaurants. Many restaurants are, of course, very happy to have local farmers deliver produce to them once or twice a week, and many growers are equally willing to oblige. If this is a marketing avenue you'd like to pursue, approach smaller, high-quality restaurants in your area and tell them what you do. Better yet, put together a list of the crops you grow, their prices, the quantities you can provide, and the dates they are available. Top-quality food, fresh from the field, is a great way to a chef's heart. Offer a few samples of any unusual vegetables or herbs, or items you have plenty of (and are particularly keen to sell). Sometimes the chefs will return the favor on your next delivery by handing you a prepared dish to take away.

It's important to develop strong relationships with restaurants, their chefs, and their food buyers. The restaurant business is a demanding one, and it's vital that you make good on your promises. If a chef is planning his or her menu around produce that you've said you will deliver, it's critical that the produce arrive on time and in good condition. If, because of some mishap or act of nature, you're not able to come through with your order in its entirety, call the restaurant and let them know in advance. Explain the reason, and suggest substitutes that you can provide.

Payment. A word of warning: Unfortunately, not all restaurants are 100 percent reliable in the payment department. Don't let yourself get burned. Tell restaurant clients that you need to receive a check each time you make a delivery and that if this doesn't happen you won't have enough money to buy seeds, or fuel for your truck, or something like that. If you're a trusting soul and decide to allow some credit, don't allow much — a couple of weeks or a month at most — otherwise you might find yourself out serious money for produce long-since eaten and forgotten. By the way, in case you haven't experienced it, chasing down people who owe you money is not a fun thing to do.

Health Food Stores

Health food stores offer a somewhat different marketing opportunity for organic growers. They are more likely than restaurants to seek out organic produce, but they tend not to want to pay as much for two reasons: One, they have to place a substantial markup on what they buy from you to make a profit; and two, if they don't sell it, they'll lose money on the deal. The problem is greatest with perishable items such as lettuces or other leafy greens because there is a small window of opportunity for selling the produce. They are more likely to go for storage crops such as winter squash or garlic and shallots.

Another problem with health food stores is that most of them don't have sufficient refrigerated display space —if they have any at all — to accommodate many bulky perishable items. High-end grocery stores that cater to an affluent clientele are more likely to buy your perishable vegetables, but the prices they offer you may be disappointing.

Food Co-ops

Increasingly, food co-ops are purchasing local food when they can. These businesses are usually worker or customer owned and might take the form of retail stores or buying clubs. They seek high-quality food of good value for their members. Food co-ops tend to be very community oriented and supportive of local and organic agriculture.

Large food co-ops, like the Park Slope Food Coop in Brooklyn, New York, which has more than 12,000 members, go through an enormous amount of fresh produce and other grocery items every week. They need and rely on farmers and other food producers who can supply them with high volume and consistency. Like colleges, schools, and other large institutions, they present marketing opportunities that are more suited to large and experienced growers.

Smaller food co-ops may present opportunities for smaller growers, but growers still need to meet quality and consistency requirements. The Coop Directory Service and LocalHarvest (see Resources) lists food co-ops and other cooperative resources throughout the United States.

CHAPTER 17 RECAP

Major Marketing Opportunities

▶ Farmers' markets
▶ CSAs
▶ Farm stands
▶ College campuses, schools, and other institutions
▶ Restaurants and health food stores
▶ Food co-ops

Chapter 18

....

More on Farmers' Markets

W E'VE BEEN SELLING OUR PRODUCE at farmers' markets for 26 years. Both my employees (at least most of them) and I look forward to going to market; usually, it's just 1 day a week for each of us. The market is a highly social event and a nice break from farm work. It's always gratifying to see the smiles on our customers' faces and hear how much they appreciate our efforts to bring them fresh, nutritious food. The markets also provide valuable feedback. We get to see firsthand what shoppers want and the type of presentations and displays they respond to.

This chapter will take the form of a Farmers' Market 101 course. I'll cover what you need to get started at a market, how to design a stand and set prices, and the virtues of bunching vegetables and herbs and maintaining good customer relations.

What You Need to Get Started

Here, I'll describe the basic elements we need to have in place on our farm before setting out for a day of selling in New York City. Different growers in different markets in different parts of the country may come up with a somewhat different list. But I'm confident that at least the bigger items — a vehicle, a shelter, display tables, a scale, bags, and signage — are almost universal requirements. Before I set out for the first market of a new season, I make a checklist of all the hardware and paraphernalia that we'll need. Many of the items are pretty obvious, but it helps to have them listed on a piece of paper in front of me, just in case I forget one.

A Vehicle

Many small growers start out with a pickup truck or van. As their business grows and there is more to sell, they might acquire a trailer, but in time, most growers prefer a box truck with a box big enough to stand up in. A box truck will protect your produce from the elements and enable you to organize it for easy access. It will also spare you the trouble of tie-downs and tarps, which you will need if you are using an open-back pickup.

Most growers should have ample space in a standard 16-foot box, but smaller and larger boxes are readily available. Depending on the state you live in, you might need a commercial driver's license (CDL) to drive a truck with a box larger than 16 feet. Before buying a box truck, make sure there is room for it at the market you plan to attend.

Refrigeration. Should you look for a refrigerated or non-refrigerated truck? Farmers have different opinions on this subject. Refrigerated trucks need to be insulated, and the insulation takes up a fair amount of space. Consider the climate zone in which you farm and the distance you travel to market. Most farmers' market managers don't want you to leave your truck running while at market, for obvious reasons. This means that once you arrive at your space and the day begins to heat up, the refrigeration unit has to be turned off unless you have access to an electrical power source, which is unlikely. So it comes down to a question of how much you gain from cooling your produce on the way to market. We've done okay without refrigeration for 26 years.

Shelter and Furnishings

Most farmers' markets are outside events, so you need a canopy to protect yourself, your customers, and your produce from sun, wind, and rain. These days, many growers use pop-up or E-Z Up–type tents. Depending on the size of your space, you might need more than one. They're a little bulky to transport, but one reasonably strong person can handle the setup and breakdown very quickly.

Another option is to construct your own canopy to fit exactly the space you have. To do this, you'll need the following:

- 1-inch (interior diameter) EMT conduit, which can be purchased in 10-foot lengths from any large hardware store

- Canopy corner and angle fittings

- Grommeted tarp of whatever size you choose. Silver tarps are best for reflecting heat and don't cast unusual colors on your produce.

- Bag of bungees to attach the tarp to the EMT conduit frame (see Resources for online sources for canopy corners, tarps, and bungees)

Pop-up-style canopy

It will take more time and expertise to erect this type of shelter, and you might need more than one person, but it will be less expensive, will last longer, and can be configured to fit your space more exactly.

Shade tarps. Direct sun withers leafy greens and is not good for most other vegetables. Black-mesh shade tarps do a decent job of protecting the goods on the sunny side of your stand. They can be attached to the canopy poles with bungees and moved as the sun moves. But be careful how you use them — they can restrict customer flow.

Tie-down weights. In a strong wind, an unsecured pop-up tent or any other type of canopy can become airborne and pose a serious danger to you and your customers. Tie down each corner of your tent to a heavy weight, such as a sandbag or a feed sack with 30 or 40 pounds of gravel in it. If your vehicle is adjacent to or behind your stand, tie down to it as well. Do this as a matter of course, not just on windy days — you never know when a serious gust is coming your way.

Folding tables. These come in very handy at a farmers' market, especially for holding digital scales, cash boxes, and the like. They also provide good surfaces on which to display your wares. Folding sawhorses and sheets of plywood or hollow doors (they're lighter than plywood) will also do the job. GAP rules require that all produce be at least several inches off the ground at all times. Heavy-duty vinyl tablecloths that can be sponged off after each market will enhance your table displays and give your stand a cleaner look. Try a colorful, checkered pattern that doesn't show the dirt.

Shelving. If space at your stand is limited, you might try exploiting the vertical dimension. You can buy or build your own collapsible two-, three-, or four-shelf display stand. The shelves should pitch forward slightly, and the structure should be sturdy but easy to dismantle and lay flat so it doesn't take up too much space on your truck.

Cash box. A metal or wooden cash box, or two, will help you organize bills of different denominations for quick access. Drill a small hole through the bottom of your cash boxes and run a knotted piece of string through them. Tie the other end of the string to a leg of your folding table. If anyone tries to run off with a box, they'll have to take the table with them. A compartmentalized tray for coins will speed up the change-making process. If you buy a metal cash box, it will usually come with a coin tray.

Money belts. Another, and perhaps more secure, way to organize bills and coins is to wear a money belt or apron with pockets. A money belt enables you to move around your stand and make "roving" sales without having to return to the cash box to get change. You can buy ready-made money belts or make your own. Look for belts with three or four pockets so that you can separate different denominations of bills and loose change.

Ultimately, the safest place for the cash that comes your way, especially the bigger bills, is on your own person. I recommend a pair of loose-fitting pants with big, deep thigh pockets. If you're uncomfortable with a lot of money in your pockets, you might funnel cash into a small metal safe bolted to the inside of your truck. If you go this route, make sure the safe is securely fastened to the truck and always obscured from sight.

Digital scale. Unless you plan to sell everything by the bunch or the piece, you'll need one or more scales. Hanging scales do an okay job and have a nice farm stand look, but to use them, you need to do a lot more arithmetic. Digital scales cost a little more, but they'll make life easier and move things along faster when you have a line of customers. You can buy models that run on regular or rechargeable batteries. From time to time, check to see that your scales are reading accurately. If they are not, you might anger your customers or get fined by the local weights and measures authority. Some growers keep an item of known weight (for example, a solid object that weighs exactly 1 pound) at the stand so that they can periodically verify the accuracy of their scales and/or satisfy suspicious customers.

Display containers. These can be baskets, wooden crates, coolers, bread trays, or the plastic lugs you harvest directly into. Different display containers are suited to different items. For example, potatoes or carrots look good in a deep container, but tomatoes should be in a shallow tray (just one or two layers) so they don't squash each other. Bunches of herbs do well in small baskets. Pumpkins look good simply sitting on a table or upturned crate. Salad mix should be presented in a very clean cooler.

The type of containers you choose will help define the "look" of your stand. An observant walk around a successful farmers' market will give you plenty of ideas. And always remember: A bountiful and overflowing display catches the eye much faster than a meager one.

Bags. Plastic bags are a ubiquitous blight on the planet, but at a farmers' market or farm stand, it's hard to avoid using them. At our stand, we've tried a number of ways to wean our customers and ourselves off this earth-choking, petrochemical product. We've provided the first bag free but charged for additional bags (5 cents apiece)—you'd be surprised at how much this encourages doubling up. We've rewarded customers who bring their own bags; they really appreciate this.

We've supplied shopping baskets and encouraged people to fill them with various unbagged items that require weighing, such as tomatoes, potatoes, and winter squash. They bring the shopping basket to us; we weigh the contents and place them in one large bag—preferably a reusable shopping bag—instead of several small plastic ones. We've designed our own recycled cotton "made in the USA" shopping bag, which has been quite successful and which we sell at cost. These tote bags have our farm name along with my wife's woodcuts of a tomato and a garlic (our signature crop) printed on them. Underneath the bulb of garlic are the words, "Arguably the best garlic in the known universe." This gets a chuckle out of some of my more literate customers.

Finally, we've purchased compostable bags and encouraged customers to bring them back to us for composting because there is very little opportunity for composting in New York City. Unfortunately, these "compostable" bags got two strikes against them right off the bat: Our certification organization (NOFA–NY) decided they had an unacceptable ingredient in them; therefore, they could not be used to make compost that could be applied to our fields. Second, these bags had a slight degree of porosity, which caused them to wick moisture away from whatever was placed inside them. This wasn't a problem for potatoes and tomatoes, but leafy greens that were stored in the compostable bags wilted after a couple of days, even in a refrigerator.

(437)

Paper bags, of course, are a possibility, but their manufacture takes a toll on the environment. Also, they are more difficult to carry and can break apart when wet. Because all our greens are kept a little wet up to the point of sale to maintain freshness, paper is not a practical alternative for us.

All in all, we've significantly cut down on plastic use at our stand, and many of our customers appreciate this and have been very supportive.

Signage

A good farm sign that can be seen from a distance is a must, unless you like to keep a low profile, which I would not recommend. Designing a farm sign, like choosing a name for your farm, deserves a fair amount of thought, or perhaps a "eureka" moment. Before you proceed, it might be worth talking to a graphic designer, or at least a friend who has a good eye for things graphic and visual. A farm sign might include a simple image along with your farm name; it should have personality and express something special about your farm; and it should be sturdy, durable, easy to read, and easy to mount at your stand.

You also need signs with prices for each kind, and even variety, of vegetable and herb you sell. Heirloom tomatoes, for example, will sell better when you identify them by name — 'Brandywine', 'Cherokee Purple', and 'Aunt Ruby's German Green' are more enticing than simply Heirlooms. Your customers will appreciate a few words describing the flavor of a vegetable or ways in which it might be used. Here are some examples:

'Winterbor' Kale — High in iron and vitamins A & C. Good in
   soups or steamed. $2.50/bunch
'Hakurei' Turnips — Japanese turnip with a sweet and fruity
   flavor. Excellent raw or lightly cooked. $2.00/bunch
Red Mustard Greens — Hot and spicy flavor that mellows when
   cooked. $2.50/bunch
'Keuka Gold' Potatoes — Yellow flesh, smooth buttery texture.
   $2.60/lb
'Golden Chard' — A nutritious member of the beet family.
   Delicious steamed or stir-fried. $2.50/bunch
Escarole — Rich in vitamins and minerals. Makes a wonderful
   soup. Or enjoy it steamed or stir-fried. Extra-large heads:
   $3.00 each

Price and descriptive signs should be reusable, so you don't have to prepare new ones for each market. They can be made from different materials such as card stock, plastic sheeting, or ¼-inch plywood. You can use sharpies and assorted Magic Markers or paint markers to do the writing.

We've used laminated index cards as signs for many years. We attach them to our display containers with a nifty little clip-on sign holder that tilts in any direction. These can be obtained from a supply company called Hubert (see Resources).

Note the very handy clip-on sign holder on this basket of garlic that is on display.

Miscellaneous Items

The following are somewhat lesser items that will come in handy when you go to market. Leaving behind any one of them will very likely slow you down, compromise your efforts, and in some cases, leave you feeling cold, wet, and unloved.

Start-up cash (coins and bills). You need enough cash ($150 in 1s, 5s, and 10s would be a good amount) to make change for that first round of $20 bills and the occasional fifty that comes your way in the morning. Also bring along several rolls of quarters, dimes, and nickels.

Raingear and change of clothing. Having a good canopy doesn't necessarily mean you won't get wet on occasion. For one thing, you will need to set up and break down without the protection of the canopy, and when the canopy is in place, rain can still come in from the sides. It's a good thing to have a change of clothing, especially socks. Driving home with wet feet, after a long day at market, is no fun.

Gloves. It's wise to wear a pair of gloves when setting up and breaking down your stand, especially in cold weather. Remember, you only get one set of hands in a lifetime, and you're going to need them throughout your farming career. Learn to look after them.

Rope. Always keep a couple of lengths of rope in your truck. You'll be surprised at how often you reach for them. An example: If a strong wind develops, you might tie your canopy to a spare tire.

Photos of farm and promotional materials. Once they get to know you, your customers will want to see what your farm looks like. A nice photo or two will add a personal touch and give people an image of where the food they're eating is coming from. A copy of your organic certification statement should always be on display, and don't shy away from using any news clippings or promotional materials that mention your farm. An inexpensive laminator will enable you to reuse these types of materials.

Index cards and Magic Markers. Bring to market a supply of index cards of different sizes, a few sheets of 8½- by 11-inch card stock, and some Magic Markers. In addition to your regular signs, sometimes you'll want to make spur-of-the-moment signs to highlight special offerings or an end-of-day sale.

An inverter. This device will enable you to draw a minimal amount of power from your vehicle's battery to run a couple of floodlights at your stand. It's handy if you sell at a market that runs at night or late on a winter afternoon. The drive home should be enough to restore to your battery the small amount of juice that is lost.

Rubber bands and sundry other items. It's also good to bring string, tape, pens, scratch pad, receipt book, business cards, pocketknife, scissors, and salad tongs (if you sell washed salad mix). All of these items will come in handy at a farm stand.

Designing the Stand

The amount of space you are allocated at a farmers' market and the amount of produce you have will determine the size of your canopy and the way you lay out your stand. If your space has a depth of 10 feet or less, you might be better off with a frontal display on tables. You or your helpers can stand behind the tables to conduct transactions, but make sure you have an easy way to get out in front to clean and restock the tables.

A deeper space will allow you to take better advantage of the shade provided by your canopy and to create a layout that invites customers to come into the stand and move around islands of produce. In this type of situation, there should be enough aisle space for people to pass each other. If your stand gets busy, there should also be enough space for customers to form a line that doesn't impede the access of others who wish to browse.

Take a piece of graph paper and sketch a few different layouts for your stand. Consider which vegetables you have the most of and which you want to give prime display space to. Organize what you have into product groupings and keep related products together — for example, all the greens in one location; potatoes in another; herbs in a third location; onions, shallots, garlic in a fourth, and so on.

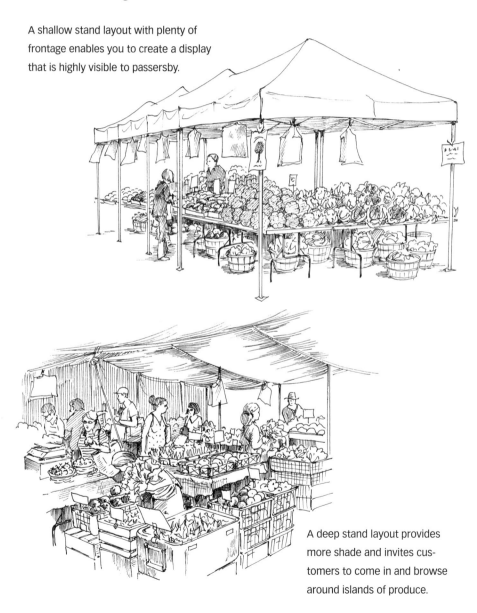

A shallow stand layout with plenty of frontage enables you to create a display that is highly visible to passersby.

A deep stand layout provides more shade and invites customers to come in and browse around islands of produce.

(441)

Take into account how customers will enter and exit your stand. Note parts of your stand that will get exposed to the sun at different times of day. These are bad places to put greens. If you have a choice, always keep your back to the sun. The layout and display choices you make are expressions of your farm, your philosophy, and your personality. Go for something that is enticing, colorful, bountiful, and customer friendly.

Once you're satisfied with your layout, stay with it for a while and try to keep items in the same place from week to week, though as the season progresses, you'll probably need to make some changes to accommodate a shifting line of vegetables.

Pricing

Some people are a little uneasy setting prices. If you're one of them, get over it as soon as possible. It's part and parcel of being a direct marketer. Bottom line: You need to come up with prices that are acceptable to a broad range of customers so your product will move off the stand at a good rate. At the same time, you need to make a reasonable profit — at least enough to keep you in business — but there's a bit more to it than this.

Seven Factors to Consider When Setting Prices

1. How much does it cost you to grow a particular vegetable? Allot some value to your land and equipment. Consider how much labor was needed, including your own, and the value of that labor. Consider the cost of seeds, amendments, irrigation supplies, water, and fuel. Finally, assign a value to your marketing efforts.

It'll take time and thought to come up with dollar estimates for all these different steps, but the answers, however approximate, will tell you a lot. Besides helping you arrive at stand prices, this type of production cost analysis will enable you to compare the returns on different vegetables and determine which are the most profitable. This knowledge can help you make better crop planning decisions in the future.

2. Do you have a lot of a specific vegetable you wish to price, or just a little? If a lot, you might set a lower price. If you have only a little, a slightly higher price makes more sense.

3. Are you pricing a vegetable that must be sold fresh, such as a head of lettuce, or a storage crop, like a potato? At the end of a slow day, you might lower the price on your lettuces but not on your potatoes or garlic. As the market draws to a close, bargain hunters often show up. Give them bargains on perishable items but not those vegetables you can bring back to another market.

4. What are other farmers in your market charging for similar items? Are they organic? Regardless of your production costs, you need to be somewhat in line with prices in the market. Pricing too low will make you unpopular with other farmers, and you will be short-changing yourself; pricing too high will reduce sales.

5. What are supermarkets and other food retailers in the area charging for similar items? You don't want to be too much out of alignment with their prices, either. But remember: Your vegetables have not traveled a thousand or more miles; they are a lot fresher, and they were grown without toxic chemicals. They are worth a little more.

6. Are any of your items uncommon? If you have an uncommon item that is in demand (for example, a few dozen persimmons off your backyard tree) and that other growers don't have, you may be able to charge a little more for it. But don't price gouge. Soon enough, customers will figure out whether you're offering them a fair deal. They appreciate fair deals and will come back. If they feel overcharged or ripped off, they'll drift away and will probably never tell you why.

7. Avoid pennies. Round numbers work better at a farmers' market, preferably to the nearest quarter.

To Bunch or Not to Bunch

Sooner or later, you'll find yourself confronted with this question. Not bunching generally means selling by weight; bunching means selling by the piece. For the sake of argument, I'd like to use the term "bunching" very loosely, so that it encompasses not only bunches held together with rubber bands or twist ties but also produce in containers, such as pints, quarts, or net or cloth bags.

In the early days at our farmers' market, it seemed as though it was a lot easier to sell most crops by weight. All we had to do was cut kale, collards, Swiss chard, and the like, and throw them into a lug. Some vegetables, such as heads of lettuce or escarole, could easily be sold by the piece, and herbs seemed to be obvious candidates for bunching; however, just about everything else we sold loose. In other words, we allowed our customers to take whatever quantity they desired and bring it to us to weigh. For a season or two, I wondered why most other farmers went to the trouble to portion out so many items into pint and quart containers and to make evenly sized bunches of greens, beets, and turnips. Why go to all that extra work? Having been a slow learner all my life, it took me a while to figure out what was going on, but eventually, I did. Now we bunch, or sell by container, whatever we can.

Three Reasons to Bunch

A bunch is a known quantity. Bunching does take longer, but it results in more sales. Most people find it easier to visualize the quantity of a particular vegetable they need, rather than try to judge the weight. They will guess that a 5-pound bag of potatoes weighs just 2 or 3 pounds, or that a handful of kale weighs 8 ounces when, in fact, it is a full pound. A few experiences like this make them uncomfortable. A bunch of kale or a quart container of small potatoes are known quantities and, perhaps more importantly, known prices. No surprises. A hanging scale for customers to use (before they get to the cash register) will be appreciated by those who are looking for a specific weight of a given item.

Bunches speed up the action. On a busy market day, weighing a lot of vegetables slows down the action, which means that customers might have to stand in line longer and cash flows into our hands more slowly. Bunches, heads, and small containers keep business moving briskly.

Bunches reduce the harm done by vegetable fondlers. Finally, there is the matter of the vegetable fondlers. This is a broad term we've come up with to describe those folks who are insecure in their ability to select vegetables, are always looking for the perfect specimen, or just like to touch everything. Some even go so far as to bring almost every vegetable in contact with their noses. Regardless of the particular strain of the pathology, any of these traits can be troublesome, especially when the subject of interest is a lug full of leafy greens. Such individuals will rifle through the container, inspecting every leaf, some perhaps more than once. They just can't make up their minds. The greens, of course, suffer from so much handling and soon show signs of stress, which disincline others to buy them.

Unusual Items

At our farm, we are always looking for unusual items that will bring in some extra dollars at the beginning and end of the season, when we don't have a lot of staple foods to offer. As holidays approach, anything that might qualify as a gift is a good bet.

Unusual items for late spring:

- Perennial herbs (tarragon, thyme, sage, mint)
- Edible weeds and wild plants (stinging nettle, lambs quarters, ramps)
- Bunches of immature alliums (garlic, onions, shallots)

Unusual items for late fall:

- Herb wreaths (made from sage, thyme, and rosemary)
- Garlic braids
- Small net or cloth bags of garlic and shallots
- Bouquets of dried grasses and berries

Just a small minority of people are vegetable fondlers, but those who are can do inordinate damage when presented with loose vegetables. With bunches, they have a limited opportunity to do harm (but sadly, bunches won't deter the sniffers).

Quality and Pricing

When you sell loose, people get to pick and choose and will naturally cull out inferior produce. When you sell by the bunch (or the pint or quart), the onus is on you to make sure that everything is of good quality. Beware of a soft potato at the bottom of a quart container, and don't let yellow or faded leaves get hidden in the middle of a bunch. They will disappoint.

Price bunches and containers of vegetables based on what you feel you need to get per pound, but limit yourself to just two or three different bunch prices. It's better to adjust the bunch size of different vegetables to reflect the per-pound price you want. For example, we sell most of our herbs for one price, but our bunches of French tarragon are always smaller than our bunches of Italian parsley. Our greens go for the same per-bunch price, but the collard bunches weigh more than the bunches of mibuna (a delicate Asian green).

Tips for a Successful Market Day

Be friendly and helpful to customers, even the more difficult ones. If they show any interest, share with them what's happening on your farm — when the new crop of potatoes will be in, when the first garlic will be harvested. Tell them if you've just had a hailstorm that has strafed your greens or a lot of rain that has caused your tomatoes to split. Don't ask for their sympathy; just help them understand that local food should not be taken for granted.

The more your customers learn about the daily workings of your farm, the more connected they will feel to you, the farmer, and the food you provide. If you do well by them, they will repay you many times over with their friendship, loyalty, and support. They don't get many smiles or much personal attention when they go into supermarkets. You have an opportunity to give them wholesome, fresh food with a human face on it and to make them feel good at the same time. Don't pass it up.

Most of the following points are no-brainers but are worth paying attention to nonetheless.

Be consistent. Try to maintain consistent prices, consistent quality, and consistent bunch sizes from week to week. Show up at the same time, and always be in a good mood — even when you're not.

Have a return policy. If someone tells you they got a bad potato the week before, or a rotten onion, tell them you're sorry and that they should take another one. If they happen to be a regular customer and would prefer a refund, don't hesitate to give it to them.

Offer samples. Giving away free samples can be a good way to make a sale, but be careful that your samples are clean. For some samples, toothpicks and napkins may be appropriate. To be on the safe side, ask your market manager or local Department of Health if there are any rules regarding ready-to-eat samples.

Keep your stand clean. Remember, you're selling food. A clean stand tells customers that you care about your produce and care about them. A dirty stand brings into question your farming practices.

Wear clean clothing. It can be old, it can be faded, but it should not be dirty or stained, especially at the beginning of the day.

Don't eat, drink, or smoke. Sure, you'll need to eat and drink at some points during the day, but don't do it while serving a customer. And definitely, never smoke at your stand.

(446)

Greet your customers by name. Depending on how many customers you have, this may be unrealistic, but make an effort to get to know the names of as many of your regulars as you can. They'll love you for it. Jot the names down on an index card that's taped to the inside of your cash box to help you remember. This kind of personal touch is what sets the farmers' market experience apart from almost all other retail transactions.

Don't argue with customers. If you're like me, this can sometimes be a challenge, but over the years, I've learned that arguing with customers is seldom worth the effort. It uses up energy and can make others who are standing nearby uncomfortable. Occasionally, you might need to take a firm stand if someone is doing something really egregious, such as stealing or verbally abusing your staff.

Remove damaged produce. Some vegetables, such as tomatoes, show signs of wear and tear when they are handled too much. Cull them out as soon as you can. Sometimes, they can be turned into seconds, but don't give seconds prime display space.

Make it easy. This applies to just about everything — reaching for produce, reaching for a bag, weighing vegetables that need to be weighed, and any other steps that are part of the shopping experience at your stand.

CHAPTER 18 RECAP

Checklist: What You Need for a Farmers' Market

- ☐ Vehicle
- ☐ Shelter
- ☐ Shade tarps
- ☐ Tie-down weights
- ☐ Folding tables
- ☐ Shelving
- ☐ Cash box
- ☐ Money belts
- ☐ Digital scale (and possibly a hanging scale)
- ☐ Display containers
- ☐ Bags
- ☐ Signage
- ☐ Startup cash (coins and bills)
- ☐ Raingear and change of clothing
- ☐ Gloves
- ☐ Rope
- ☐ Photos of farm and promotional materials
- ☐ Index cards and Magic Markers
- ☐ Inverter
- ☐ Rubber bands and sundry other items

Three Springs Farm

Owners: Emily Oakley and Mike Appel
Location: Cherokee County, Oklahoma
Year Started: 2006
Farm Size: 20 acres, with 6 acres in production
Crops: Mixed vegetables and fruits

EMILY OAKLEY AND MICHAEL APPEL studied sustainable agriculture in college and in such diverse parts of the world as Africa, Asia, Latin America, and the Middle East. They also worked on a couple of well-established organic farms in California — Eatwell Farm and Full Belly Farm — gaining practical experience that they consider invaluable.

In 2003, Emily and Mike returned to Emily's native Oklahoma. For 3 years, they farmed on leased land to get their feet wet and have time to look for a place of their own. They eventually found 20 acres of relatively inexpensive (and nutrient-poor) land outside Tulsa, and in short order established a viable small-farm business that supports the two of them.

Three Springs Farm grows a mix of some 30 annual vegetables, along with strawberries and asparagus. Tomatoes are the biggest and most profitable crop. They also grow perennial fruits, including apples, peaches, and blueberries, though not all are ready for market.

To overcome their soil's nutrient deficiencies, Emily and Mike have made use of the abundant supply of chicken manure in their vicinity. They work the manure into the ground with a chisel plow soon after spreading. They use cover crops for fertility enhancement and organic matter, preferring millet and cowpeas during the summer months and oats and field peas in winter. Emily would like to rely only on cover crops for soil improvement and believes this will soon be possible.

Equipment used on Three Springs Farm includes: vintage 35-horsepower John Deere tractor, newer 45-horsepower Massey Ferguson tractor, rototiller, bed shaper, mechanical transplanter, chisel plow, disc, row hipper, and grain drill for seeding cover crops.

Plants are started in standard 128-cell flats in a small (10- × by 20-foot) greenhouse that can accommodate up to 15,000 starts. Mike and Emily run drip irrigation from a generous well, which allows them to water 3,500 row feet of vegetables at the same time.

Farmers' Market CSA

When Mike and Emily started farming on their own, they chose two marketing models they were already familiar with: the CSA and the farmers' market. The two worked nicely together because the CSA shares were distributed on Saturday mornings at their farmers' market in Tulsa. Shareholders would simply come to the market to pick up their boxes while other customers were making their individual purchases.

One day, a customer approached Mike and Emily with a novel proposition. He said he would like to support their CSA and provide them with start-up cash in winter but feared that he would not be in town during portions of summer to receive his weekly share. Why not, he suggested, sell him a CSA share in winter and then allow him to take produce of his own choosing in summer, when he was in town. The value of the produce taken could be deducted from his CSA credit balance. Mike and Emily liked the idea and thus was born their farmers' market CSA.

Now, 5 years later, Mike and Emily still sell at the Tulsa farmers' market, but they have dropped their regular CSA and replaced it with the new farmers' market CSA, which has more than 100 shareholders. They no longer worry about packing boxes of produce for members and wondering whether all the ingredients will be equally appreciated.

Under the new model, members choose what they want. If they can't make it to a market one day, they just have more dollars to spend the next week. To receive the widest selection of produce, CSA members are encouraged to come to the market early, preferably before noon.

For signing up and paying in advance, the market CSA shareholders receive a 10 percent bonus. Shares range from $150 to $300 for a 20-week season that runs from mid-April until Labor Day. Members can add to their shares at any time during the season and still receive the 10 percent bonus. Some have gone as high as $700.

Mike and Emily use an Excel spreadsheet to keep track of each member's account and bring a printout with names and balances to market every week so that members can find out how much credit they have remaining.

The market CSA has worked out well for Mike and Emily, as well as their customers. They have a very high member-retention rate and a waiting list for new entrants. Regular farmers' market customers still account for about two-thirds of their annual income. The other third, which flows through the market CSA, arrives in winter and makes ordering seeds and supplies for the coming season a lot more agreeable.

Chapter 19

....

Marketing through the Seasons

WHAT TO GROW? HOW MUCH TO GROW? When to grow it? These are critical questions for most small, diversified farmers, especially those who are just starting out. You've probably already guessed that there are no pat answers. Each of us has to take into account the usual factors — among them, acreage, soils, irrigation, equipment, labor, and markets — and then come up with a crop plan that makes sense. We've already considered most of these variables in other chapters of this book. Here, I want to take another look at the last of them — markets — and, specifically, consider the ebb and flow and shifting demand of the market season. Once again, I'll use our farm as an example.

When preparing a crop plan, I find it useful to think of the year as having five overlapping marketing periods. These periods coincide with what we have available — both from the field and from storage — to bring to our farmers' market. They are:

- **Late spring** (late May to late June)

- **Early summer** (late June to late July)

- **Summer** (late July to late September)

- **Early fall** (late September to mid-October)

- **Fall and early winter** (mid-October to late December)

(Note: We don't, at this point, go to market through most of winter and early spring. If we did, there would be a sixth marketing period — one heavily dependent on storage crops.)

Certainly, there will be peak-season times (for many of us, mid- to late summer and fall) when we have a large supply of harvestable produce, we are functioning at full throttle, and we are bringing home lots of bacon, so to speak. There will also be times (often at the beginning and end of the marketing season) when we have a diminished supply and perhaps we're just bringing home a few sticks of lard (if you'll excuse the metaphor). This is to be expected, but the value of those sticks of lard should still be enough to cover our costs and provide at least a small profit.

If you plan poorly, you will see sudden spikes and dips in what you have to offer. Spikes and dips will cost you money and cause you grief. It doesn't do much good to have more perishable produce than is necessary for your CSA members, or more than you can sell at your farmers' market in a given week or month (unless you've had the foresight to line up other outlets — but that's a different subject). Large surpluses represent unnecessary expenditures of energy — energy that could have been better spent elsewhere.

And of course, a shortfall is even more distressing. Not having enough to satisfy your CSA shares or enough to sell to cover the cost of going to market, can put you in the red and give you a serious case of self-doubt. It will also disappoint your customers or shareholders and perhaps cause them to doubt your ability to come up with the goods.

So the challenge facing you is to maintain a steady flow of salable produce to the extent that nature and your efforts will allow. A good overall plan, a planting schedule, and careful record keeping will help enormously in this regard, as will understanding the market and its shifting demands.

In chapter 3, I went over the details of the Weekly Planting Schedule and other planning and record-keeping tools used on our farm. Here, we'll focus on the market itself. We'll consider the five marketing periods listed above — how they differ from one another and the range of produce we have available for each of them. I'll follow each period with a few comments and observations.

Late-Spring Lineup

Our first markets in late May and June are usually our weakest. We have to reestablish ourselves after our winter absence, and we don't have an abundance of fresh produce to sell in the first few weeks. Potatoes from the previous year are usually beginning to sprout and are less firm than they were, but they can still taste remarkably good — they have less moisture and more essence of potato flavor. We rub off any sprouts and tell our customers to eat them within a week or two. Shallots from the previous season, if well cured and stored, can sell well, but we seldom have many of them.

In the first couple of weeks, our lettuces may be on the small side and our bunches of chard, dandelion, and kale are not especially generous because these crops are still small in the field; nonetheless, they are appealing to city dwellers who may be starved for fresh greens. Mesclun salad mix, though labor intensive, is a good early-season seller.

Bunches of green garlic, onions, and shallots have been a staple early offering at our stand for many years. These are grown from sets (rather than seeds) and reach a marketable size fairly quickly. The idea is that people can chop up and eat the green tops, as well as the very small bulbs. They sell well enough at a time when there isn't much fresh, local food to be had, but they are not snatched up as fast as the mature garlic, onion, and shallot bulbs that come to our stand several weeks later.

As we move toward the end of June, basil from one of our high tunnels is a popular item, but we don't have a lot of it because we have only two tunnels and there are several contenders for the limited space. Rhubarb is also a good seller, but again, the quantity is limited — we don't have a big rhubarb patch. The peas, when they arrive in mid-June, are a welcome sight, both for our customers and for us. Edible weeds, especially stinging nettle and lamb's quarters, have gained favor in recent years. Our many perennial herbs, which grow vigorously in late spring, help fill out the display. They sell relatively well, but not as well as they will later in the season, when we have more substantial vegetables to go with them.

Late Spring

Basil (from tunnel)
Bunched green garlic
Bunched green shallots
Bunched set onions
Edible weeds (stinging nettle, lamb's quarters)
Italian dandelion
Lettuces
Mesclun
Perennial herbs (anise hyssop, catnip, chives, lavender, lovage, mint, oregano, rosemary, sage, tarragon, thyme, winter savory)
Potatoes, shallots, and storage onions from previous year, if we have any left
Red Russian kale
Rhubarb
Sorrel
Sugar snap and snow peas
Swiss chard

Bottom line: At this time of year, we have to scratch around to put together a modest offering of eclectic items. More would be better. We're always on the lookout for new early-season crops to supplement our somewhat meager offerings and generate more income.

Though we may not have a lot to offer at our early markets, it's important for us to be there. Our regular customers understand that the growing season is a trajectory — that we are at the beginning of it and so are they.

Early-Summer Lineup

During the first half of this period, from late June to mid-July, we continue to sell bunches of green garlic, but by the middle of July, we switch to individual bulbs, which are always a big seller for us. Also, by mid-July, we begin to phase out the bunches of green onions and shallots. For a few weeks, garlic scapes fill a nice niche — they are a highly seasonal item that attracts plenty of attention, and we have lots of them.

Toward the middle of July, tomatoes and peppers from our high tunnels and some new potatoes from the field make their first appearance. These always excite our customers and invariably sell out. Squash, zucchini, and cucumbers liven up the stand with their interesting shapes and colors, but almost every vegetable grower in the market has an abundance of them. Lettuces and bunched greens are now of respectable size and continue to sell well. So do the herbs, especially basil, which is now coming from the field. Business is picking up.

Early Summer

Basil
Bok choys
Carrots (small)
Cucumbers
Edible weeds
Garlic (bulbs)
Garlic scapes
Green garlic
Green onions and shallots
Italian dandelion
Lettuces
Mesclun
Parsley
Peppers
Perennial herbs
Potatoes
Red Russian kale
Rhubarb
Sorrel
Spinach
Sugar snap and snow peas
Summer squash
Swiss chard
Tomatoes
Zucchini

Summer Lineup

The big money makers during this period from late July to late September are tomatoes, tomatoes, tomatoes — all shapes, sizes, and colors. Everyone wants them — local, vine-ripened tomatoes are so much better than their industrial counterparts. But there are many other items in demand as well. These include both perennial and annual herbs (especially basil, parsley, and mint), salad onions, green beans, zucchini, squash, cucumbers, peppers, and eggplant — all popular summer foods.

Lettuces, mesclun salad, and any other greens we can muster continue to sell well, but growing them can be a challenge — summer heat causes many of the greens to bolt. The turnaround time (from seeding to harvest and sale) needs to be short. Hot market days and direct sun also take their toll on fresh greens, causing them to heat up and wilt while on display. A covered truck (rather than an open-back pickup), shade tarps, and misters can help alleviate this condition.

Garlic and salad onions remain strong sellers. Potatoes, cooking onions, and shallots sell reasonably well, but they are not yet at peak demand. Their day will come.

The important thing to note about our *summer* marketing period is that the volume of sales for certain items goes up substantially, even exponentially, and we have to be ready for this. In the *early summer* period we might sell 100 to 200 pounds of tomatoes at a market because the fruits are just beginning to ripen and this is all we have to offer. In August and September, we routinely harvest 800 or more pounds for a single market and sell most of them. Basil follows suit, with as many as 200 bunches harvested for our Saturday market. And when tomatoes and basil are selling in large quantities, there are more people in the market and more customers at our stand, which means that we sell more of everything else. Business picks up substantially.

Summer

Apples
Basil
Cilantro
Cucumbers
Eggplant
Garlic
Green beans
Italian dandelion
Lettuces
Marjoram
Mesclun
Onions
Parsley
Peaches
Peppers
Perennial herbs
Potatoes
Shallots
Summer savory
Summer squash
Swiss chard
Tomatoes
Zucchini

The challenge now is to have enough good food to meet the increased demand, and there's no better way to make this happen than to have a well-thought-out and well-informed plan.

Early-Fall Lineup

For us, early fall is the best-selling period of all. People are back from their vacations; schools and colleges are in session; the hot, muggy days of summer are now only a memory. Most folks are in a good mood. At the market, there's a tantalizing scent of fresh vegetables, herbs, and fruit in the air. Our regular customers are out in full force. Many other city residents, who seldom visit a farmers' market, are now ready and eager to partake of the local harvest. It's almost as though their inner clocks are ticking, telling them to go out and stock up with good fare before the long, dark winter is upon them.

Almost everything we had to offer in the *summer* marketing period, with the possible exception of summer cucurbits (which often succumb to disease), is still available and in substantial quantities. Tomatoes continue to hold the day, but the basil goes into decline as we approach the end of September and cooler nights.

The demand for potatoes, cooking onions, and shallots takes a jump, as does interest in perennial herbs — especially rosemary, sage, and

A busy day at the market

thyme — because people are preparing more hot dishes. Then, there are the fall crops that many customers have been waiting for — several different kinds of kale, broccoli, mustard greens, collards, Asian greens, carrots, celeriac, winter squash, turnips, radishes. Now we have the makings of a real party.

There are certain challenges that come with the surging demand and sales of fall. One is having enough labor to harvest and prepare everything that is available as the days shorten. Imagine the time you need to pick a couple of hundred pints of cherry tomatoes, or harvest and wash a dozen coolers of arugula, or cut 75 bunches each of sage, thyme, and rosemary — all of uniform size. Add to this every other fresh item that needs to be harvested in a single day, and you can have a very full plate. At this time of year, if we have extra hands on the farm, they will almost always pay for themselves and more. At the market, too, we can benefit from extra help, but it pays to line this up in advance.

The capacity of our truck, with its 16-foot box, is another consideration —we might have to pack certain items more tightly to ensure there is room on board for everything we have. And with so much to sell, display space at the stand is at a premium. If space on either side of us were available, we would gladly pay for it, but that is unlikely, since most other producers at the market are also experiencing the fall bonanza. One solution is to set up shelving to take advantage of unused vertical space. Another is to narrow access isles, but this can lead to congestion and toppling signs, crates, and coolers.

Again, the point I'm trying to make is that a good plan, which takes the many changing

Early Fall

Apples
Arugula
Asian greens
Basil
Broccoli
Carrots
Celeriac
Cilantro
Cucumbers
Eggplants
Garlic
Green beans
Italian dandelion
Kales (assorted)
Lettuces
Marjoram
Mesclun
Onions
Parsley
Peppers
Perennial herbs
Persimmons
Potatoes
Pumpkins
Quinces
Radishes
Raspberries
Red and green
  mustard
Shallots
Summer savory
Summer squash
Swiss chard
Tomatoes
Turnips
Winter squash
Zucchini

variables into account, will win the day. Forewarned is forearmed. True, if you are a beginning farmer it might be asking a lot to be in full alignment with the changing seasons, shifting demand, and other vagaries of the market. And indeed it is. But don't feel bad. Rome wasn't built in a day, and neither will your farm business be. Look at each year as a learning experience, an opportunity to observe, record, and incorporate new information into your plan. In time, you will be able to anticipate and prepare for what is coming your way, and that is both empowering and lucrative.

Fall and Early-Winter Lineup

The days, and especially the nights, are getting chilly in mid-October. By now we've almost certainly had a frost. The basil and beans are finished, and so are the tomatoes and peppers, except for any late bloomers that were picked while still green and brought inside to ripen. The hardy fall greens continue to both look and taste good. They sell extremely well, but their growth has slowed down dramatically. Lettuces, some of them now under row cover, are holding up and also selling well. The days of Italian dandelion and Swiss chard are numbered — by November, they are usually gone.

Garlic, potatoes, carrots, and celeriac continue to sell well. Pumpkins are a good item in the weeks leading up to Halloween and Thanksgiving (for pumpkin pies). Winter squash, sage, thyme, and rosemary are good sellers throughout the fall and terrific before Thanksgiving; the trick is to have enough of them on hand and know when to bring them to market. (Looking back at Harvest and Sales Records from previous years keeps us on top of these sudden surges in demand.)

Fall & Early Winter

Arugula
Asian greens
Broccoli
Carrots
Celeriac
Dandelion
Garlic
Italian dandelion
Kales (assorted)
Lettuces
Mesclun
Onions
Parsley
Perennial herbs
Potatoes
Pumpkins
Radishes
Red and green mustard
Shallots
Swiss chard
Turnips
Winter squash

The goal: a steady flow. Our goal is to have an adequate supply of produce to sell during each of the above periods. We want to cover our operating costs, the foremost of which is labor, and justify our marketing efforts, which in our case, involve driving a truck to New York City, setting up a stand, and staying there all day — an undertaking that has a significant price tag associated with it. We also want to make a profit.

I suspect that most small, diversified growers — regardless of whether they choose to operate their own farm stand or CSA, deliver to a restaurant or food co-op, or sell at a farmers' market — have similar objectives. As business people, we need to figure out how to have a steady flow of produce for as long as possible each year so that we can keep our customers well fed and the dollars coming in. You don't need me to tell you that the bills come in regardless of cash flow.

PART

VI

COMPETING FORCES

Chapter 20

Regarding Weeds

VEN AFTER EXERTING CONSIDERABLE EFFORT to control weeds, most organic growers end up with a seemingly unlimited supply of them. Note that I say *most* growers, not all — there are some who, with great perseverance and long-range strategies, have devised systems for keeping weed numbers very low. Anne and Eric Nordell at Beech Grove Farm in Pennsylvania are shining examples of this (their articles in the *Small Farmer's Journal* and their booklet *Weed the Soil, Not the Crop* describe the methods they use), but the reality for most of us is that serious competition from weeds is an ongoing fact of life. We expect to be always engaged in a holding action and to have little chance of ever being able to declare a decisive or long-lasting victory.

My computer's dictionary program defines a weed as "a wild plant growing where it is not wanted and in competition with cultivated plants." That's not a bad definition. After you've been farming for a year or two, you'll understand just how accurate it is, and you might wish that all the weeds would go away forever. But that would not be a wise thing to wish for — at least, not at a planetary level.

Weeds are a vital part of the earth's natural cover. They thrive on disturbed ground that has lost its natural vegetation — be that grassland

Important Cross-References

We've already touched on weed-control implements, methods, and strategies in a few other parts of this book, most notably in:

Chapter 4, Tractors and Tractor Implements
Chapter 5, Small Equipment and Tools
Chapter 10, Cover Crops and Green Manures
Chapter 11, Crop Rotation
Chapter 13, Our Most Profitable Crops

or forest. They are the first plants to come in and stabilize the soil after it has been violated and laid bare at the hands of man or the elements. After the rapid-action weed squad has secured the site, slower-growing and more long-lived species gradually move in, and a more stable ecology is established. Without weeds a lot more of the planet's topsoil would have washed into the ocean and we would have a much harder time making a living.

Any type of farming that involves repeatedly tilling the land is, perhaps, the ultimate form of soil disturbance. Most of us vegetable growers do this in one form or another (excluding those who engage in no-till practices, but that's another subject altogether). Every time we till the land, weeds will attempt to correct the damage that has been done. More often than not, they are already in the soil waiting for an opportunity. When that opportunity comes, they do their job with remarkable efficiency, and therein lies the problem for farmers: Weeds are so good at what they do that the great majority of them will outcompete the vegetables we plant every day of the week.

In this chapter, we'll take a more general look at weeds, then come up with some guidelines for controlling them or, at least, coexisting with them as best we can.

Get to Know Your Weeds

First of all, get to know the common weeds on your farm and learn to identify them at different stages of growth. Learn their common and scientific names. Find out their preferences — do they flourish in compacted soil, poorly drained soil, fertile soil, or soil that is alkaline or acidic? How fast do they grow? Do they spread laterally or grow straight up? Are they annual, perennial, or biennial? Do they propagate themselves vegetatively or by seed or by both means? And perhaps most important of all, how long does it take for them to flower and produce viable seed?

I strongly recommend that you buy a good weed-identification guide for your region. Get one that has clear illustrations or photographs of each weed's leaves, stem, flowers, and other distinguishing features. It should provide information on the weed's range, preferred conditions, and life cycle. Also helpful would be strategies for elimination or control.

ANNUAL WEEDS

Redroot pigweed

Lamb's quarters

Hairy galinsoga

Purslane

Henbit or
dead nettle

Chickweed

PERENNIAL WEEDS

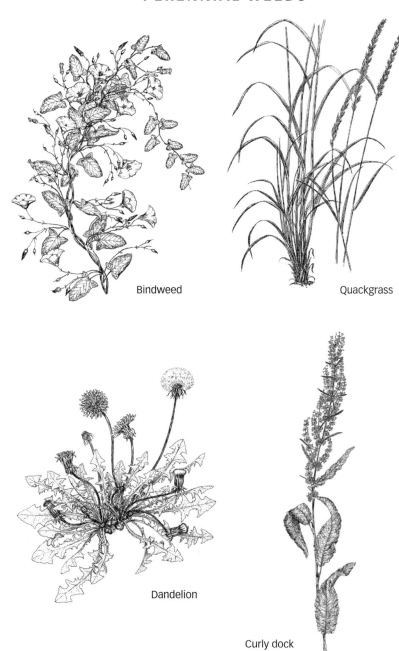

Bindweed

Quackgrass

Dandelion

Curly dock

Annual Weeds

The majority of weeds that farmers deal with are summer annuals. Annuals germinate, grow to maturity, produce seed, and die — all in just one season. They do not reproduce vegetatively (from bulbs, corms, rhizomes, stolons, or severed pieces of root), as do many perennials. The annuals compensate for this shortcoming by growing rapidly and setting huge numbers of seed — often tens of thousands per plant — in the limited time they have. Take redroot pigweed (*Amaranthus retroflexus*), for example. A single plant can reach a height of 6 feet or more and easily produce 100,000 seeds during its short life. Other common and prolific seed-producing annuals are lamb's-quarters, purslane, and hairy galinsoga. The seeds of most annual weeds live in the soil for 5 to 7 years. After this, they are unlikely to germinate.

Summer annuals are active during the warmer months of the year. They withdraw from the scene when cold weather arrives. Winter annuals are a little different; they germinate and grow much later in the season. When winter comes, they go into dormancy, but they come back to life in spring and rapidly proceed to grow, flower, set seed, and then die by early summer. Chickweed and henbit (dead nettle) are winter annuals.

Perennial Weeds

Perennial weeds live for 3 years or more. Many of them reproduce both by seed and vegetatively. Some, such as dandelion and curly dock, have very deep taproots that are difficult to extract completely. Even a small piece of root left in the ground can regenerate the plant. Others, such as bindweed and Canada thistle, have extensive creeping root systems that spread out laterally, conquering new ground as they go.

Perennial grasses, such as quackgrass and Johnsongrass, reproduce from seed or from underground stems called rhizomes. The rhizomes can move many feet away from the parent to begin a new plant. Once established in your fields, perennials are the most difficult weeds to get rid of. When you chop one up with a rototiller, you may have just multiplied your problem. The best strategy for controlling perennials is to exhaust them with frequent low cutting. This reduces their ability to accumulate and store the energy they have gained from photosynthesis. It is also critical that you mow or cut them before they get a chance to produce viable seed in their reproductive phase.

Biennial Weeds

Biennial weeds live for two seasons but don't flower or set seed until their second year. Like the perennials, most of them propagate both by seed and vegetatively. Common biennials are Queen Anne's lace, burdock, and mullein.

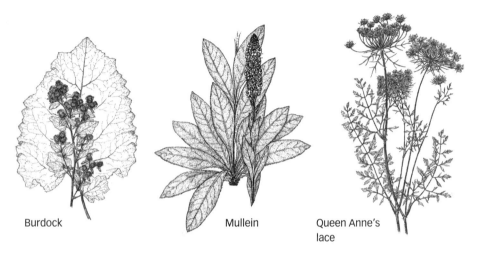

Burdock   Mullein   Queen Anne's
                                                lace

Twelve Weed-Control Strategies

Once you've identified the major weeds on your farm and learned a little about their life cycles and methods of reproduction, you'll be in a far better position to exercise some control over them. Many of the field choices you make — the crops you grow and where, when, and how you grow them — will be influenced by your knowledge of weeds and the resources and techniques in your weed-control arsenal.

    1. Don't let weeds go to seed. This is rule number one and the reason you need to know a little about the life cycle of each of your major adversaries — especially when they will develop flowers and when those flowers will have viable seed. When you incorporate weeds into the soil after they've gone to seed, your fields will look good for a week or two, but soon you'll have thousands of their offspring poking their noses at you. The most challenging weed in this respect is hairy galinsoga, commonly known as quickweed or shaggy soldier. This summer annual can flower as early as June and set seed just days later. At the first sign of this hairy demon, remove it from your land entirely.

2. Dispatch weeds while they are still small. "A stitch in time saves nine" is an old saying from the days, not so long ago, when people still mended clothing that had come apart at a seam or darned a sock with a hole in it. It was good advice. Waiting for an entire seam to unravel just meant you'd have a lot more work on your hands. The same is true with weeds. While still small, they are a lot easier to dispatch with a hoe or a cultivator. As they gain size, they become more difficult to behead or uproot and even when uprooted can reestablish themselves. A big part of successful weed management is good timing.

3. Avoid deep cultivation. Many weeds need sunlight to germinate. Others need to be very near the soil surface. Deep cultivation (more than a few inches down) may knock out existing weeds, but it will also bring buried weed seed closer to the surface and into the germination zone, which is another good reason to hoe or cultivate when the weeds are still small. As weeds get bigger, you have no choice but to cultivate more deeply. (See chapter 5, page 118, for a discussion of wheel hoes and hand hoes that are designed for shallow cultivation.)

The Most Troublesome Weeds on Our Farm

ANNUALS

- Common chickweed (*Stellaria media*)
- Common lamb's quarters (*Chenopodium album*)
- Common purslane (*Portulaca oleracea*)
- Common quickweed (*Galinsoga ciliata*)
- Pennsylvania smartweed (*Polygonum pensylvanicum*)
- Redroot pigweed (*Amaranthus retroflexus*)
- Velvetleaf (*Abutilon theophrasti*)

PERENNIALS

- Curly dock (*Rumex crispus*)
- Canada thistle (*Cirsium arvense*)
- Field bindweed (*Convolvulus arvensis*)
- Quackgrass (*Elytrigia repens*)

BIENNIALS

- Common burdock (*Arctium minus*)

4. Rotate crops. Rotating between cash crops and cover crops, between early- and late-season crops, and between crops that compete well with weeds and those that don't are all good strategies for disrupting weed cycles. (For more details, see chapter 11, page 273.)

5. Choose the right time of day. Whenever possible, hoe, cultivate, or hand-pull your weeds in the earlier part of a hot, dry, and preferably windy day. You want the roots to dry out as quickly as possible. When this happens, the weeds are less likely to reroot themselves. Weeding late in the day or in damp, overcast weather is much less effective.

6. Use a smother crop during a summer fallow period. A fast-growing, broadleaf cover crop, such as buckwheat, should be able to outcompete the weeds at their own game. To get a good stand, seed heavily, preferably before rain, unless you can sprinkler-irrigate soon after seeding. Be sure to incorporate the smother crop before it sets its own seed; otherwise, it might morph into a weed.

7. Use stale seedbed or false seedbed techniques on fallow ground. With these two methods, which are discussed on page 275, you actually encourage weeds to germinate, then dispatch them with a cultivator or flame weeder. The purpose is to deplete the seed bank, or supply of weed seeds, in the soil.

8. Use plastic and organic mulches to smother weeds. Black plastic and other infrared transmitting (IRT) mulches are excellent weed barriers and work well with heat-loving plants such as tomatoes and peppers. Organic mulches, such as straw or wood shavings, can also do a good job of blocking annual weeds but are not so effective against already established perennials. If perennials come through the mulch, you need to uproot them by hand, maybe with the assistance of a dandelion weeder.

9. Use a flame weeder. A flame weeder (either backpack or tractor-mounted) will knock out small annual weeds that appear after you have direct-seeded a slow germinating crop such as carrots or parsley; however, the timing is critical. You need to use the flame weeder a couple of days before the carrots or parsley begins germinating (see page 308).

10. Plant and mow living mulches. Seed living mulches, such as perennial rye or white clover, in the access paths between beds of crop plants. In their early growth stage, living mulches will suffer serious competition from weeds. Regular mowing, though, will favor the mulches and exhaust the weeds. After the first month or so of mowing, the mulch plant (or plants — you can use more than one) should gain full control. The living mulch itself then can be mowed less frequently.

11. Transplant crops whenever possible. When you transplant seedlings into a clean bed, you gain a head start of a couple of weeks over the weeds. You'll have to do more early weeding of those crops that must be directly seeded (carrots, turnips, peas, beans, and radishes). Try using a flame weeder or the stale (or false) seedbed technique with them.

12. Space rows to allow for easy cultivation. One of the main reasons for planting cash crops in rows is to make them easier to cultivate or hoe. Make sure you leave enough space between rows so your equipment can get in without damaging the crop plants. Many growers like to use a standard row spacing (perhaps 12, 15, or 18 inches) whenever they can so they can employ the same cultivating equipment. Sometimes, fast-growing broadleaf crops are planted in close rows with the expectation that they will establish themselves soon enough to shade out the weeds in between.

The "Up" Side of Weeds

As noted at the beginning of this chapter, weeds play an important role in the ecology of our planet. They can also do some good for a farmer, as long as they are managed knowledgeably and not allowed to go to seed.

Weeds as cover crops. If the weather or soil conditions prevent you from sowing a winter cover crop (and this can happen quite often), it's far better to have a stand of mowed weeds rather than no cover at all. The weeds will hold the soil together and prevent nutrient leaching. When they are turned in the following spring, before they can produce seed, they will contribute organic matter to the soil, just as any cover crop or green manure would.

Catch crops and nutrient scavengers. Weeds hold on to minerals and plant nutrients that might otherwise leach out of the soil in winter or during periods of heavy rain. Deep-rooted perennial weeds are adept at capturing and retaining nutrients from the subsoil. When they eventually die and decompose, the nutrients they have mined from below become available to crop plants rooted in the topsoil. There's some poetic justice to be found in allowing cultivated weeds to become compost for your next cash crop. Just be sure they haven't already gone to seed.

Indicator species. Because different weeds flourish under different conditions, they can tell you a lot about the state of your soil. Lamb's quarters and wild mustards, for example, prefer alkaline soils. Chickweed, plantain, and dock favor soils that are acidic. Horse nettle and quackgrass tolerate compacted and crusty soils. If you have a field with lots of pigweed, purslane, and lamb's quarters, you can be fairly confident, at least, that the soil's fertility level is high.

Tasty and nutritious food. Edible weeds can make good eating and have very high nutritional value. We've been selling purslane by the pound and bunches of wild stinging nettle for many years. Lamb's quarters is another good one, but only when the plants are small.

Guardians of genetic knowledge. Weeds, like all wild plants, are a rich repository of evolutionary and genetic knowledge accumulated over great reaches of time. They know more about resisting pests and diseases, or surviving drought and flooding, than virtually all of our vegetables. This knowledge is often put to use by plant breeders looking to improve a trait of a vegetable. And of course, we should not forget that most of the vegetables we enjoy today have very weedy ancestors in their not-so-distant lineage.

Food for others. Last, but by no means least, some weeds provide food and habitat for wildlife, birds, and beneficial insects — many of which play a positive role on an organic farm. This is a good reason to tolerate the weeds growing in your hedgerows and on patches of idle or nontillable land.

Twelve Weed-Control Strategies

1. Don't let weeds go to seed.
2. Dispatch weeds while they are still small.
3. Avoid deep cultivation.
4. Rotate crops.
5. Choose the right time of day.
6. Use a smother crop during a summer fallow period.
7. Use stale seedbed or false seedbed techniques on fallow ground.
8. Use plastic and organic mulches to smother weeds.
9. Use a flame weeder.
10. Plant and mow living mulches.
11. Transplant crops whenever possible.
12. Space rows to allow for easy cultivation.

Chapter 21

....

The Four-Legged Competition

NATURE THRIVES ON DIVERSITY, struggle, and interconnectedness. It is at once inclusive and highly competitive. It is constantly in flux, yet always seeking a rough equilibrium. At times, the natural world appears sweetly tender and innocent, but it can also sport a brutally harsh side. In nature, there is not much time for sentimentality or self-doubt. We humans occupy an odd spot in this grand venture, and figuring out exactly how to comport ourselves is no easy matter.

The relatively new science of ecology attempts to understand how different forms of life function together and complement each other. Ecology, along with good stewardship, is a big part of what organic farming is about. On a healthy organic farm, there will be a lot more than just humans and vegetables. There will be, or there should be, an entire panoply of species — microbial, plant, insect, and animal — that compete, interact, and coexist with each other. When all goes well, there is a variable, give-and-take kind of balance. This is the premise we start with and the goal we aspire to.

The reality, however, is a bit murkier. There will be times when the competition, if allowed full rein, could run us out of business. The challenge we then face as farmers becomes how to maintain diversity with its inherent healthfulness and, at the same time, produce vegetables for market.

For me, one of the great joys of being a farmer is that I get to interact daily with so many different living creatures. Each contributes its unique presence and is part of the larger whole. Sometimes, our immediate interests conflict, but that doesn't mean we can't share the same piece of land within reasonable limits. Birds sample the ripe peaches and raspberries on our back lawn. Rabbits dine on the lettuces and broccoli seedlings on the edges of the fields. Voles nibble on the carrots where they emerge from the soil. All can be irksome, and now and then, we may need to take some action, but none of them poses quite as serious a problem as our two major adversaries: woodchucks and deer.

In this chapter we'll consider some of the steps you might take to protect a crop. Much of the advice I'll offer involves different kinds of fencing to exclude herbivorous animals. I'll also have a few words to say about the use of repellents and patrolling dogs. Then, there are the more aggressive and violent actions one might resort to. To some people, these may be distasteful and offensive. Even so, there will very likely come a time in the life of a farmer when the more passive controls are not working and lethal action becomes part of the job description.

For most of this chapter, I'll discuss ways to prevent or minimize the damage done by woodchucks and deer. Then I'll quickly mention a few other critters that can also give farmers a headache.

Deer

In many parts of North America, deer numbers are on the rise, and this can spell big trouble for vegetable growers. A few deer in your fields at night can wipe out a month's income and maybe even your business. Deer will eat your palatable crops at any time of the year, but they are more likely to do so in winter, early spring, and late fall, when there's a shortage of natural browse in the woods and they are especially hungry. There are several options for dealing with this plentiful and rather beautiful animal, and none of them is easy.

Fencing

Good fencing is probably the best way to keep deer out of your fields, but depending on the severity of the deer problem you are facing and the type of fencing you choose, it can be an expensive undertaking. Read about one couple's deer fence in the profile on Waterpenny Farm (page 482).

Permanent Woven-Wire Barrier Fence

If your deer problem is severe, a permanent barrier fence, at least 8 feet high, may be your best bet. These fences typically have 12-foot posts of cedar or yellow pine sunk 3 or 4 feet into the ground (depending on how deep the freeze line goes) and spaced 15 to 20 feet apart. A common option for the barrier itself is 12.5 gauge, fixed-knot, galvanized wire mesh. You'll need to brace corners of the fence and construct gates for entry and exit.

Cost. Fences of this sort can cost $6 or more per linear foot, installed. How much, then, might such a fence set you back? Well, a square acre is about 208 feet on a side. That means it has a perimeter of 832 linear feet. Multiply 832 by $6 and you arrive at $4,992. If you were to fence in 4 square acres, at this rate, the perimeter distance per acre would come down by one-half and you would end up with a price tag of just under $10,000. A square 9 acres would cost around $15,000. Rectangles and irregularly shaped areas will cost more because they have more linear feet of perimeter than squares of equal acreage.

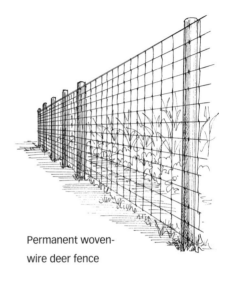

Whatever size fence you put up, leave 10 to 15 feet inside the enclosed area, on all sides, for tractors (or horses) and equipment to turn around and people to move about. In other words, if you enclose exactly 4 acres, your actual planting area will be somewhat smaller.

Permanent woven-wire deer fence

Maintenance. Permanent fences of this sort need periodic maintenance. As an organic grower, you are not going to be using herbicides, so maintenance entails cutting or removing grasses and weeds that grow at the base of the fence, where a brush hog or lawn mower cannot quite reach. If you don't let this vegetation get too big, a weed whacker should do the job. You've also got to watch out for trees that might fall on the fence and damage it.

Putting up a permanent barrier fence is no small undertaking, and the approach outlined above is by no means the only one. Consult with other farmers and at least a few companies that sell fencing supplies before you forge ahead. If you have a tractor, a three-point-hitch posthole digger, and a good crew, you might take on the job yourself and reduce the cost considerably.

Permanent Electric Fence

A permanent electric fence should also be at least 8 feet high and would most likely use ¾- to 1-inch fiberglass rods for posts and tightly stretched, high-tensile wire lines spaced about 12 inches apart. You might need sturdier corner posts. Power can be supplied through a charger or energizer that plugs into an electric outlet or, if such a power source is not available, through a solar-powered energizer located in the field. For an electric fence to work properly, it must be grounded to a ground rod that is inserted several feet into soil with some moisture in it. This type of fence is not as sturdy as a permanent barrier fence, but it is easier to erect and should cost far less.

Temporary and Movable Electric Fences

These might be adequate if deer are not a major problem in your area. The simplest temporary fence will have just one line of non-rigid, poly-wire twine, about 3½ feet off the ground, running between fiberglass rods.

With this type of fence, use a "bait and shock" strategy. Tightly wrap small ribbons of tinfoil around the electrified line, every 10 feet or so, and smear the foil with peanut butter. You want browsing deer to pick up the scent of the peanut butter and touch the tinfoil with their moist noses for the shock to be a sufficient deterrent.

For it to be effective, this type of fence should be put in place before deer have had a chance to sample your crops. If you do it after they have enjoyed a meal or two, they might just decide to jump over this rather minimal barrier.

More advanced fences. You might choose to include two electrified lines, instead of one — the first 2 feet above the ground and the second about 4 feet. Another approach is to run two fences — one about 5 feet inside the other. Deer are excellent high jumpers and don't need to think twice about clearing a 4-foot fence. They are not, however, confident long jumpers and can be put off by a double fence, assuming they can see both outside and inside lines (you can wrap pieces of surveyors tape or old row cover around the electric lines to help in this respect).

Quick setup and dismantle time. Once the fundamentals are in place — namely, a charger, access to an electrical outlet, a ground rod, fence posts (we use 1-inch-diameter posts on the corners and ½- or ⅜-inch posts in between), and an insulated wire to carry the charge to your fence lines — you can quickly put up a temporary electric fence. In a remote location, where there is no electrical outlet, a solar-powered fence charger will do the job.

You will need to make holes for the 1-inch corner posts, but it should be possible to push the smaller posts into the ground by hand. It's easy to roll the polywire off a spool and string it around your fence posts. Simple spring clips fit over the posts and hold the wire at the level or levels you choose.

The advantage of temporary fences is that you can erect them when and where you need them. Usually, you would do this soon

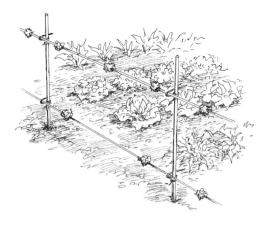

Temporary electric fence showing fiberglass posts, polywire, and spring clips. Note tinfoil and peanut butter bait.

after you've planted a crop that you know deer would like to eat. When the crop is harvested or no longer vulnerable, you can quickly dismantle the fence. With the fence gone, it's a lot easier to come in with a tractor and do field preparation when, for example, you decide to seed a cover crop.

Temporary Barrier-Net Fence

The heavy-duty, permanent barrier fence mentioned above uses woven-wire mesh and 4- to 6-inch-diameter wooden posts. A far less expensive, and also less sturdy, version of a barrier fence uses 8-foot-tall plastic netting with a 2- by 2-inch mesh and 10-foot-long fiberglass sucker rods (sucker rods are left over from oil drilling projects; they can be purchased used and cut to whatever length you wish). One-inch-diameter rods are suitable for the corners and ¾-inch-diameter rods for all posts in between. The corner posts may need to be secured with guy ropes to prevent them from bending too much in the direction that the fence is pulling. The plastic net is attached to the fiberglass rods with spring clips and cable ties.

This type of fence can be erected fairly quickly and used as a temporary barrier. Once they know what they are doing and have a system in place, two or three people should be able to fence in an acre of land in less than half a day. It is important to stretch the fence as tightly as

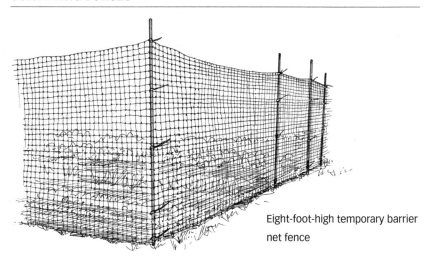

Eight-foot-high temporary barrier net fence

possible between the posts, which is why an extra pair of hands can be useful. Some growers run an electrified line around the outside of this type of fence as an additional deterrent.

Shooting

If you can prove that deer are damaging your crops, most state fish and wildlife agencies will issue a nuisance permit allowing you to kill a certain number of deer or turn the permit over to hunters who will do the work for you. If there are plenty of deer in your area and you want to make a real impression, you'll need to kill a lot of them. Shooting just two or three probably won't make much difference. Others will quickly come in to take their places. The wholesale killing of deer that have discovered the bounty of your vegetable fields is not for everybody. Moreover, it is a violent, bloody, and quite difficult thing to do.

Other Strategies for Minimizing Deer Damage

Other approaches to preventing deer damage include planting unpalatable crops, using dogs, and using natural repellents. These strategies might be employed in combination with temporary low fences and possibly occasional shooting.

Unpalatable crops. Plant crops deer don't eat in areas that you don't want to fence, or where deer pressure is greatest. We've used garlic, onions, shallots, and most herbs in this way for many years.

Dogs. If you can train two or three hardy dogs to patrol your vegetable fields and chase off intruders, especially at night, deer will tend to stay away. Unfortunately, most dogs like to sleep at night. We've used an Italian breed, called Maremma, with limited success. These beautiful, even-tempered, stoical dogs (they look a lot like Great Pyrenees) are good for protecting sheep or goats from wolves, coyotes, and other predators, but they don't always get the idea that you want them to protect vegetables from herbivores.

Repellents. There are numerous concoctions on the market for repelling deer and other animals. Some can be sprayed directly on crops, others on a ribbon fence around the perimeter of your fields. Most of them probably work to some extent, some of the time. You might try them in conjunction with other measures, but it would be a mistake to rely solely on them.

Keep them guessing. Don't underestimate a deer's capacity to learn (the same goes for many other vertebrate animals that eat our crops). Just because a low, baited electric fence has worked well for one season is no guarantee that it will work well the next. A hungry deer can figure out a way to get through, over, or under to satisfy its appetite. By combining and varying your preventive measures, you'll have better odds of protecting your crops.

Woodchucks (a.k.a. Groundhogs)

This burrowing rodent is common throughout the eastern United States and much of Canada. It is a vegetarian with broad tastes, though it will also eat grubs, grasshoppers, and other large insects. If your farm is within its range and has any type of cover where a woodchuck can dig a hole, you'll be in competition with this fellow soon enough. Like most other herbivorous mammals, woodchucks don't just settle down and dine on a whole head of lettuce or a single broccoli plant. They like to sample several different ones, or a dozen or more at a sitting, usually taking the new growth in the center. This quickly becomes a problem that you have to deal with one way or another.

Woodchucks are prolific breeders and will need to be controlled every year. You might clear out all the woodchuck holes in the vicinity of your vegetable fields, but in all likelihood, there will be a new set of residents to greet you the following season.

Keep the grass well mowed around your vegetable fields. Many small mammals, but especially woodchucks, don't like to venture too far from their holes or burrows, especially when they don't have much cover. A clear 15-foot strip between any cover (such as a hedgerow) and your vegetables will discourage them. An access route for tractors and other vehicles will serve just fine as a clear strip.

Here are some other control options.

Traps. Havahart-type traps, baited with a piece of apple, catch woodchucks alive. They work best when you leave a trail of small apple pieces leading into the trap. If that doesn't work, try smearing some peanut butter on a few of the apple pieces. But don't expect to catch your 'chuck on the first night. If using a Havahart or other live trap, be sure to check it regularly. You're not doing a woodchuck or any other animal that gets trapped any favors if you leave it to die from dehydration or heat stress.

Once you have a woodchuck in a live trap, you have to decide what to do with it. Personally, shooting a trapped animal, unless it is seriously injured, does not appeal to me and drowning seems very cruel. I or someone else on the farm usually transports trapped woodchucks several miles away, preferably on the other side of a large body of water, but I have a feeling the local authorities do not approve of this. (Indeed, it is illegal in many areas to transport or release trapped wildlife, so check into your local laws before doing so.)

Another type of trap is the Conibear. These must be set at the entrance to the woodchuck hole. If all goes according to plan, they spring closed around the animal's neck or head, killing it, but not always quickly. They are unpleasant devices that can kill other animals such as opossums, raccoons, or possibly the neighbor's small dog or cat.

Electric fence. An electric fence needs to be very low to control woodchucks. Two lines — 3 and 6 inches above the ground — might do the job. It's essential that the animal's nose, face, or paws make contact with the live fence. As an additional visual deterrent, you might consider putting up a 3- or 4-foot-high net barrier fence about 12 inches inside the electric fence. If the woodchuck doesn't catch a shock on the way in, it might not be so lucky as it tries to dig under the interior fence.

To deter woodchucks with electricity, it is important to use an energizer powerful enough to handle some vegetative interference. If you don't, you will have to diligently keep down the grasses and weeds below the electric line. Any plants that make contact with the line will cause it to short out, greatly reducing the charge. If you've spent any time mowing lawns, you'll know that grass, after a nice soaking rain, can grow an inch or more in just a couple of days, so keeping it down could be an endless undertaking. One solution would be to place a strip of weed barrier around the perimeter of your fields and set the electric fence into it. These measures might be effective on a garden scale but seem impractical for anyone growing acres of crops.

A dog. In my experience, this is the best defense against woodchucks. A tough, fast, fit hunting dog, preferably with some hound genes, will be an effective control officer. Such a dog will take real pleasure in his work and may even make a meal out of his quarry. We currently have a free-ranging male Black Mouth Cur named Kobe. His breed was featured in the Disney movie *Old Yeller*. Kobe lives in an outside doghouse year round but spends most of his time hunting in his designated territory. He's contained within about 30 acres with an invisible fence line and a collar around his neck that holds a 9-volt battery. Within this area, only the wiliest of woodchucks can escape his fast and formidable jaws.

Gas cartridges or woodchuck bombs. These put woodchucks to sleep, permanently. They require that you locate and block all entrances to the hole, except the one that you're going to throw the ignited cartridge into, and then quickly block that one, too. Obviously, they also require that the woodchuck be at home. We've had mixed success with gas cartridges. There can be as many as five entrances to a woodchuck hole, with subterranean tunnels as long as 50 feet. On our farm, the multiple entrances are usually strategically located along almost impenetrable hedgerows. Finding all of them is a challenge.

Exhaust fumes from a tractor or truck. This is done by placing one end of a long piece of metallic tubing over your vehicle's exhaust stack or pipe, while the engine is running, and the other end into the woodchuck's hole. As with a gas cartridge, all entrances to the hole need to blocked with soil, so that the carbon monoxide stays inside.

Rifle or shotgun. This is only an option if you're willing to shoot animals and know how to do it and live in an area where the discharging of firearms is not frowned on. Also, shooting woodchucks isn't that easy and can take a lot of time. They seem to have an uncanny ability to know when you're out looking for them with intent to kill. Local hunters might be recruited to thin down your woodchuck population.

Other Sometimes Pesky Critters

There are a multitude of other plant-, seed-, and grain-eating animals living in North America. Most won't pose much of a problem, but some might. Every farmer should become familiar with the troublesome ones common in his or her area and bone up on damage prevention and control methods. Here are a few other sometimes pesky critters you might run into.

Rabbits. They like to nibble on many vegetables, especially young peas and beans and leafy greens. Dogs and shotguns are probably the easiest way to keep rabbits' numbers down, or at least to keep them away from your fields.

Voles. This small rodent, which looks a bit like a snub-nosed mouse, can do significant damage to carrots and other vegetable crops. On our farm, voles nest amongst the larger carrots and eat the top part of the carrot root at soil level. Cats, mousetraps, and scent hounds (dogs with an excellent sense of smell) will make a dent in the vole population, as will barn owls and kestrels — both good birds to have on your farm. You might look into ways to create nesting sites for these birds of prey.

Chipmunks. These little critters may occasionally eat seedlings and large seeds of vegetables such as pumpkins and squash. They are likely to operate only on the edges of fields, close to trees and other cover. Chipmunks are rarely more than a minor nuisance. If they become a more serious problem, they can be fairly easily trapped or shot.

Raccoons. These highly intelligent, omnivorous animals are very fond of sweet corn but are seldom satisfied with a single ear. If you wake up one morning and discover a lot of broken plants and partially eaten ears, it's very likely that a raccoon and his friends have paid you a visit in the night. A hunting dog on the loose will definitely dissuade raccoons, as will a low electric fence. By the way, raccoons like chickens, too.

Meadow vole

Gophers. There are several different species of gophers in North America. They occur in western, central, and southwestern regions of the continent. Gophers spend almost all their time underground in narrow tunnels that can be anywhere from 3 inches to 2 feet deep. They dine on roots, tubers, bulbs, and seeds and can do significant damage to vegetable crops. They also like alfalfa. If you can find the several entrances to a gopher's tunnel system, fumigation is probably the best way to deal with this pest.

CHAPTER 21 RECAP

Fencing Options

► **Permanent woven-wire barrier fence.** This does the best job of keeping out large and medium-sized herbivores but is expensive and requires periodic maintenance.
► **Permanent electric fence.** This type of fence is easier to erect and less expensive than a permanent barrier fence but is not as sturdy.
► **Temporary and movable electric fence.** This is easy to set up and dismantle and can be moved where you need it.
► **Temporary barrier net fence.** This type of fence is also relatively easy to erect and dismantle, and you can add an electrified line around the outside for extra protection.

Other Methods of Controlling Critters

► Plant unpalatable crops.
► Employ dogs.
► Use repellants.
► Set up traps.
► Use a rifle or shotgun.
► Use gas cartridges, woodchuck bombs, or exhaust fumes directed into tunnels.

Waterpenny Farm

Proprietors: Rachel Bynum and Eric Plaksin

Location: Sperryville, Virginia

Year Started: 2000

Farm Size: 27 acres, with 10 acres in production

Crops: Mixed vegetables, flowers, chickens

WHEN RACHEL BYNUM AND ERIC PLAKSIN started Water-penny Farm, they were in their mid-20s, but they already had plenty of experience under their belts, including several seasons with Chip and Susan Planck at Wheatland Vegetable Farms in northern Virginia, a place where many young farmers have learned their trade.

Rachel and Eric felt they had the knowledge and drive to operate their own farm, but they lacked the capital to buy the high-priced land in the region of Virginia where they had their hearts set on farming. They overcame this obstacle by acquiring a long-term lease (40 years) on 27 acres of bottomland that is part of a much larger farm. The details of their lease arrangement with the landowner are nicely spelled out in an article Rachel and Eric wrote for *New Farm* magazine. This article can be found on the farm's website under "Press" (see Resources).

Equipment at Waterpenny Farm includes a Kubota M6800 65-horsepower tractor, a 6-foot Imants rotary spader, a water-wheel transplanter, a chisel plow, a plastic mulch layer, and a 35-horsepower Deutz 4507 tractor. The smaller tractor is used mainly for laying plastic, and the larger one operates the spader. The farm is divided into 40 rotation blocks, each 300 feet by 30 feet, with 12-foot grass and clover pathways in between.

Rachel and Eric are big believers in using mulch to control weeds. They use black plastic to grow tomatoes, peppers, eggplant, melons, squash, cucumbers, onions, scallions, chard, herbs, beans, winter squash, pumpkins, and flowers. White plastic, which doesn't heat up so much, is used for lettuces, kale, cabbages, and broccoli. The farm is gradually working more with biodegradable mulches. The pathways between the beds of vegetables are mulched with hay from round bales, a technique that Rachel and Eric learned at Wheatland Farm.

Rachel and Eric use pelleted, composted chicken manures and regular foliar applications of hydrolyzed fish and kelp, along with winter cover crops and summer rotations, to maintain fertility. The farm's two hundred egg-laying chickens are rotated behind harvested vegetable crops. At night, they are housed in a couple of old trailers that move on wheels. A tractor or truck is used to reposition the trailers in the fields.

Waterpenny Farm offers a 19-week CSA share and sells at three farmers' markets in the Washington, D.C., and Takoma Park, Maryland, areas. Five to seven interns work and live on the farm each year.

Deer Fence

Deer damage to crops at Waterpenny Farm was getting out of hand. The hungry ungulates seemed to be adding different vegetables to their diet each season and passing the newly acquired tastes on to their offspring. The deer learned to paw through row cover to get at the choicest morsels and break open melons with their hooves. Finally, Rachel and Eric decided to bite the bullet and spring for a serious deer fence. In spring 2007, they enclosed 18 acres of relatively level ground with 4,400 feet of 8-foot-high woven-wire mesh fence. They hired a local fence builder to do the job. At $6 a foot, the price tag was $26,000. This included two small, people-size gates and nine other gates that can accommodate tractors and farm equipment. The fence builder used 6-inch-diameter, 10-foot-long, treated wooden posts and sunk them 2 feet into the ground. The posts were spaced 20 feet apart.

Rachel and Eric are happy with their fence and feel that it was worth the expense, though Eric notes that any grower contemplating such a fence should have in place the production capacity and markets to justify the considerable cost. These days, the only deer that get into the vegetable fields at Waterpenny Farm are those that occasionally wander through an open gate.

Rachel and Eric say they didn't recognize the extent of the deer damage they were experiencing until they fenced the critters out. They believe the extra harvest paid for the fence within 3 years. They expect the structure to have a lifespan of 30 to 40 years, with minimal upkeep, although some maintenance is necessary. One person mows around the perimeter once or twice a year, and a four-person crew spends about a day each year cutting down vines and saplings that have taken root at the base of the fence.

Insects and Diseases

THE MOST I CAN HOPE TO DO in this chapter is introduce you to the subject matter, for two reasons: First, the number of insect pests and diseases that afflict vegetable crops throughout North America is large — larger than can be addressed in a book of this sort — and second, I'm far from being an expert on these matters. Rather than try to itemize and describe all the individual pests and diseases — though I will list some of them — I'd like to concentrate on general strategies for avoiding trouble in the first place. This approach has served our farm pretty well for the past 25 years. We've had our problems, but we've never had a year in which our business was severely threatened by a pest or disease.

Let's start with some definitions and basic information.

Insects

An insect is a small invertebrate with three pairs of legs and either one or two pairs of wings. During their adult phase, insects have three segmented body parts (head, thorax, and abdomen) and an exoskeleton that functions a little like body armor. All insects pass through some form of metamorphosis. For insects with complete metamorphosis (e.g., caterpillars and moths), the immature and adult phases bear little resemblance to one another. For insects with incomplete metamorphosis (such as plant bugs and leafhoppers), the immature form resembles the adult but

Squash vine borer in larval (left)
and adult stages

doesn't have wings. There are estimated to be more than 1 million different insect species in the world. Only about 1 percent of them are pests.

Beneficial Insects

The vast majority of insects found in North America (99 percent of them) are, from a farmer's point of view, either beneficial or benign. The beneficials, or "good guys," can be divided into two groups: predators and parasitoids. Beneficials make up about 25 percent of all insects.

Predators. Predatory insects, quite simply, eat their victims. The victims are mostly, but not always, pest insects. During their life span, predators can eat a lot of prey. The predator is usually most voracious in the larval stage. In some species, it is only the larvae that are predatory.

Parasitoids. These beneficials lay their eggs on, or in, their prey, which is frequently caterpillars or other insects in their soft-bodied phase. Often, they paralyze their victims first. When the eggs hatch, which is usually in a relatively short period of time, the parasitoid larvae have a ready source of food. Unlike predators, a parasitoid insect requires only one host to reach maturity. After it is selected, the unfortunate host gets eaten from the inside out. It usually doesn't take very long.

PREDATORS

Lady beetle in adult (left)
and larval stages

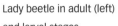

Adult lacewing

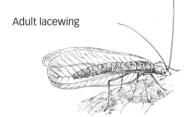

Lacewing larva
eating aphids

Adult assassin bug

Adult spined
soldier bug

Common Predatory Insects and Their Prey

Predator	Prey
Assassin bugs	Larvae, and sometimes adult forms, of many pest insects
Big-eyed bug	Aphids, leafhoppers, spider mites, and other immature bugs
Damsel bugs	Aphids, leafhoppers, spider mites, caterpillar eggs, and insect larvae
Lacewings	Aphids, as well as larvae and eggs of many insects
Ladybugs	Aphids, thrips, spider mites, and other soft-bodied insects and insect eggs
Pirate bugs	Thrips, aphids, spider mites, small caterpillars, and insect eggs
Praying mantids	Many insects, including each other
Spined soldier bugs	Larvae of Colorado potato beetles, fall armyworm, and many caterpillars and grubs
Syrphid flies	Aphids, leafhoppers, thrips, and other soft-bodied insects

Note that parasitoids are different from parasites, which have a definite interest in keeping their hosts alive. As adults, many parasitoids feed on nectar and are excellent pollinators of flowering plants. The most common and valuable parasitoids are wasps and flies.

Common Parasitoids and Their Hosts

Parasitoid	Host
Braconid wasps	Larvae of various caterpillars and aphids
Ichneumon wasps	Corn borers, cutworms, caterpillars, and other larvae
Tachinid flies	Many caterpillars, plant bugs, adult beetles, grasshoppers, and other insects
Trichogramma wasps	Eggs of moths and butterflies

PARASITOIDS

Braconid wasp

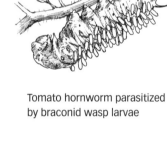

Tomato hornworm parasitized
by braconid wasp larvae

Tachinid fly

"Pests"

Many of the "bad" insects that damage our crops from time to time have a definite role to play on an ecologically sound farm. They provide an essential food source for predators and parasitoids, both of which are valuable friends for a farmer. If you destroy all the problem insects, you run the risk of upsetting nature's balance. The "good guys," many of which are important crop pollinators, will have nothing left to eat and no suitable host for their eggs. You force them to look elsewhere.

Some Common Pests of Vegetable Crops

The list below is limited to some of the common insect pests we are familiar with in our region and the crops that attract them — just to give you an idea of what one farm might have to contend with. The lineup would be different if we were farming in the Southwest or the Southeast or the Pacific Northwest, though I'm sure some of the same culprits would keep reappearing. Wherever you happen to be farming, you shouldn't have any trouble finding out what insects you'll be up against. It's likely they'll introduce themselves to you.

Pest Insect	Host Plant
Aphids	Many leafy greens and peppers; frequently a problem in greenhouses
Cabbage looper	Most brassicas and many other vegetables
Colorado potato beetles	Potatoes, eggplant, tomatoes, and peppers
Cucumber beetles	Cucurbits (spotted and striped)
Cutworm (several species)	Most vegetables, especially in seedling stage
Flea beetles	Most brassicas (especially the "spicy" ones), potatoes, eggplant, sweet corn, and tomatoes (each of these crop families has a different species of flea beetle, or group of species, that attack them)
Imported cabbageworm	Brassicas, especially cabbage, kale, and collards
Potato leafhopper	Potatoes, beans, celery, eggplant, and many other crops
Squash bug	Cucurbits
Squash vine borer	Cucurbits
Tarnished plant bug	Many vegetables, cover crops, and weeds
Thrips	Onions, shallots, garlic, cabbage
Tomato hornworm	Nightshades

PESTS

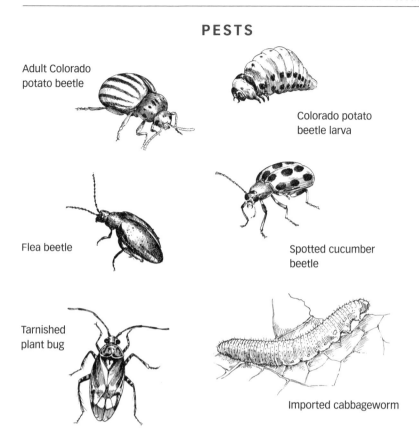

Adult Colorado potato beetle

Colorado potato beetle larva

Flea beetle

Spotted cucumber beetle

Tarnished plant bug

Imported cabbageworm

Diseases

The first challenge a farmer faces with a vegetable disease is identifying it. In this respect, diseases are a lot more difficult to pin down than insect pests. The latter shouldn't give you too much trouble if you have a good ID book, a hand lens, a specimen, and a little time on your hands. The problem with diseases is that you can't see the identifying characteristics (often spores) without a compound microscope. And to compound the problem, many diseases share similar symptoms.

Sometimes symptoms that suggest a disease are, in fact, caused by an abiotic problem, such as a particular nutrient deficiency. Or a disease may be secondary to some form of plant tissue damage, such as sunscald.

To be sure you've got the right disease, you may need a profession-al's help. With plant diseases, prevention is always the best policy. Once you've got a disease, the most you can hope to do is minimize the symptoms. For organic growers, in most instances, eradication is either difficult or impossible.

There are three categories of plant diseases: fungal, bacterial, and viral.

Fungal Diseases

Most of the common plant diseases are caused by fungi. Fungi are simple, threadlike plants that lack stems, roots, and leaves. They also lack chlorophyll and are therefore unable to produce food for themselves. Instead, they parasitize vegetable crops and other plants, using the stored energy these higher plants have gained, through photosynthesis, to nourish themselves. The minute "seeds" that fungi produce are called spores. Overcast, damp, and wet conditions facilitate the growth and spread of many fungal spores.

Those fungi known as saprophytes are often beneficial to a farmer because they feed on dead plant matter and, in so doing, advance the processes of decomposition and recycling of minerals and nutrients. Others, unfortunately for us, feed on both organic matter and living plants.

Symptoms. Common symptoms of fungal diseases of vegetables include: mildews (such as downy and powdery mildew); lesions on the leaves, fruits, and roots; and leaf rusts. If conditions are favorable, fungal

Common Fungal Diseases in North America

Disease	Affected Crops
Anthracnose	Beans, peas, cucumbers, mustards, turnips, tomatoes, and peppers
Botrytis gray mold	Most vegetables
Damping-off	Almost all vegetables in the seedling stage
Downy mildew	Cucurbits, crucifers, beets, peas, lettuces, and onions
Early blight	Tomatoes and potatoes
Fusarium wilt	Beets, crucifers, tomatoes, peppers, onions, chard, and spinach
Late blight	Potatoes and tomatoes
Powdery mildew	Cucurbits, crucifers, peas, beans, and peppers
Rusts	Beans, peas, carrots, corn, eggplant, lettuce, onions, and spinach
Verticillium wilt	Most vegetables, but especially tomatoes, potatoes, and cucurbits

Caution: The common names of fungal diseases (listed above) are not always reliable identifiers. The same name might be used to refer to slightly different diseases on different plants and even in different regions of the world. There are several species of downy mildew, for example, that affect different plants. To clearly identify a disease, for management purposes it is best to use its scientific name.

diseases can spread quickly over large portions of a plant and, before long, an entire crop. It's important to identify them in their early stages and take rapid action, assuming that there is an effective remedy.

Common fungal diseases. Some fungal diseases, such as early blight of tomatoes, are very common and not necessarily the end of the world. Plants can often tolerate and even outgrow low levels of infection. Others, such as late blight of potatoes and tomatoes, must be aggressively managed as soon as they appear, as an entire planting can be lost when weather conditions permit the disease to spread.

Bacterial Diseases

Bacteria are microscopic, single-celled organisms that can easily mutate in response to changing environmental conditions.

Symptoms. Some bacteria cause various wilting diseases by gumming up a plant's vascular system. Others result in abnormal growth, such as galls, and still others cause spotting or rotting of plant tissue. Any wound or natural opening in a plant's epidermis is a possible entry site for bacteria.

Transmission. Bacterial diseases are most often transmitted through infested seed or soil. Soilborne transmission commonly occurs when heavy rain splashes off the soil and onto a plant. Bacterial diseases can be moved from one field to another through mud or soil on a tractor tire or a farmer's boots. A small amount of soil on your hands or gloves could be the cause of transmission. Bacterial diseases can even be transmitted on tools or clothing when no soil is present. So if you know you have a bacterial disease, you need to take serious precautions to avoid moving it around on your farm.

Common Bacterial Diseases in North America

Bacterial Disease	Affected Crops	Bacterial Disease	Affected Crops
Bacterial blights	Peas and beans	Black rot	Crucifers
Bacterial spot	Peppers and tomatoes	Potato ring rot	Potatoes
Bacterial wilt	Cucurbits	Scab	Potatoes
Black leg	Potatoes	Soft rot	Most vegetables

A few bacterial diseases have insect vectors. For example, bacterial wilt of cucurbits, a serious problem for organic growers, overwinters in, and is transmitted by, striped or spotted cucumber beetles when they dine on members of the cucurbit family.

Disposal. Bacterial diseases are not curable. If you know you have a bacterial disease, uproot and dispose of infected plants as soon as possible to prevent the disease from spreading to other plants. The best methods of disposal are burning or burying in a deep hole, or possibly trucking infected plants to a landfill. Do not try to compost them, even in a well-heated compost pile. Bacteria can survive high temperatures.

Viral Diseases

Viruses are made up of genetic material in a protein sheath. They replicate inside infected cells and disrupt cellular processes. Like fungi and bacteria, they are essentially parasitic, but they are obligate parasites, which means they depend on a living host to reproduce or replicate themselves.

Symptoms. Vegetables infected by viruses are usually stunted, mottled, yellowed, or malformed.

Transmission. Viruses are often transmitted by insect vectors, such as aphids, but they can also be passed along on fungi, nematodes, certain weeds, and infected seeds. It's also possible to transmit them through infected soil or plant material on equipment or a farmer's hands or boots. Viruses have a tendency to mutate in response to changing environmental conditions, and some of them can remain alive on dead plant material for many years.

Common Plant Viruses

Viral Disease	Affected Crops
Bean mosaic	Beans
Cucumber mosaic	Cucurbits, tomatoes, peppers, celery, and spinach
Potato leaf roll	Potatoes and other nightshades
Tobacco mosaic	All nightshades but especially tomatoes, spinach, beets, and some crucifers
Tobacco ring spot	Cucurbits, beets, nightshades, and spinach

Nematodes

Though usually grouped in with plant diseases, nematodes are not really diseases themselves. Rather, they are tiny wormlike creatures that are almost invisible to the naked eye. Of the thousands of different species of nematodes that exist, only a small percentage are harmful to plants. The harmful few most often attack plant roots. They may be vectors for various bacterial and viral diseases, but even when they are not carrying diseases, the microscopic wounds they leave behind weaken plants and serve as entry points for other damaging organisms.

Symptoms. Affected plants usually show stunting, wilting, and yellowing. Plant roots often exhibit swelling and gall-like growths. Nematode damage can render root crops unmarketable.

The root-knot nematode causes the most familiar nematode-related symptoms, including thickening and galling of roots. Many plants, including potatoes, tomatoes, carrots, and beans, are vulnerable to this disease.

Root-lesion nematode is also relatively common. It causes brown lesions on roots.

Transmission. Nematodes are mainly transmitted by moving infected plants or nematode-infested soil on farm equipment.

How to Minimize Damage

The first step, which may take several years, is to familiarize yourself with the insect pests and diseases that are common in your region. There are many websites, books, leaflets, and fact sheets that will help with identification and provide information on such matters as:

- Range of the pest or disease
- Life cycle and when it is an active threat
- Host plants
- Feeding habits (for insects)
- Transmission strategies (for diseases)
- Methods for prevention and control

In the beginning, don't feel that you have to memorize all this stuff. Just know it exists and how to access it. As you gain firsthand experience with individual pests and diseases, you will accumulate knowledge. When you do encounter a problem, the most important thing is to find out what is causing it. Then you can focus on specific details.

If you are unable to make a positive identification on your own, your local Extension agent or educator should be able to help. If the agent is not sure, he or she will send a specimen of your plant or insect to a university diagnostic lab or specialist. Identifying the culprit is the first step, and it will enable you to develop an appropriate course of action.

Rotate Crops

The subjects of diversity and crop rotation keep cropping up (no pun intended) again and again in this book because they really do embrace so much of what organic farming is about. Large plantings of the same vegetable, year after year, on the same piece of land, will cause diseases to stay around and pest insects to get out of control. When this happens, naturally enough, the disease or pest population surges to take advantage of a vast and readily available food supply. See pages 270 and 272 for some of the numerous benefits gained from moving crops around.

Rotate by family. Remember, many pest insects, and to a lesser extent diseases, focus their attention on a particular family of vegetables and avoid plant species from other families. The imported cabbage worm, for instance, likes to eat cabbages (no surprise there) and similar cole crops such as collard greens and kale. Crucifer flea beetles prefer the mustardy-flavored brassicas, such as arugula, mizuna, tatsoi, and red and green mustard. Colorado potato beetles like potato foliage but will also enjoy the leaves of other members of the solanaceous family (peppers and eggplant). None of these pests is likely to switch to another family, even when their food of choice is not available. So when it comes to moving crops around, it's best to rotate families, not just individual vegetable types.

Unfortunately, this strategy doesn't work in all cases. Some pests — most notably, aphids and thrips — are quite happy to cross from one vegetable family to another. Many insects are mobile and able to move considerable distances to find a host, so it's necessary to rotate as far as possible from an overwintering site to delay movement to the next year's crop.

Include alliums. The *Allium* family (onions, shallots, scallions, and garlic) are believed to repel many insects; therefore, including alliums in your crop rotation plan is a natural pest-control measure.

Provide Habitat for Beneficials

It is crucial to maintain suitable, year-round habitat for a broad range of beneficial insects. Patches of wild and flowering perennials are the best way to do this. Such wild plants might be growing on nearby idle land, along hedgerows that separate fields, or in permanent strips between beds of crops. Winter cover crops are another good place for the beneficials to hang out in the cold months.

Many herbs — especially perennials such as oregano, mint, thyme, and sage — are attractive to beneficial insects, both for the cover they offer and the nectar they provide during their flowering stage. The greater the diversity of vegetation on your farm, the greater will be the diversity of your beneficials.

Adjust Planting Times

Another good strategy for dodging troublesome insects and diseases is to adjust planting times so they don't coincide with a pest's peak population. This requires understanding the pest's life cycle and perhaps analyzing some recent climate data, such as growing degree days in your area. (For a discussion of growing degree days, see page 79.) Much of this information is available online, in books, and in Cooperative Extension handouts, if you're willing to do a little poking around.

In the Northeast, Penn State University's website (see Resources) is a good real-time resource. It tracks the emergence and movement of multiple vegetable pests and diseases. Other useful sites include the IPM PIPE (IPM stands for Integrated Pest Management) and USAblight, which will keep you informed on the incidence of late blight in tomatoes and potatoes in the United States. See Resources for their web addresses.

You might also ask your local Cooperative Extension office about other pest-watch services in your area.

Employ Good Field Hygiene

Don't leave weeds and plant debris, especially debris from diseased plants, lying around in your fields at the end of the season. Such material can harbor pests and diseases that can come back to bite you the following spring or summer. If you know that a particular crop had a disease or pest problem, it's best to turn it into the soil before cold weather sets in and there is still time for biological decomposition to take place. Or you might uproot affected plants and burn them, or bury them in a deep hole. It is generally not a good idea to compost diseased plants. Many plant diseases, especially those that are bacterial in nature, can survive the heat of a compost pile.

Some pests deposit larvae and pupae in the top few inches of the soil before cold weather sets in. Tilling early in spring, to a depth of 6 to 8 inches, will expose many of these malefactors to the elements, causing them to die.

Maintain Good Air Circulation

By maintaining good air circulation around plants, you can discourage many diseases. Densely packed plants and heavy weed growth at the base of plants create damp microenvironments in which diseases are more likely to take hold and proliferate. Moist conditions and stagnant air around plants also favor certain insect pests, especially slugs.

Plant Resistant Varieties and Use Clean Seed

Some hybrid plant varieties have resistance to various diseases bred into them. If you know you have a particular disease problem, check the seed catalogs for resistant varieties. They may not solve your problems altogether, but they could reduce them. Over time, experiment with resistant varieties and see if you can determine which, if any, make a noticeable difference.

Be sure the seed you plant is clean and of high quality. Reputable seed companies should test for seed-borne diseases when there is even a small probability that they could be present. Don't be shy about asking them how much of each seed lot they test and what other precautions they take to avoid passing on plant diseases. If you decide to save your own seed, find out the best way to process it to decrease the chance of transmitting disease. Be especially careful with plants that have large seed pieces, such as potatoes and garlic.

Use Protective Barriers

Organic growers often use lightweight floating row cover or fine-mesh gauze to keep insect pests away from their crops. These protective barriers are especially valuable when the crop plant is small, succulent, and more vulnerable to pests. In many cases, when the protected plants put on some size, the barrier can be removed. Larger, stronger plants can often stay a few steps ahead of whatever insect depredations they might have to endure.

Bring in Your Own Beneficials

Many of the beneficials we've discussed in this chapter can be purchased from insectaries and released on your farm. This can be costly and does not carry any guarantee of success because there is no way to ensure that the insects you buy — which are usually in the egg or larval stage — will stay where you want them and do the job you have in mind. After they hatch, they might choose to fly off to the neighbor's place.

Releasing beneficials into a greenhouse or high tunnel is another story. In these cases, there is a much better chance that they will stay put, but timing is critical. The beneficials must be released preventively or at the first sign of insect infestation. On several occasions, we've had reasonable success releasing wasps and lady beetles into our greenhouse to control aphids. In general, the best candidates for this type of biocontrol are those beneficials that attack pest insects during their larval stage.

If you want to learn more about or purchase beneficials, try IPM Labs (see Resources). They have a reputation for offering good products and good advice.

Try Organically Approved Sprays

If you've followed the above practices and still have persistent pest or disease problems, there are a number of organically approved pesticides on the market. Some are quite pest-specific and some are broad-spectrum and capable of harming many different insects. You should take a close and critical look before using any of them. Information on some products allowed for organic production can be found in the Resource Guide for Organic Insect and Disease Management, which can be accessed online (see Resources for website).

In my early days of farming, rotenone, pyrethrum, and sabadilla — all botanicals (i.e., derived from plants) — were the most common pesticides available to organic growers and were used fairly widely. They did a reasonable job of killing insect pests but, unfortunately, were just as deadly against benign insects and beneficials. These broad-spectrum insecticides are still available, but most organic growers now give them a wide pass or use them only in extreme situations.

Bacillus thuringiensis (Bt). Sold under different trade names, Bt is a biological insecticide that is much more environmentally friendly than most of the botanicals. It is a naturally occurring bacteria that, when consumed, is toxic to a certain pest insect during its larval stage. We use the strain *Bacillus thuringiensis* subspecies *kurstaki*, to control imported cabbage worms and cabbage loopers on our brassicas. Other strains of Bt are available for other pests.

Spinosad. Sold under the labels Entrust and Monterey Garden Spray, spinosad is derived from the fermentation of the naturally occurring actinomycete (*Saccharopolyspora spinosa*). It is effective against several pest insects (among them caterpillars, leafminers, thrips, flea beetles, and Colorado potato beetles) but has low toxicity to most beneficials. It is quite expensive. Some growers alternate between spinosad and Bt.

Insecticidal soaps ("Safer" is a common brand). These are made from potassium salts and are highly biodegradable and relatively safe to use. When sprayed directly onto the pest insect, they act as a smothering agent or dessicant. They work best on soft-bodied insects such as aphids, whiteflies, and spider mites. They are best used early or late in the day, or when the sun is not shining brightly.

The Trouble with Broad-Spectrum Sprays

Broad-spectrum pesticides frequently kill pest insects and beneficials alike, and often they are more toxic to the beneficials than the pests. A further problem is that pest insects are often able to quickly repopulate and launch another assault, while the beneficials invariably take much longer to regroup. The farmer who has used a broad-spectrum spray, whether it be a conventional or an organic product, often has little choice but to spray again because any possibility of natural control has been greatly reduced or eliminated.

Is It on the OMRI Products List?

The Organic Materials Review Institute (OMRI) publishes an annual *Products List* that contains all the materials that are allowed for use in organic production, processing, and handling (see Resources for website). The list changes frequently as new products are added, and sometimes old products are removed. If you're a certified organic grower, you need to be sure that any pesticide — or any other product, for that matter — that you plan to use is currently on the list. Not all certifiers accept all the products on the OMRI list, so again, if you're a certified grower, you should always check with your certifying agency before using a new product. Also, note that not all formulations of the handful of products listed on pages 498–499 are necessarily approved for organic use.

Trichoderma harzianum (**T-22 Planter Box**). This is a beneficial fungus that colonizes the roots of vegetable plants and protects them against certain soil borne root diseases, including fusarium, rhizoctonia, and pythium. It is expensive but does the job. We use it as an inoculant on potato seed pieces before planting. It can be used on many other vegetable seeds.

Bacillus subtilis (**Serenade**). This biological fungicide is a bacterium that works against some fungal diseases, such as powdery mildew and early blight.

Neem. This botanical insecticide is derived from the seed of the neem tree, which grows in India. It is used to control many pests of vegetable crops, including aphids, whiteflies, thrips, hornworms, and cabbage loopers. Like rotenone, pyrethrum, and sabadilla, it is a broad-spectrum insecticide, but because it works more as a growth regulator, it is somewhat less toxic to beneficials than these three.

Sulfur. This is used in the form of copper sulfate or elemental sulfur, primarily as a fungicide. When applied at the first sign of disease, sulfur can be very effective against powdery mildew but hard on some beneficial mites, which can lead to spider mite outbreaks.

Copper. Different formulations of copper and copper hydroxide have fungicidal and bacteriostatic properties. Like sulfur products, copper products should be used early in the progression of a fungal disease.

Keep a Positive Attitude

Try not to let the rather daunting subject of plant pests and diseases bring on a bout of anxiety that could dampen your enthusiasm for farming. If you follow good practices, particularly organic methods, pests and diseases should not be a major impediment to your success. Another thing: Plant diseases cannot be transmitted to humans, so you've got nothing to worry about there.

Plants are tough. Many plants have developed their own defenses, especially against diseases. A healthy plant might release substances to ward off a disease in an infected area, or it might cause a portion of its tissue to die to deprive the disease of further nutrition. In some cases, plants create physical barriers to halt the spread of a disease.

Most organic farmers believe that vegetables growing in healthy, organically managed soil are better able to withstand plant diseases and might be less palatable to insect pests than crops growing in soils that are subjected to agricultural chemicals.

Customers might not mind. Just as many healthy plants can withstand some amount of insect damage and low levels of disease without being severely compromised, humans, too, can be reeducated to accept minor flaws in the fresh produce they buy. More and more, I'm finding that people who seek local, organic food are not deterred by a small spot or ragged hole that suggests some other creature has taken a bite. Some even see such imperfections as the organic stamp of legitimacy and regard visually flawless food as possibly suspect. They've had their fill of the cosmetically perfect stuff from the supermarket produce aisle. Its bland flavor has disappointed so many times that they are ready for something different.

One of our customers once left a note for me that I've never forgotten. It read: "I found a caterpillar on something I bought at your stand last week and it just thrilled me to see it."

Now, I wouldn't push this point too far. Organically grown food can and should look vibrant and healthy, but a minor blemish here

SAR Pesticides

New pesticides on the market actually attempt to boost or mimic plants' immune responses. These products are called systemic acquired resistance (SAR) response pesticides. Some might be approved for organic use.

or there might be quite acceptable. Diseases and pests become real problems only when they begin to have a significant impact on the growth and quality of a crop. In some cases, no action is the best action.

Plant insects and diseases represent two very large and fascinating subjects. You could spend your whole life studying one or the other. Here, we've only touched lightly on them. There is so much more to learn and understand—for all of us.

CHAPTER 22 RECAP

How to Minimize Insect and Disease Damage

- ▶ Rotate crops.
- ▶ Provide habitat for beneficials.
- ▶ Adjust planting times to avoid a pest's peak population.
- ▶ Employ good field hygiene.
- ▶ Maintain good air circulation.
- ▶ Plant resistant varieties and use clean seed.
- ▶ Use protective barriers.
- ▶ Bring in your own beneficials.
- ▶ Try organically approved sprays.
- ▶ Keep a positive attitude.

PART VII

TAKING CARE OF BUSINESS

....

Chapter 23

Running a Business

A SMALL FARM IS A BUSINESS, like any other, and should be managed as such. There will be records to keep, bills to pay, income to report, and a multitude of minor headaches to endure. It's the lot of the small businessperson in this day and age. Your best defense is to have a stoic attitude, a good bookkeeping system, and a habit of making timely bill and tax payments.

In addition to the above three lines of defense, look for a good accountant who has some knowledge of farming matters, will stay with you over the years, is willing to become familiar with your business, and will guide you through the tedious and treacherous regulatory waters. A good farm accountant can advise you on such matters as how to structure your business, when to make capital improvements, when to purchase new equipment, how to calculate the depreciation of existing equipment, and the like, as well as keep you up to date on all tax filing and payments.

Shop around until you find an individual who will work with you on all of these fronts. Don't be afraid to quiz potential accountants on their knowledge of the tax code with respect to agriculture. A competent accountant will know the right moves and get the right numbers faster than you ever will. This will leave you more time to do the actual work of farming. A local insurance agent, who can bundle together different kinds of insurance, will also make your life a lot easier.

Sure, you might start out filing your own taxes and buying different types of insurance on the open market, but as time passes and your business grows, you'll benefit greatly from professional help. Certainly, there will be fees to pay, but you can deduct these, along with all other business expenses, from your gross income. Try to see all the record keeping, filing, and tax reporting as tools to help you manage your farm and chart a course for the future. This sometimes onerous work is as important to your success as a good crop plan and a working tractor. Viewed in this light, it will be a lot more palatable.

(503)

Ten Mile Farm

Proprietors: Kevin Toomey and Christina Carter
Location: Candler, North Carolina
Year Started: 2005
Size: 3 acres
Crops: Mixed vegetables

K EVIN TOOMEY CAME TO OUR FARM IN 1999 after studying agricultural communities for his degree in anthropology. He liked what he had learned about agrarian cultures and wanted to experience small-scale, diversified farming for himself. After his season with us, Kevin moved to Asheville, North Carolina, and explored other elements of what he calls the "food chain." He cooked for a vegetarian restaurant and worked for a gardening landscaper. He put in some time at a Japanese sushi bar. On the side, he tended a large garden and sold specialty vegetables to one of his restaurant employers. After 5 years, he was ready to give farming a serious try.

In 2004, Kevin teamed up with Christina Carter (the two are now married), who also had a strong interest in growing plants and herbs. A year later, the young couple decided to launch their own farm business. Because they couldn't afford to buy a farm in the Asheville area, they opted to lease land and continue living in the city. The two parcels of fertile bottomland that they found are approximately 10 miles from their home in Asheville — hence the name, "Ten Mile Farm." During the growing season, they commute to work every day.

Kevin and Christina grow more than 40 varieties of vegetables and herbs and sell their wares at two farmers' markets in Asheville. They also operate a small CSA and sell to restaurants. They started out growing on just 1 acre, using a BCS walk-behind tractor with a tiller, brush mower, rotary plow, and sprayer, to help them get the work done. As their operation has expanded (they now farm 3 acres), so has their equipment list. They've added a Kubota 5040 tractor with a front-end loader, a Celli spader, a disc, a flail mower, and an Allis Chalmers G cultivating tractor. Kevin is especially enamored of the Celli spader and the Allis Chalmers G.

Ten Mile Farm now has its own 14- by 48-foot greenhouse for seed starting. It uses a radiant-style heating system that runs off a standard hot-water heater. The hot water is circulated under the benches that hold the seedling flats. This system, like that of heat mats, keeps the potting soil at an optimum temperature for germination and plant growth. Crops at Ten Mile Farm are irrigated with drip lines and overhead sprinklers. The water source is a nearby creek.

Kevin and Christina make most of their own compost using biodynamic preparations. They further enhance their soil's fertility with bone meal, blood meal, azomite, humic acid, and mycorrhizal fungi. Foliar applications of fish and kelp and compost tea are applied directly to growing plants. Christina and Kevin started out doing all their own growing and selling, with just occasional help from individuals willing to work in exchange for food and knowledge. Their first year with a full-time employee was 2011, and they expect to hire more help in the future. Ten Mile Farm follows organic and biodynamic methods but is not certified.

For Christina and Kevin, leasing land and living off-farm has both good and bad features. Packing lunch and snacks and making sure they have everything necessary to do their work is a daily challenge. It's a 20-mile round trip to fetch a forgotten seed packet, tool, or item of clothing. Not being able to step inside for lunch and a nap on a hot day, or dodge a heavy rainstorm, makes life a bit more rigorous. And there are times when the greenhouse might benefit from watering while they are still packing their van for the day. No doubt about it: Farming 10 miles from where you live has its challenges.

On the positive side, Christina and Kevin are adding to their farming knowledge every day and learning how to run a business, without having the monthly expense of a mortgage hanging over their heads like the sword of Damocles. Each year, they are growing and selling more vegetables, attracting new customers, and gaining confidence that they are able to support themselves as farmers. The Asheville market for local food is strong.

Now, after 5 years, Christina and Kevin have saved some money and are seriously looking for a place of their own. They know what they need and what they want. Fifteen to 20 acres, most of it tillable land, would be perfect. A patch of woods and some pasture would be a plus. Kevin would like to plant an orchard and incorporate animals into their future farm. He is also keen to venture into some high-tunnel growing. In both his and Christina's eyes, it is just a matter of time.

A Bookkeeping System

Bookkeeping is essentially the recording of financial transactions such as sales, purchases, payments, and income. It is something that only you or someone working closely with you can do, and it is often done on an almost daily basis. Usually, this information is periodically turned over to an accountant who prepares the necessary reports and submits them to the appropriate government agencies.

Key Elements

There are three key elements in a basic bookkeeping system:

- **Income.** This is recorded by type — such as cash sales, restaurant and CSA share payments (often by check), credit card sales, Food Stamps, and Farmers' Market Nutrition Program coupons.

- **Expenses.** These should be broken down into categories relevant to your farm business. Keep receipts of all expenditures and invoices and method of payment. More on this in a moment.

- **Cash flow into and out of business.** This includes bank deposits and withdrawals and any funds transferred from a personal to a business account, or vice versa.

These days, most farmers opt for an electronic bookkeeping system, though some begin by keeping manual records, then switch to a computerized system later on when they have a better idea of their needs. There are numerous farm management and bookkeeping programs to choose from. Some provide more than basic bookkeeping software. Features may include tax reporting, invoicing, payroll management and check writing, cost-of-production analysis, inventory management, and other farm support functions. My advice would be to start out with something simple and let your accountant handle the more complicated matters.

Just remember, if you're relying on a computer to store and organize your business data, you need to make frequent backups and, preferably, have a fallback manual system in place in case of a power outage or computer failure. If you use more advanced web-based software, your data should never disappear, even if your computer crashes. Also, your accountant will be able to access the data at any time.

A Filing System

You'll also need a system for filing invoices and receipts, credit card statements, cancelled checks, and bank statements. Solid metal filing cabinets with hanging manila folders are still in vogue and will serve you well. Every deduction on your tax return should be verifiable. If you get audited and cannot document an expense, the IRS will simply disallow that deduction.

Expense Categories

Depending on the size and nature of your farm operation, you'll want to create your own list of expense categories and set up a system for entering data under these categories. Your system might be as simple as a manual ledger with rows and columns, in which case you'll need columns for the following:

Date of Purchase

Method of Payment (check #, credit card, cash, automated or online payment)

Paid To (include brief description of item purchased or service rendered)

Amount Paid for All Purchases

Amount Paid for Individual Items (in a separate column for each category of purchase or service)

A step up from the manual ledger would be a bookkeeping software package tailored to small farmers or a Microsoft Excel spreadsheet. Both will tally up numbers for you so you can keep track of how much you're spending in each expense category.

To get you thinking along these lines, here's a list of expense categories we use on our farm:

Expense Categories Used on Our Farm

Supplies (e.g., plastic bags, berry baskets, rubber bands)

Repair and Maintenance:

      Trucks

      Tractors

      Other Equipment

Professional Fees (lawyer, accountant)

Advertising (any promotional materials, employment ads, etc.)

Entertainment (if you take a job applicant, a client, or an employee to dinner)

Office Supplies (computer, software, printer, fax, copier, stationery, etc.)

Farmers' Market Fees

Organic Certification Fees

Fuel, Oil, Grease (for tractors, vehicles, and other equipment)

Heating Fuel (propane for greenhouse, oil for worker housing)

Veterinary Bills (for any animals that support the farm business)

Animal Food (if your dog is a working dog who keeps away predators or kills woodchucks, his food can be deducted)

Postage and Shipping (stamps and shipping fees for business-related mail and shipping)

Conferences, Dues, Publications (e.g., organic farming conferences, subscription and membership dues, books on farming)

Hardware and Small Equipment — Under $300 (e.g., small tools, nuts, bolts, nails, etc.)

Hardware and Large Equipment — $300 and above (e.g., an air compressor, a tractor, a brush hog)

Potting Mix and Soil Amendments (e.g., off-farm compost, manure, fish emulsion, rock phosphate, greensand)

Seeds and Plants

Crop Expenses (e.g., irrigation supplies, plastic mulch, floating row cover and hoops, organic sprays)

Land, Building, and Fence Repair (e.g., gravel on driveway, coat of paint on barn)

New Construction (e.g., a new pole barn, a high tunnel, worker housing)

Taxes:
    School and Property
    Federal
    State
    Other

Insurance:
    Farm Buildings and Equipment (fire, theft, etc.)
    Liability (both on farm and at market or for a CSA)

Vehicles (commercial insurance will cost more but give you
better protection against lawsuits)
Employment-Related Expenses (Workers Comp., Disability,
Unemployment)
Land Rental and Lease (in the event you rent land)
Mortgage and Loan Expenses (note principal and interest paid)
Electric Utilities (allot some of utility bill to personal living, some
to farm business)
Telephone and Internet Service
Miscellaneous Expenses

Payroll Data

If you have employees, you will need to have them fill out the federal
Form W-4 and federal Form I-9 (Employment Eligibility and Verifi-
cation), and possibly other forms, depending on what state you are in.
You will also need to keep payroll information. This should include each
employee's name, address, Social Security number, and the number of
deductions he or she is claiming, along with the following weekly pay-
roll data:
Pay Period
Gross Pay
Social Security Withholding
Medicare Withholding
Federal Income Taxes
State Income Taxes
Disability Insurance
Net Pay

If you prefer hard copy, this data can be manually entered on a quarterly payroll sheet for each employee. If you're using a computer, a basic
payroll program will save you a lot of time and arithmetic once you've
mastered it. Your accountant should be able to advise you on these
matters.

Ways to Structure a Farm Business

There are a few different ways to structure a farm business, and each has its pros and cons. What is most appropriate in the early years of your farm may be less suitable later on. Most small growers begin as sole proprietors of their farm business, or as partners if there is more than one individual involved. This is the simplest way to get started; however, as the business grows and there are more assets to protect, many farmers consider incorporation.

The information given below is general and intended only to get you thinking about these rather complicated matters. There's more to setting up a business than is presented here. Your personal financial situation and plans for the future are highly relevant. Unless you are a business or accounting major, you would be wise to seek professional advice before deciding how to structure your farm business. On the other hand, if you know what you want and think you can do it on your own, there are online services (see Resources) that will enable you to forge ahead.

Sole Proprietorship

In this arrangement, the business is owned by one person who has full control over, and responsibility for, everything that happens. The business can hire employees, sell to the public, offer services, and do almost anything that you might want to do as a farmer. Profits go directly into your personal or business account, and you file a personal tax return. It's worth noting that, since you are self-employed, you pay Social Security and Medicare both as employer and as employee, in addition to federal and state taxes. This is known as the self-employment tax. It essentially doubles the Social Security and Medicare payments.

A sole proprietorship is the simplest type of business to set up, and it will require the least amount of paperwork. Its main disadvantage is that it does not protect your personal assets. This means that if you get sued, you could lose everything you own, including your farm. In the most extreme and unlikely case, someone could come onto your land, trip over a rock, break his or her neck, then sue you personally. In a sole proprietorship, you also have personal liability for all debts incurred. When you die, the business is essentially over.

Partnership

A partnership is similar to a sole proprietorship but is owned by two or more people. The partners contribute funds and assets and share in profits and losses according to prearranged agreements. Such arrangements should be spelled out on paper and signed by all parties in the presence of an attorney. As with a sole proprietorship, the owners of a partnership share unlimited liability for debts incurred and can be sued for personal assets. Each partner, or entity, is liable for the actions of other partners. If one partner dies, the business transfers to the other partner unless a will designates otherwise.

A "limited partner" is more like an investor. Such an individual does not usually participate in the daily operations of the farm. A limited partner's liability for debt or lawsuits is usually limited to the amount he or she has invested.

S Corporation

In the United States, corporations are legally formed and controlled at the state level. Once you have formed a corporation, regardless of the type, you can elect to be taxed under Subchapter S of the Internal Revenue Code. This means that the corporation itself will not pay federal income taxes, though it will file its own tax return (Form 1120S). Profits or losses are passed on to the owner of the S corporation, or its shareholders, who report them on their individual tax returns.

There is no minimum on the number of shareholders in an S corporation. There can be just one — you, the owner and president of the business; or two — a married couple, or father and son, or mother and daughter; or any assortment of individuals who have come together to form a partnership.

An S corporation's profits can be paid to the owner or shareholders as salaries; distributions; or, more often, as a combination of the two. If all of the profits are paid as a salary, the federal government will collect Social Security and Medicare in addition to federal tax. For Social Security and Medicare purposes, the payroll is treated as self-employment. As is the case with a sole proprietorship, this means you pay twice.

On the other hand, if some portion of the profits is paid to the owner or shareholders as distributions, rather than as wages, no Social Security will be assessed on that portion. This can save you a significant amount of money, though your Social Security account will end up with less in it. It should be noted that the IRS generally requires that a reasonable amount of the profits be paid as a salary, based on industry standards, and they might ask you to justify your allocation. So you can't call all of the profits a distribution. An S corporation doesn't necessarily have to show a profit for the principal or principals to collect a salary.

Another thing to bear in mind is that salaries paid to members of an S corporation are treated as payroll. Social Security, federal, and state deductions must be made from each paycheck issued; then these deductions are turned over to the relevant government agencies, usually on a monthly basis. If you are the sole employee, this can mean some additional paperwork.

An S corporation is a separate legal entity from its one or more shareholders and, as such, provides significant liability protection as well as protection from creditors should your business fall on hard times. Generally, the shareholders retain personal ownership of the major assets, namely the farm, its buildings, and its equipment. These are usually leased back to the corporation. Unless gross negligence on the part of the owners can be proven, any lawsuit would be filed against the corporation and covered by the corporation's liability insurance. So if something goes seriously wrong, you are less likely to lose your farm and everything you own.

There are a few minor drawbacks to being incorporated. These include:

- A few hundred dollars in start-up fees

- An annual fee of about $100 (this may vary depending on the state in which you are incorporated)

- More paperwork and higher accounting fees

Limited Liability Company

In the United States, a limited liability company, or LLC, is another way to avoid or limit personal liability for debts and lawsuits filed against the company. LLCs share some of the characteristics of corporations, partnerships, and sole proprietorships.

Like an S corporation, an LLC does not pay federal income tax on profits. Profits are passed on to the owner or owners, who declare them

Pros and Cons of Business Structures

Sole Proprietorships and Partnerships. Easiest to set up. Least amount of record keeping and paperwork. No protection of personal assets from creditors and lawsuits. Must pay self-employment tax.

S Corporation. More costly to set up and maintain. More paperwork and record keeping. Corporation does not pay tax. Profits or losses passed along to owners. Owners pay self-employment tax only on salaries, not on distributions. Protection from creditors and lawsuits, except in cases of gross negligence. Less likely to get audited than a sole proprietorship.

LLC. More setup and paperwork than a sole proprietorship but less than an S corporation. More flexible than an S corporation. LLC does not pay tax. Profits or losses passed along to owners who must pay self-employment tax on all net income. Protection from creditors and lawsuits except in cases of gross negligence.

C Corporation. Least suitable for small farmers. C corporations pay tax on profits. Shareholders pay tax on dividends. Liability protection and protection from creditors.

on their individual tax returns. Multiple owners of an LLC are usually called members.

In general, an LLC is more flexible and requires less paperwork and recordkeeping than an S corporation. An LLC can file a sole proprietorship, a partnership, or a corporation tax return. Profits from an LLC can be distributed as the owners see fit. Profits from an S corporation must be allocated according to the amount of stock each shareholder owns. This difference between the two entities is not relevant if there is just one owner. Laws governing LLCs can vary somewhat from state to state.

The owner or owners of an LLC are considered to be self-employed; therefore, they are subject to the self-employment tax, meaning they must pay Social Security and Medicare both as employer and employee. Unlike an S corporation, the entire net income of the LLC is subject to the self-employment tax, but this tax is paid just once a year.

C Corporation

The C corporation is the standard for large publicly held corporations. It is not likely to be a suitable business structure for a small farm. C corporations are taxed on their earnings, and shareholders pay taxes on the dividends they receive. This is referred to as double taxation.

A Workforce

Many growers start out farming alone or with a partner or spouse and do not aspire to become employers. That's how it was for me, but after a couple of years of working alone, I began to see that life might get a little easier and the work get done more efficiently if I had some help. When two or more people work together, there's often an energy created that is hard to summon when you're working alone. This is especially true when you are confronted with a long and tedious job.

There are also economies of scale that come with an expanded operation. Once you've done any or all of the following: built a greenhouse; installed an irrigation system; purchased a tractor, a truck, and other necessary equipment; and decided to go to a farmers' market, how much better it is if you can grow enough vegetables to make sales of $3,000 in a day instead of $2,000. The problem is that you'll be hard pressed to produce $3,000 worth of vegetables every week on your own, unless you're a superhuman. Even $2,000 might be a big challenge.

Interns

Interns are usually young, often college-educated Americans who want to give farming a try. Most are unhappy with the industrial food system and the health and environmental issues it has spawned. Many are disenchanted with other aspects of modern corporate consumer culture as well. These individuals — and there are an increasing number of them — believe small, organic farms are a way to create a viable, alternative, and sustainable food system. They also see working on such farms as a way to bring purpose into their lives.

Selecting the Right Ones

Many prospective interns have seldom, if ever, been on a farm and have little in the way of practical skills or experience with physical labor. Although their hearts and their politics may be in the right place, they are not always quite up to the job. Some are highly motivated, and energetic, and take to the work quite well; others disappoint. The challenge is to sort out the former from the latter, preferably before signing them on. Slow and physically incompetent interns will drag you down, and they are not easy to get rid of. Of course, if you can find someone who has prior farm experience (and a good recommendation from another farmer), you can be a lot more confident that you're getting the kind of person you need.

The best way to select suitable interns is to establish a lengthy application and interview process in which you find out, in advance, as much as you can about them. We require a cover letter, résumé, and references, and an on-farm interview that usually lasts a few hours. Even better than an interview is having the applicant stay and help out on the farm for a day or two. The idea is to find people who are genuinely interested in organic farming, are suited to the work, and will be glad they came to your place.

Determining Wages

Farmers who offer internships usually provide housing and allow interns to eat whatever is grown on the farm. In most cases, there is a weekly stipend, though this may not always be equivalent to minimum wage. For the most part, the sometimes-less-than-minimum-wage aspect of farm internships has gone under the radar. But in some instances, state authorities have disapproved and meted out fines that have put farmers out of business. It's a touchy subject that can be argued convincingly from both sides.

If you decide to take on interns and feel you cannot afford minimum wage, it's very important to draw up a "Farmer Intern Agreement" that spells out exactly what you are offering and what you expect in return. Both parties should sign the agreement, and you, especially, should be careful to come through with your side of the bargain, and a bit more if you can. You want to keep your workers happy and productive. You definitely don't want a disgruntled employee going to the state labor department and complaining about your operation. Before forging ahead with an intern program, it might be a good idea to speak with one of your state's Department of Labor agricultural specialists. There might be some major dos and don'ts you should know about.

Providing Learning Opportunities

In any internship, there is an expectation that interns will have plenty of opportunity to learn and will be involved in most aspects of the farm operation. They don't need to receive a formal education, but the farmer should be willing to explain why he or she is taking a particular action and be ready to answer questions when they come up. An experienced person should work alongside interns whenever possible. Also, it is best to give interns a variety of tasks to perform in a day, rather than relegate them to a job such as weeding carrots for 8 hours at a stretch.

Personally, I like to run a fairly tight ship, so that everyone, myself included, knows exactly what is expected of him or her and when. We start on time, break for lunch on time, and finish on time, unless there are extraordinary circumstances.

Dealing with Seasonal Employment

A major problem with interns is that they usually stay with you for just one season. They are young, adventurous, and still testing the many waters that seem to lie before them. In most cases, after they've experienced your farm, they are ready to move on and try something else. This means that soon after they develop some good skills, you lose them. It also means you end up training a lot of new people every year, which can get very tiring.

The challenge of training new people each year is not confined to farms that take on interns, though it is far more common with interns. Because much of North America gets a fairly serious winter, most agriculture is seasonal, which makes it very difficult to keep employees throughout the year. It's hard to justify keeping a workforce when there's not much work to do and little or no money coming in. Some farmers go to great lengths to generate enough income in the winter and off-season months to retain and pay a crew, even though there may be no profit in it for them.

Finding Interns

A good place to advertise for interns is the National Center for Appropriate Technology (NCAT). Their Appropriate Technology Transfer for Rural Areas (ATTRA) project has an online listing (with detailed descriptions) of well over one thousand farms across the United States and Canada that offer apprenticeships and on-the-job learning opportunities (see Resources for website).

Interns

IN MY THIRD YEAR OF FARMING, I took on a 19-year-old college student as a summer intern. We basically worked side by side all day long, and he lived in an old trailer that I picked up at a bargain price. The experience was a good one. The following year, I hired a young couple who had just returned from the Peace Corps. A couple of years later, I moved up to three people, then four and five — all born and bred Americans.

Today, we typically hire eight seasonal workers, all interns, though those that return for a second or third year are more like employees — they take on more responsibility and receive greater compensation. We provide clean, well-furnished housing. We pay for utilities, telephone, and Internet service and offer a fairly generous stipend, even for first-year interns. Workers prepare their own food and can eat anything grown on the farm. The value of the whole package, when prorated to an hourly basis, is more than minimum wage.

The New England Small Farm Institute in Belchertown, Massachusetts, also compiles an annual list of farms in the northeastern United States that are looking for interns. Search for "northeast workers on organic farms." Many organic farming organizations across North America maintain their own online lists of internship opportunities.

Migrant Workers

Most of the farm work in North America is done by migrants from Mexico, Central America, and other parts of the world. Migrants are usually accustomed to field work and are skilled at it. In most cases, they will get the job done in much less time than would a crew of North American interns. Of course, there are exceptions. Watch out for those migrants who have grown up in cities, rather than rural areas. Their willingness and ability to do long hours of field work might be limited.

Migrants are usually paid by the hour at the minimum wage rate or higher. They may or may not receive housing. Most agricultural migrants want to work as many hours as they can to maximize their earnings. Putting in 70 or 80 hours a week is not uncommon. Federal law does not require that employers pay overtime to agricultural employees, as long as their duties are limited to agricultural work performed on the farm.

With migrant workers, you have the issues of language and communication. Many migrants speak little English. If you don't speak their language, you will have difficulty explaining what you want done. This can be a minor issue on large farms that practice monoculture, but on a small, highly diversified vegetable farm, where there are multiple tasks to do every day, it can present a problem. Poor English skills also make many migrants less able to deal with customers at a farmers' market or CSA distribution point, unless, of course, some of those customers speak the same language as your migrant workers.

With migrants, there is also the thorny question of whether they are legally entitled to work in the United States. Their papers are often not entirely legitimate. You need to be careful that you are covered on this matter, in case Immigration and Customs Enforcement comes knocking.

Many of the larger farmers I know use migrant workers and are very happy with them. In many cases, they have the same people — often extended families — come back year after year.

Local Workers

Local residents in your area are another source for farm labor. If they have experience and an interest in the work, they can be desirable employees. If they are just looking for a job, they will often disappoint. Generally, local residents will ask for and expect significantly more than minimum wage, and they are not likely to want to work more than 40 or 50 hours per week.

Unfortunately, these days, working-class people in North America are not drawn to farm work as a way to make a living — partly because of the physical nature of the work and partly because of the low earnings potential. Most would sooner gravitate toward a Walmart or McDonald's. This might change if farms become more profitable and are able to offer better wages. Already, small, organic farms are being seen more clearly for what they are — attractive and valuable assets in their community. This is a step in the right direction.

Chapter 24

....

Looking After Number One

ORGANIC FARMING HAS A LOT TO DO WITH STEWARDSHIP. This involves the care and management of land, soil, crops, animals, water, and other resources, so that all remain productive, in good health, and in a state of balance with regard to one another. A good farmer strives toward these goals, even as he or she knows their full realization is always a little beyond reach. There is one other goal that is equally important and more readily attainable, yet ironically, this one often gets overlooked. It is the care of your own body and spirit.

To farm — to actually get out there and do the work that needs to be done, day after day — you need a healthy body and a good attitude. Assuming we can make a go of it, most of us take up farming for the long haul. We don't set out to farm for a few years before moving on to some other preferred line of work. Consider the superathlete who performs wonderfully for a decade, then starts to physically unravel. A farmer cannot afford to have this happen. The time line is so much longer — maybe the rest of your life.

If you want to have a successful career as a farmer, looking after your physical body, observing safe practices, and paying attention to your mental and emotional well-being should be critical priorities. Of course, one could direct this advice toward virtually any career or pursuit in life, but with farming it is especially appropriate. Your body is the most important apparatus you have. It needs attention, care, and respect. At the most basic level, this means eating well, getting enough rest, and learning how to pace yourself and manage stress, but there's more to it than that. In this chapter, I will discuss some of the dangers — large and small — that lurk in the shadows of the farming life and how to reduce your chances of falling victim to one or more of them.

Not just you. Without a doubt, the following advice applies not only to you but also to your spouse, partner, employees, or any other individuals you work with. If you are the farm owner or operator, you're the one who's responsible for providing a safe work environment and relevant information and protective equipment, when needed. Ultimately, though, each of us must take responsibility for looking after ourselves.

Physical Health and Safe Practices

Our bodies are made up of many parts. Damage to any one of them can affect our well-being and ability to do the physical work that farming entails. Here, I'll identify some of a farmer's more vulnerable body parts and organs and offer basic advice on how best to look after them.

Skin Protection

All people who work outdoors and leave parts of their body exposed to the sun for more than short periods of time are at risk of developing skin cancer, the most prevalent of all cancers. Skin cancer rates among farmers are alarmingly high. The disease comes in three forms:

- **Basal cell carcinoma.** The most common and least likely to metastasize (spread).

- **Squamous cell carcinoma.** Less common but more likely to metastasize.

- **Malignant melanoma.** Least common but most virulent; can metastasize rapidly to the body's lymph system and internal organs. Unless detected early and treated, can be fatal.

Symptoms. It's a smart idea to learn how to recognize the early stages of the various forms of skin cancer. Unusual fleshy bumps, scaly patches, and discoloration of the skin should all be reasons for concern. Watch out, especially, for skin blemishes larger than a pea that are asymmetrical and change, over time, in shape, color, or size. If you can manage it, an annual visit to a dermatologist is a good idea.

Treatment. Early treatment of all forms of skin cancer is usually successful. The American Academy of Dermatology's website (see Resources) provides information that will help you recognize and guard against this largely preventable disease. Fair-skinned people are most vulnerable to skin cancer, but those with darker skin are by no means immune.

Protection. In the summer, you can best protect yourself against skin cancer by wearing loose-fitting long pants, a long-sleeved shirt, and wide-brimmed hat. The hat should have a gauge that is heavy enough and a weave that is tight enough to stop the sun completely. Day after day, year after year, flimsy, threadbare hats won't do a good enough job. Some companies make hats with high sun protection factor (SPF) ratings. Coolibar is one of them.

It's also important to wear suntan lotion on your face, neck, and the backs of your hands, but be aware that the protection offered by most suntan lotions is not as long-lasting as advertised. A few applications per day might be necessary, especially if you're sweating and throwing water on your hands and face. Choose lotions with an SPF of 30 or higher. In winter, when you are likely to be wearing gloves, it is mostly your face that needs protection.

Healthy Hands

Most farmers develop strong hands because they use them a lot. This doesn't mean those hands should ever be employed as battering rams or in place of a mallet. Even the toughest hands are made of flesh, blood, and bone and as such are vulnerable to accidents and abuse. As with the rest of your body, it pays to start looking after your hands at a young age. You're going to need them for many years.

Protection. Use insulated gloves to protect hands from severe cold, which can result in arthritis later in life. Keep a pair of waterproof, insulated gloves for when it's both wet and cold. Avoid exposing hands to toxic substances by wearing nitrile gloves when necessary. Use suntan lotion or gloves to protect your hands from the sun and thereby reduce the risk of developing one of the forms of skin cancer mentioned above.

Many farm supply catalogs (such as Gempler's and Galeton) offer a wide selection of gloves for diverse applications. It's a good idea to try several different gloves to find out which are most suitable for you and your workers. Then buy them in bulk.

Preventing cuts. Most vegetable growers use knives or clippers to harvest many crops, and cut fingers are common. On our farm, I insist that my workers wear a glove on the non-cutting hand — the one that reaches down and holds the head of lettuce or bunch of Swiss chard while the other hand bears down with a knife.

Sooner or later even the most experienced knife wielder is bound to miscalculate, for whatever reason — carelessness, distractedness, tiredness, emotional distress — it really doesn't matter. If he or she is wearing a glove on the non-cutting hand, the result is rarely more than a slice in the glove and possibly a shallow nick in the skin of an index finger, which a Band-Aid can easily take care of. When the person doesn't wear a glove and a mishap occurs, he or she may need to go to a hospital emergency room for stitches.

To convince my workers of the importance of protecting the non-cutting hand, I keep a supply of old gloves that have served to protect former wearers. All of them show cuts or slices in the fabric of the glove, where an index finger once lived. A close inspection of this exhibit usually makes the point better than I can.

Ear Protection

Farm machinery, such as tractors, mowers, and chain saws, can be very loud and can damage the internal workings of your ears. Farmers tend to be exposed to loud noises far more than other people and therefore are especially vulnerable. Hearing impairment is an insidious condition that can creep up on you without your realizing it. At some point, you find yourself complaining that people are not talking loudly enough, or a spouse or friend notices that you're saying "What?" a lot more than you used to. That's when you know something is wrong, but by then, it's too late to turn back the clock.

It's never too late to use hearing protection when working with noisy equipment. If you don't, your hearing is likely to gradually deteriorate. Hearing aids are expensive; often uncomfortable; seldom covered by insurance; and, on top of all this, are not likely to bring your hearing back

Protection with Chain Saws

When using a chain saw, always wear a head protector with a hard hat, earmuffs, and a face screen. Gloves that permit good manual dexterity and chain-saw chaps, which are designed to protect your legs from the moving chain, are also highly recommended. Just one mistake with a chain saw can change your life. Because chain saws are particularly dangerous, it's a good idea to get some guidance and instruction from a professional before you start using one. Even then, it's wise not to work alone or, if you must, at least carry a cell phone in case something goes wrong.

to where it once was. I know this because I wear hearing aids. If I'd taken protective measures 25 years ago, maybe I wouldn't need them.

These days, I keep a pair of industrial-grade earmuffs hanging on the steering wheel of each of our tractors and lawn mowers. They stay with the machines. This way, there's always one there when we need it. (A single floating pair was not enough — no one was ever quite sure where it was.) When buying earmuffs, choose ones with a noise reduction rating (NRR) of 28 or higher. Disposable earplugs that fit right into your ears will also do the job. Some farmers wear earplugs all day long.

Eye Protection

If you plan to use power tools of any sort (and you almost certainly will), eye protection is recommended. It doesn't take much for a splinter of wood or metal to find its way into an eye. We keep a pair of protective goggles, hanging in a plastic bag, in all workplaces where power tools reside.

Some goggles are designed to protect your eyes from dust. These come in handy when mowing or doing other tractor work in dry conditions. Certain organically approved pesticides, such as copper, can be extremely dangerous to the eyes, especially when they are being mixed with water. Splash goggles, which provide full (rather than just frontal) protection, should always be worn when mixing up and using any pesticidal brew. Just because a substance is permitted under organic standards doesn't mean it can't hurt you.

Keep an eye bath on hand in case some foreign liquid or material does find its way into one of your eyes. And of course, always read pesticide labels carefully — they'll tell you what precautions you should be taking.

For those hard-core do-it-yourself farmers who take up welding so that they can repair broken steel implements and even fashion new ones, proper eye protection is essential; that is, unless you don't mind losing your sight.

Maintaining Healthy Lungs

You might think that country living and clean air are good for your lungs, and you'd be right — as long as you're wise enough to protect yourself in certain situations. Here again, I speak from experience. After 20 years of farming, much to my surprise, I developed a mild case of chronic obstructive pulmonary disease (COPD) or emphysema. Looking back, I can see that I made two big mistakes that might have caused the onset of this malady.

Exhaust fumes. Mistake number one was to allow leaky exhaust systems on two tractors to go unrepaired for several years. In both cases, the leaks directed exhaust fumes toward my mouth and nose and from there, into my lungs. If you spend a couple of hours on an old tractor, which I frequently did, you're likely to inhale a fair amount of diesel fumes. When I stepped off the tractor, I'd be dizzy from imbibing so much diesel. Still, I didn't get around to doing what needed to be done. Not very smart.

Dust. The other mistake involved curing garlic and other alliums such as onions and shallots. Harvested alliums usually carry some field dirt with them, especially around their roots. The problem is worse when they are pulled out of wet soil. We bring these crops into our barn and other sheds and use fans to dry them down and cure them for long-term storage. The dirt on the roots soon turns to dust, and the fans keep the dust agitated and in the air. Over time, this can amount to a lot of dust entering the lungs.

Protection. Having learned the hard way, I now keep plenty of dust masks on hand and wear one whenever working with drying alliums in a confined space or, for that matter, in any dusty environment. I encourage my workers to do the same.

The exhaust systems on my tractors have been repaired, but not the damage to my lungs. These days, I wear a respirator every time I do tractor work to prevent even a small amount of diesel fumes from entering my respiratory system.

Respirator

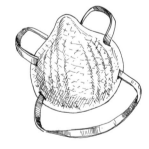

Dust mask

These are just a couple of examples of minor oversights or mistakes that, over time, gave birth to real problems. No doubt, there are many other seemingly innocuous perils. Every farmer who wants to stay on the job should be on the lookout for such dangers and take precautionary or preventive action before it's too late. Think of your body as the most important piece of machinery you have. Like a tractor or pickup truck, if it receives good maintenance and is not subjected to abuse, it will most likely keep running and serve you well for a long time.

Back Care

When you're doing a lot of heavy lifting — as in moving around crates of winter squash or potatoes — it's easy to throw your back out. One careless move can put you out of commission for a week. Wearing a back support belt is a good way to minimize this sort of injury. Back belts promote better posture and better lifting techniques and provide crucial support for the lower back. Have a few on hand for yourself and your crew members.

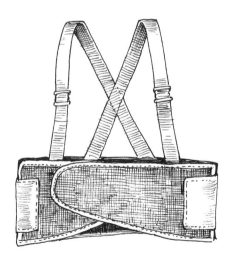
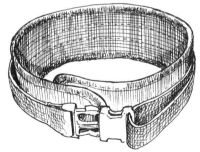

Back support belts

Preventing Dehydration and Heat Stress

Problems arising from dehydration and overexposure to heat are common on farms, especially among inexperienced field workers, and they can be serious. Fortunately, there are simple precautions to take to steer clear of these conditions.

Dehydration. Water plays an important role in regulating many body functions and helps maintain peak-level performance. We all need it, but those of us who spend a lot of time outside engaged in physical work have an even greater need. It's wise to get into the habit of drinking two or three glasses of water in the morning and several more glasses throughout the workday.

Eight 8-ounce glasses of water is a reasonable daily quota for most people, but in hot, dry weather, when you're perspiring, you might need more.

Don't rely on thirst to tell you when you need a drink. Under certain conditions you can be seriously dehydrated and not feel especially thirsty. Develop a routine for drinking your water, whether thirsty or not, and stick with it.

Heat stress. Heat stress occurs when your body has absorbed more heat than it can cope with. It is a serious and sometimes even life-threatening condition.

Lightweight, light-colored, loose-fitting clothing over your entire body — this means avoiding T-shirts and shorts — will greatly reduce your chance of suffering from heat stress. Also, wear a broad-brimmed hat or at least a baseball cap. But even with these precautions, there can be very hot days when it's wise to limit your amount of outdoor physical activity until you have a good idea of what your body can handle.

Even experienced farmers should be on their guard against heat stress when they are operating under extreme conditions. A good example would be when they wear protective clothing during pesticide applications on a hot day. The personal protective equipment or PPE (such as chemical resistant and nonbreathable

Symptoms of Heat Stress

- ▸ Dizziness
- ▸ Nausea
- ▸ Fatigue
- ▸ Impaired coordination
- ▸ Headache
- ▸ Irrational behavior
- ▸ Dry mouth
- ▸ Extreme thirst

How to Deal with Heat Stress

If you or someone you're working with exhibits any of the symptoms of heat stress, there are a few steps that should be taken immediately:

- Get to a cool, shaded place.
- Drink plenty of water.
- Remove any clothing that might be contributing to the problem.
- Cool off with a wet cloth or sponge.
- Lie down and rest.
- Seek medical help.

overalls, rubber boots and gloves, goggles, and respirator) might do a good job of keeping chemicals out, but they also trap heat inside.

Beginning organic farmers might think this warning does not apply to them. That would be a mistake. As already noted, there are an increasing number of quite effective pesticides that are approved for organic use. They can be botanical or biological formulations, or naturally occuring elements. Just because they are not synthetic chemicals doesn't mean they will not cause you harm. At some point, you might find yourself donning protective equipment and going out to the field with a backpack sprayer. If you do, be aware of the danger of heat stress.

Other Physical Hazards

I don't want to scare you off, but you should know that farming ranks as one of the most dangerous occupations in America — right up there with ocean fishing, mining, and logging. There are three major reasons for this: agricultural chemicals, large animals, and farm equipment.

Exposure to Agricultural Chemicals

Conventional farmers use a wide array of herbicides, insecticides, fungicides, and other synthetic concoctions to control whatever ails their crops. All are designed to kill one form of life or another. It doesn't take much imagination to realize that many of these "magic bullets" are also harmful to human health. Even with the best protective gear, accidents can occur, and long-term exposure to small amounts of pesticides can lead to serious health problems.

PPE (Personal Protective Equipment)

The term PPE refers to special clothing and other equipment that are designed to protect the wearer from a broad range of occupational and health hazards that go well beyond the field of agriculture. Items on a farmer's PPE list are likely to include dust masks, respirators, splash-proof goggles, dust and impact goggles, earmuffs and earplugs, nitrile and neoprene gloves, and protective coveralls and suits.

Pesticides approved for organic agriculture — should you choose to use them — can also be hazardous to your health. When using any pesticide, whether it be organically approved, all precautionary instructions, and especially those concerning the use of PPE, should be closely adhered to. It is, in fact, the law.

Danger with Large Animals

If a cow or horse wants to push you over and stomp on you, it can do so with ease. We humans rely on the goodwill of these jumbo-sized creatures, but even an inadvertent head butt from a friendly cow can knock a person off his or her feet.

This is relevant only to organic vegetable growers who use animal traction (namely, draft horses or oxen to work their fields) or wish to incorporate large livestock into their farm plan.

Safety around Tractors and Other Farm Machinery

Big machines with moving parts are inherently dangerous. Tractor rollovers have ended many a farming career; power take-off shafts have separated more than a few limbs from their owners. And too many human bodies have been crushed or injured when large pieces of equipment are coupled together.

After a few close calls (including both a rollover and a runaway tractor), I've trained myself to slow down and take a methodical approach whenever working with a tractor, and not just when driving one. It's equally important to be careful when attaching large implements such as rototillers, chisel plows, or brush hogs. My rear-mount rototiller weighs 3,000 pounds. That's a lot of unyielding metal to come down on an arm or a leg.

Many Cooperative Extension Services and organic certification organizations conduct or host tractor workshops. Inexperienced growers who do not have a personal tractor mentor should plan on attending a few of these workshops or classes before embarking on extensive tractor use. They usu-

ROPS

If you're farming hilly or uneven ground, your tractor should have a roll bar or rollover protection system (ROPS), and you should wear a seat belt. These are basic and necessary precautions.

ally focus on tractor safety and maintenance, and will sometimes offer guidance on the right size and type of tractor for your operation; all very handy information.

Tractor with rollover protection system (note seat belt)

Mental and Emotional Health

If you're suited to it, farming can be a good life, a full life, a life you'll be glad you chose. It will also be demanding and sometimes stressful. When you're running your own business, failures or mishaps, when they occur, will rest squarely on your shoulders. This is something you've got to get used to.

Some days, there'll be too much to do and not enough time in which to do it. You may have to practice some form of triage. And of course, things don't always turn out the way we would like. These and many other troubles — a difficult employee, a devastating hailstorm, economic hardship, an extended period of drought, an aching back, a case of the blight — are likely to surface sooner or later. When they do, we suffer — sometimes a little and sometimes a lot.

But most problems are not as bad as they seem at first. Many can be solved quite easily once you stand back and take a clear-eyed, objective look at them. Some troubles simply resolve themselves with the passage of time and little or no intervention on our part. Others are more intractable and may take considerable effort to overcome. Unfortunately, all of them, even small setbacks, can cause stress if we let them. And stress can affect our physical and emotional well-being, which, in turn, can affect our ability to farm safely and effectively.

So what's a farmer to do? Well, as stated earlier in this chapter, a healthy diet with three square meals a day, along with adequate rest at night, are a good place to start. But let me add a few other things you can do to stay on an even keel.

Develop a Work Schedule

Establish regular work hours and good work habits for yourself and your crew. Try to begin and end work at the same time each day, and allow for some downtime at night before going to bed. On our farm, we stick to a strict schedule on what we call "farm work" days — 8:00 AM to 6:00 PM with a 1-hour break for lunch or, in the heat of summer, 7:00 AM to 6:00 PM with a 2-hour lunch break. During the peak of the season, when there's a lot to harvest and sell, we might work for an extra 1 to 1½ hours on "pick days" (of which there are two each week) to put together a load for the next day's market. As the days shorten and there is less to harvest, we usually finish pick days earlier.

The important thing here is regularity. Our bodies and our psyches like to know what is expected of them and when they will be rewarded with rest, food, or simply something different to do.

Stretch Daily and Exercise

Some form of daily exercise that focuses on stretching and flexibility will have a calming effect, dissipate stress, and keep your body in good form. Consider taking up yoga, chi gong, tai chi, or some other gentle practice with a meditative underpinning. You'll find this is especially valuable as you get older. Our bodies really do appreciate being loosened, stretched, and warmed up before a day's work. Encourage your workers to incorporate stretching into their daily routine. You might even try stretching with them for 5 or 10 minutes each morning.

Don't Sweep Stress under the Rug

When you face a particularly vexing problem, talk to a friend or family member. Often telling a trusted person will ease the load, provide perspective, and even shine light on a solution.

Small amounts of stress are part and parcel of life. They usually go away on their own or after a night's rest. The important thing is to recognize when you're experiencing stress and, if possible, find some way to alleviate it. Not acknowledging stress, sweeping it under the rug, or allowing it to build up, without relief, is a recipe for trouble.

NYCAMH

The New York State Center for Agricultural Medicine and Health (NYCAMH) has plenty to say about preventing and treating occupational injuries and illnesses common to farming. This organization offers on-farm safety surveys and on-farm safety training to all New York State farmers, free of charge. You can learn about them on their website (see Resources). Ask your state agriculture department if there are similar programs in the state or province where you live.

Don't Rush In

When a difficult situation develops, it's often wise to step back and let a little time pass, rather than try to tackle it while in an agitated state.

This is especially true with issues involving employees or virtually any other human being. I've learned over the years to avoid confronting someone while in the heat of anger. Sometimes, a day later, I realize that my anger or irritation had more to do with me and the mood I was in, or other problems I was facing, than with anything the employee said or did.

Organization Helps

Good planning and good organization will reduce stress and keep unwanted surprises to a minimum. These subjects are covered in detail in chapter 3.

On the other hand, after a short hiatus and some reflection, I might decide that there still is a legitimate reason for my displeasure and that it is necessary to address the issue that concerns me. When this is the case, it's generally better for me to wait until I'm in a calm and clear frame of mind before speaking to the person. That doesn't mean you should never express anger — there might be times when it is necessary — but when you do, it should be anger that you have full control over.

Take Time Off

Allow yourself a day off each week — a day on which you don't spend much time thinking about farming. On this day, read a book or magazine, go fishing, watch a movie, socialize with friends or family, take a walk in the woods, enjoy some of life's simple pleasures. You'll find it pays off. The next day, when you go back to work, you'll have more energy and more clarity of purpose. You'll be ready to take on what needs to be done.

A Few More Words of Advice

I'll end with a few more snippets of advice, some of which have already come up, in one form or another, elsewhere in this book.

- Divide tedious or difficult jobs into more manageable parts.

- Schedule tasks so that you are using your body in different ways throughout the day.

- Go easy on your back — when you have heavy lifting to do, approach it with caution. Always look for the smart way.

- Keep an eye on the weather so you can minimize outdoor work when it's raining or extremely hot or cold.

- Drink plenty of water, especially on those hot summer days.

- Get a tetanus shot every 10 years (which reminds me, I'm due for one).

Most of this advice is simple common sense, but it's surprisingly easy to let common sense go by the wayside when we feel pressure or stress or are just plain tired. That's when an accident is most likely to occur. The wise farmer keeps a measured pace and stays on guard at all times.

There's an art to being a good farmer, just as there is an art to leading an engaged and meaningful life. It is not learned in a week or a month or a year. You have to keep working on it day after day, until the very end. Never forget that looking after yourself is an important part of the epic journey you are on.

CHAPTER 24 RECAP

Tips for Physical and Mental Well-Being

- ▶ Wear sunscreen, loose-fitting pants, a long-sleeved shirt, and a wide-brimmed hat.
- ▶ Wear the appropriate type of gloves when necessary.
- ▶ Use earmuffs when operating machines.
- ▶ Wear goggles when using any pesticides and doing welding.
- ▶ Wear a dust mask when working in a dusty environment.

- ▶ Wear a back support belt when lifting heavy objects.
- ▶ Get into the habit of drinking water — at least eight 8-ounce glasses a day — whether you are thirsty or not.
- ▶ Develop a work schedule.
- ▶ Stretch daily and exercise.
- ▶ Don't sweep stress under the rug.
- ▶ Don't rush into a difficult situation.
- ▶ Take time off.

Afterword

In his 1960s counterculture song "Subterranean Homesick Blues," Bob Dylan wrote, "You don't need a weatherman / To know which way the wind blows." Today, I would say, you don't need a clever man to know which way the world goes. For those who care to look, the evidence is rather plain to see: rising global temperatures; melting ice caps; shrinking rain forests; expanding deserts; soaring rates of plant and animal extinction; ever more frequent floods, droughts, heat waves, and famines. Even the once-teeming life of the oceans is now reduced to a fraction of what it was a few generations ago.

In short, we have some serious crises and moments of truth coming our way, and it's high time we took our heads out of the sand and started looking, without bias, into the future we are creating for ourselves. Every year there are more of us to feed, more of us seeking a larger share of the shrinking earthly pie. The old model of endless economic growth and expansion, endless consumption and extraction of resources, is no longer serving us well. It's time for a new model—one that focuses on planetary health and long-term sustainability. The earth is our legitimate home, and it is a beautiful place. We must learn to look after it.

There is a great deal of restorative work to do on many fronts. All of it is good and worthy work. Because farming is what I know best, the focus of my attention is on building a sustainable food system—one that nurtures the land rather than depletes it, one that recognizes that we are connected to the earth and its soils in profound ways, that our well-being and the well-being of the planet and its many and varied inhabitants are inseparable.

From where I stand, a sustainable food system would be one with millions of small farmers and food producers scattered throughout our rural, suburban, and even urban landscapes, rather than a handful of giant industrial entities concentrated in just a few regions. It would be predominantly, but not exclusively, a local food system. It would be one that practices diversity rather than monoculture. It would be a food system that does not depend on an endless stream of synthetic pesticides and fertilizers that are destructive to the soil and the life within it, and very likely harmful to humans as well.

Such a food system would bring the concepts of husbandry and stewardship back into agriculture. It would bring consumers back into contact with producers and foster relationships of trust and accountability. It would focus on the quality of food grown and raised, rather than merely the quantity produced at the lowest price, with the least number of people involved. The food system that I envision would be inherently more adaptive, resilient, and resourceful and more able to withstand the vagaries of climate and weather. It would not be held hostage to the global marketplace and commodities exchanges. Nor would any single part of it be too big to fail or dependent on massive taxpayer subsidies.

Thankfully, the kind of foodways that I am espousing do not just live in my imagination. Though disparaged and devalued, they have, in fact, been with us all along and now, in the twenty-first century, are undergoing a wonderful grass-roots renaissance. I refer, of course, to the organic and small-farm movement practiced at the local level. This movement toward a better agricultural tomorrow is indeed alive and well. But in the larger scheme of things, it is still very small. To bring about really positive change, we need a lot more of it. That's all.

Contrary to popular belief, the notion that such a decentralized, local, and ecological food system could never feed the world is patently false. It is a myth perpetrated by the conglomerates who have close to a monopoly over growing and distributing the food we eat, along with immense influence over our government. It hardly need be said that these giant companies profit mightily from keeping things the way they are.

Each year, new evidence shows that small, diversified growers with a grounding in sustainable practices, a plot of well-managed land, and a stake in what they do will often outproduce the agribusiness giants acre for acre and not rob the land of its health in the process. True, more hands are needed on these small farms, but this should not be seen as a negative. Young and often not-so-young people all over the world are rediscovering what humans have known for a long time but almost forgotten: Putting down roots in one place, becoming part of a community, getting to know and look after a good piece of land, and growing decent food for a living are valid and worthwhile pursuits at every age. And I'm pleased to say that, for more than 20 years, many individuals with such purposes in mind have gotten their feet wet, in more ways than one, on our farm.

But enough of food politics. I started this book on a personal note, so I might as well finish on one. In 1986, when my wife (who teaches art at a college in New York City) and I bought the farm we live on, I didn't have much of a game plan laid out, and my knowledge of growing vegetables was scant. What I wanted and needed was a change of life. I hankered to get back to a world that made more sense to me, one in which the sights and sounds and smells of nature would be a part of my daily routine, a world in which I could use my body as well as my brain, a world in which I could live more modestly and in accord with my environmental inclinations.

Becoming a farmer, it turned out, offered all this and more. It offered a life I could truly embrace. It offered a life I have been glad to live and happy to write about.

Keith Stewart

RESOURCES

The following list contains many of the books and online resources that our farm has found useful over the past 25 years. It is far from exhaustive. There's lots more out there.

Books

GENERAL
Maynard, Donald N., and George J. Hochmuth. *Knott's Handbook for Vegetable Growers*, 5th ed. Wiley, 2007.

COVER CROPS
Baldwin, Keith R., and Nancy G. Creamer. *Cover Crops for Organic Farms.* North Carolina Cooperative Extension Service, 2006.

Clark, Andy, ed. *Managing Cover Crops Profitably*, 3rd ed. Sustainable Agriculture Network, 2007.

Sarrantonio, Marianne. *Northeast Cover Crop Handbook.* Rodale, 1996.

Sullivan, Preston. *Overview of Cover Crops and Green Manures.* Appropriate Technology Transfer for Rural Areas, 2003.

CROP ROTATION
Coleman, Eliot. *The New Organic Grower: A Master's Manual of Tools and Techniques for the Home and Market Gardener*, 2nd ed. Chelsea Green, 1995.

Mohler, Charles L, and Sue Ellen Johnson, eds. *Crop Rotation on Organic Farms: A Planning Manual.* Natural Resource, Agriculture, and Engineering Service, 2009.

GARLIC
Bachmann, Janet, and Tammy Hinman. *Garlic: Organic Production,* rev ed. National Center for Appropriate Technology, 2008.

Crawford, Stanley. *A Garlic Testament: Seasons on a Small New Mexico Farm.* University of New Mexico Press, 1998.

Engeland, Ron L. *Growing Great Garlic.* Filaree Productions, 1991.

HIGH TUNNELS
Byczynski, Lynn, ed. *The Hoophouse Handbook.* Fairplain Publications, 2006.

SEED STARTING

Bubel, Nancy. *The New Seed Starters Handbook.* Rodale, 1988

DeBaggio, Thomas. *Growing Herbs from Seed, Cutting & Root,* 2nd ed. Interweave Press, 2000.

SOILS

Gershuny, Grace, and Joe Smillie. *The Soul of the Soil: A Soil-Building Guide for Master Gardeners and Farmers,* 4th ed. Chelsea Green, 1999.

Logan, William Bryant. *Dirt: The Ecstatic Skin of the Earth.* W. W. Norton, 2007.

Magdoff, Fred, and Harold van Es. *Building Soils for Better Crops,* 3rd ed. Sustainable Agriculture Research and Education, 2010.

Wolfe, David W. *Tales From The Underground: A Natural History of Subterranean Life.* Basic Books, 2001.

WEEDS AND WEED CONTROL

Bowman, Greg, ed. *Steel in the Field: A Farmer's Guide to Weed Management Tools, rev ed.* Sustainable Agriculture Network, 2002.

Uva, Richard H., Joseph C. Neal, and Joseph M. DiTomaso. *Weeds of the Northeast.* Cornell University Press, 1997.

Online Resources

COMPOST CALCULATORS

Klamath Green Welcome Wagon
www.greenwelcomewagon.com
Click on the "Tool Box" for the compost calculator.

Solid Waste Department
Klickitat County
www.klickitatcounty.org/solidwaste

HIGH TUNNELS

Cornell High Tunnels
Cornell University, Department of Horticulture
www.hort.cornell.edu/hightunnel

Crop Tunnels and Plastic Mulch
Mount Vernon Northwestern Washington Research and Extension Center, Washington State University
http://mtvernon.wsu.edu/hightunnels

High Tunnel Crop Production Project
Mississippi State University Extension Service
http://msucares.com/crops/hightunnels

High Tunnel Project
Rutgers Cooperative Research and Extension
http://aesop.rutgers.edu/~horteng/hightunnels.htm

Hightunnels.org
www.hightunnels.org
A group of Extension specialists, professors, growers, technicians, and students sharing information about high tunnels

High Tunnels: Using Low Cost Technology to Increase Yields, Improve Quality, and Extend the Growing Season
University of Vermont Center for Sustainable Agriculture
www.uvm.edu/ sustainableagriculture/ hightunnels.html
Written by Ted Blomgren, Tracy Frisch, and Steve Moore

Ledgewood Farm Greenhouse Frames
http://ledgewoodfarm.com

National Sustainable Agriculture Information Service — ATTRA
National Center for Appropriate Technology
https://attra.ncat.org

The Samuel Roberts Noble Foundation, Inc.
www.noble.org

What Is Plasticulture?
Penn State Extension
http://extension.psu.edu/ plasticulture

PEST MANAGEMENT

IPM Laboratories, Inc.
www.ipmlabs.com

ipmPIPE
Southern Region Integrated Pest Management Center
http://cdm.ipmpipe.org

PestWatch
Penn State University
www.pestwatch.psu.edu

Resource Guide for Organic Insect and Disease Management
Cornell University
http://web.pppmb.cals.cornell.edu/ resourceguide
Written by Brian Caldwell, Emily Brown, Rosen, Eric Sideman, Anthony M. Shelton, and Christine D. Smart, 2005.

USAblight
http://usablight.org

SEED COMPANIES
(MY FAVORITES IN THE NORTHEAST)

Fedco Seeds
www.fedcoseeds.com

High Mowing Organic Seeds
www.highmowingseeds.com

Johnny's Selected Seeds
www.johnnyseeds.com

SOIL TESTING AND MAPPING

Cornell Nutrient Analysis Laboratory
Cornell University
http://cnal.cals.cornell.edu

Cornell Soil Health
College of Agriculture and Life Sciences, Cornell University
http://soilhealth.cals.cornell.edu

Web Soil Survey
Natural Resources Conservation Service, United States Department of Agriculture
http://websoilsurvey.nrcs.usda.gov

SUPPLIES

Belle Terre Irrigation
www.dripsupply.com
Drip irrigation design and supplies

CT Farm & Country
www.ctfarmonline.com
Valu-Bilt tractor and equipment parts

DripWorks, Inc.
www.dripworks.com
Drip irrigation design and supplies

FarmTek
www.farmtek.com
Hoop structure and farm supplies

Gempler's
www.gemplers.com
General farm supplies

Harmony Farm Supply & Nursery
www.harmonyfarm.com
Tools and organic farm supplies

Hubert Company
www.hubert.com
Marketing supplies

Kencove Farm Fence Supplies
www.kencove.com
Farm fence supplies

Market Farm Implement
www.marketfarm.com
Vegetable crop equipment

Peaceful Valley Farm Supply, Inc.
www.groworganic.com
Tools and organic farm supplies

SharpeningSupplies.com
www.sharpeningsupplies.com
Knife sharpening supplies

Store It Cold, LLC
www.storeitcold.com
Manufactures the CoolBot, Ron Khosla's invention for turning an air conditioning unit into a turbo-charged chiller.

Tractor Supply Company
www.tractorsupply.com
General farm supplies

TRICKLE-EEZ Company
www.trickl-eez.com
Irrigation design and supplies

SUPPORT FOR NEW FARMERS

International Farm Transition Network
www.farmtransition.org
This organization holds seminars and publishes materials with the goal of fostering the next generation of farmers and assisting those who are making the transition away from farming. See their website to find Land Link and Farm Link programs that participate in their network.

National Center for Appropriate Technology (NCAT)
www.ncat.org
Listing of well over one thousand farms across the United States and Canada that offer apprenticeships and on-the-job learning opportunities.

New England Small Farm Institute
www.smallfarm.org
Provides information for organic farmers and publishes a list of farms in the Northeast that offer apprenticeships (Northeast Workers on Organic Farms).

Northeast Beginning Farmers Project
Cornell University
http://nebeginningfarmers.org

MISCELLANEOUS WEBSITES

American Academy of Dermatology
www.aad.org
    More information about skin cancer

The Commerical Storage of Fruits, Vegetables, and Florist and Nursery Stocks
www.ba.ars.usda.gov/hb66
    Written by Ken Gross. Agriculture Handbook, 66. Agriculture Research Service, 2004. Contains a wealth of information and advice on the post-harvest handling of many different fruits and vegetables

Coolibar
www.coolibar.com
    Sun protection hats and clothing

Coop Directory Service
www.coopdirectory.org
    Lists food co-ops and other cooperative resources throughout the United States

Garlic Seed Foundation
www.garlicseedfoundation.info

LocalHarvest, Inc.
www.localharvest.org

MyCorporation
www.mycorporation.com
    Information on how to set up your new business

New York Center for Agricultural Medicine and Health (NYCAMH)
www.NYCAMH.com

Organic Materials Review Institute (OMRI)
www.omri.org
    Publishes the annual *OMRI Products List* that contains all the materials that are allowed for use in organic production, processing, and handling

Waterpenny Farm
www.waterpennyfarm.com

INDEX

Page numbers in *italic* indicate illustrations;
those in **bold** indicate charts.

cultivation, 106–9. *See also* weeds/
weeding
between-row cultivators, *108*, 109
blind cultivators, 107–8, *108*
in-row cultivators, 109
irrigation following, 177
meaning of, 106
row spacing and, 468
weed control and, 106, 107, 466
customers. *See also* business and institutional customers
insect damage and, 500–501

D

daily to-do list, 74–77. *See also* to-do
lists
morning meeting, 76
sample page, **75**
scheduling considerations, 74–77
suggestions, not imperatives, 76–77
dairy farmers, 33
damage minimization, 493–501
air circulation and, 496
beneficial insects and, 495, 497
crop rotation and, 494–95
field hygiene and, 495–96
planting times and, 495
positive attitude toward, 500–501
protective barriers and, 496–97
seeds/seed condition and, 496
sprays, approved, 497–99
damping off, *157*, 157–58
day length, changing, 78–80
deer. *See also* fencing
damage minimization and, 476–77
shooting, 476
dehydration/heat stress, 526–27
Diaz, Jacob, 202–3
digging fork, 122
direct sales, 397
direct-seeding
allelopathic crops and, 233
crop rotation and, 275
soil temperature and, 54
disc harrow, drag-type, 96, *96*
disease(s), 489–93. *See also* damage
minimization

bacterial, **491**, 491–92
crop rotation and, 60, 270–71
fungal, **490**, 490–91
intercropping and, 239
nematodes and, 223, *224*, *225*, 493
untreated water and, 183
viral, **492**, 492
diversified farms, 32, 37, 76, 216. *See
also* farm stories; small farms
diversity
beneficial insects and, 238
crop, 117
intercropping and, 239
sustainability and, 535
dogs, animal pests and, 477, 479
dolomitic lime, 208, 214, 215
"double-covered low tunnels," 171, 172
draft animals, 20, 114–16
CONS, 115–16, 528
PROS, 115
drag-type disc harrow, *96*, 96
drainage. *See* lay of the land
drip irrigation, 179–82
clean water for, 181
drip lines, bare soil and, 180, *180*
drip lines, plastic mulch and,
179–80
experts on, 181–82
how long to water, 188–89
pressure regulator for, 181
record keeping for, 181
rodents and, 181
schedule, 190

E

Eagle Street Rooftop Farm, 322–23
ear protection, 522–23
earthworms, 223, 225, *225*, 227
education, formal, 17
educational institutions, 428–30
Engeland, Ron, 355
English, Jean, 171
equipment, 49. *See also* small equipment and tools; tools; tractor implements; tractor(s); workshop/work-
shop tools

U

Umbelliferae plant family, 283
unfenced areas, 282
USAblight, 495
USDA National Cooperative Soil Survey, 197
USDA NRCS, 28, 53, 202–3
USDA Plant Hardiness Zone Map, 257–59, *258*
USDA's Web Soil Survey, 25, 39, 40, 206
utilities, 139

V

vegetables. *See also* specific vegetable
    families of, 282–83, 494
    flowering/fruiting, irrigation and, 176
    for high tunnels, 167
    open-pollinated varieties, 425
    pick list, **62**
    varieties, greenhouses and, 130
vertical space, 166–67, *167*
viral diseases, 157, 492, **492**

W

walk-behind brush mower, 113, *113*
walk-in cooler, 412–13
Warner, Bill, 174–75
washing vegetables, 405–8, *408*
wasps, parasitic, 157
watering, greenhouses and, 153–56
    allowing flats to dry, 154
    basic variables, 154
    bottom watering, 155–56
    even watering, 155
    flexibility and, 155
    one person to water, 155
    organizing flats, 155
    pH of water, 153
    schedule and, 155
    stage of growth and, 154
    time of day and, 154–55
    weights of flats, checking, 155
watering specifics, 185–91

how deeply to water, 186–87
how long to water, 187–89
overwatering, 187
when to water, 186
waterlogged soil, 200
water-loving plants, 282
water tank, 34
water/water supply. *See also* irrigation;
    rainfall; watering specifics
    for drip irrigation, 181
    land acquisition and, 30–31
    pH of water, 153
    potable water, test for, 406
    rainfall, 184–85
    safeguarding of, 234
    untreated, plant diseases and, 183
water-wheel transplanter, 104–5, *105*, 185
weather. *See also* zone map
    base reflectivity maps, 73
    forecasts, 63, 73
    frost-free days, 259–60
    greenhouses and, 131–32
    harvest and sales record, 68
    plant hardiness zones and, 257–59
    weekly to-do list and, 72–73
    wet conditions, 162
websites
    Local Harvest, 422
    online research, 20
    for pests and diseases, 495
    USDA's Web Soil Survey, 25, 39, 40, 206
weed-control strategies, 465–68, 470
weed seed, 291
weeds/weeding. *See also* cultivation
    annual weeds, *462*, 464
    biennial weeds, 465, *465*
    crop rotation and, 273–76
    cultivation and, 106, 107
    edible weeds, 469
    field hygiene and, 495–96
    flame weeding, 368
    garlic, 349–50
    getting to know weeds, 461–65
    greenhouses and, 129
    hairy vetch and, 255

Other Storey Titles You Will Enjoy

The Complete Compost Gardening Guide, by Barbara Pleasant & Deborah L. Martin.
Everything a gardener needs to know to produce the best nourishment for abundant, flavorful vegetables.
320 pages. Paper. ISBN 978-1-58017-702-3.
Hardcover. ISBN 978-1-58017-703-0.

Greenhorns, edited by Zoë Ida Bradbury, Severine von Tscharner Fleming, and Paula Manalo.
Fifty original essays written by a new generation of farmers.
256 pages. Paper. ISBN 978-1-60342-772-2.

The Organic Farming Manual, by Ann Larkin Hansen.
A comprehensive guide to starting and running, or transitioning to, a certified organic farm.
448 pages. Paper. ISBN 978-1-60342-479-0.
Hardcover. ISBN 978-1-60342-480-6.

Making Your Small Farm Profitable, by Ron Macher.
An indispensable guide, full of advice on planning, marketing, and farming, to make your small farm equal big pay.
288 pages. Paper. ISBN 978-1-58017-161-8.

Reclaiming Our Food, by Tanya Denckla Cobb.
Stories of more than 50 groups across America that are finding innovative ways to provide local food to their communities.
320 pages. Paper. ISBN 978-1-60342-799-9.

Starting & Running Your Own Small Farm Business, by Sarah Beth Aubrey.
A business-savvy reference that covers everything from writing a business plan and applying for loans to marketing your farm-fresh goods.
176 pages. Paper. ISBN 978-1-58017-697-2.

These and other books from Storey Publishing are available wherever quality books are sold or by calling 1-800-441-5700.
Visit us at *www.storey.com.*